ENGINEERING SURVEYING
TECHNOLOGY

edited by
T.J.M. KENNIE
Lecturer in Engineering Surveying
University of Surrey

and

G. PETRIE
Professor of Topographic Science
University of Glasgow

Blackie
Glasgow and London

Halsted Press, a division of
John Wiley & Sons, Inc.
New York

Blackie and Son Ltd
Bishopbriggs, Glasgow G64 2NZ
7 Leicester Place, London WC2H 7BP

Distributed in the USA by
John Wiley & Sons, Inc.
Orders from the USA only should be sent to
John Wiley & Sons, 1 Wiley Drive,
Somerset, New Jersey 08873

© 1990 Blackie and Son Ltd
First published 1990

*All rights reserved.
No part of this publication may be reproduced,
stored in a retrieval system, or transmitted,
in any form or by any means,
electronic, mechanical, recording or otherwise,
without prior permission of the Publishers.*

British Library Cataloguing in Publication Data

Engineering surveying technology
1. Surveying
I. Kennie, T.J.M.   II. Petrie, G.
526.9

ISBN 0-216-92482-0

Library of Congress Cataloging-in-Publication Data

Engineering surveying technology   edited by T.J.M.
Kennie, G. Petrie.
   p.   cm.
Bibliography: p.
includes index.
ISBN 0-470-21212-8
   1. Surveying—Instruments. 2. Surveying—Data processing.
I. Kennie, T.J.M.   II. Petrie, G.
TA562.E54 1989   88-21880
526.9—dc19   CIP

Phototypesetting by Thomson Press (India) Limited, New Delhi.
Printed in Great Britain by Thomson Litho Ltd, Scotland.

EARTH-
—T-SL

Preface

The rate of technological development in the equipment and processing techniques used in field surveying, photogrammetry and remote sensing during the past five to ten years has been phenomenal, and there is urgent need for a book which can provide a comprehensive and authoritative overview of the present state-of-the-art.

The objective of this book is to examine the major changes which have influenced the measurement and processing of topographic and non-topographic spatial data. The emphasis throughout is on new and emerging technology and its applications, although fundamental principles are introduced where appropriate. The reader will not find in this volume many of the topics normally associated with textbooks on engineering surveying. Thus, for example, descriptions of standard field survey instruments such as levels and optical theodolites are not included, nor are simple computational methods such as traverse reductions, or basic techniques of setting out.

The book is structured into three parts. Part A considers developments in field instrumentation and processing techniques; Part B examines developments in remote sensing and photogrammetric technology; and Part C covers developments in computer-based mapping, terrain modelling and land information systems. These divisions are somewhat arbitrary, since the boundaries between them are becoming increasingly blurred. A short introduction preceding each Part highlights common issues affecting these overlapping technologies.

We wish to thank a number of people who have been involved in the preparation of this book. First, we thank the contributing authors for their invited contributions and acknowledge their patience and understanding. Secondly, we thank those colleagues who kindly acted as independent referees—in particular, Professor Peter Dale (North East London Polytechnic), Mrs Veronica Brown (University of Surrey), Dr John Uren (University of Leeds) and Mr William Price (Portsmouth Polytechnic). Thirdly, we thank the many instrument manufacturers and system suppliers who kindly provided illustrative material. Further, we express our gratitude to the technical support staff in the Department of Geography and Topographic Science at the University of Glasgow, particularly Ian Gerrard, Yvonne Wilson and Les Hill for their help in the production of diagrams and photographs. Thanks are also due to the publishers for their unending patience. Finally, special thanks must be given to our wives, Sue Kennie and Kari Petrie, for coping with numerous interruptions to family life during the writing of this book.

TJMK
GP

Contributors

Paul A. Cross Professor of Surveying
Department of Surveying
University of Newcastle upon Tyne
Old Brewery Building
Newcastle upon Tyne NE1 7RU, UK

Alan H. Dodson Reader in Geodesy
Institute of Engineering Surveying
and Space Geodesy
University Park
University of Nottingham
Nottingham NG7 2RD, UK

Tom J.M. Kennie Lecturer in Engineering Surveying
Department of Civil Engineeing
University of Surrey
Guildford GU2 5XH, UK

David Parker Lecturer in Surveying
Department of Surveying
University of Newcastle upon Tyne
Old Brewery Building
Newcastle upon Tyne NE1 7RU, UK

Gordon Petrie Professor of Topographic Science
Department of Geography and
Topographic Science
University of Glasgow
Glasgow G12 8QQ, UK

Donald M. Stirling Lecturer in Surveying and Photogrammetry
Department of Civil Engineering
City University
Northampton Square
London EC1V 4PB, UK

David A. Tait Lecturer in Surveying and Photogrammetry
Department of Geography and
Topographic Science
University of Glasgow
Glasgow G12 8QQ, UK

Contents

PART A: DEVELOPMENTS IN FIELD INSTRUMENTATION AND PROCESSING TECHNIQUES

1 Electronic angle and distance measurement   7
   T.J.M. KENNIE
1.1 Introduction   7
1.2 Principles of electronic angle measurement   7
   1.2.1 Electronic circle measuring systems   8
   1.2.2 Electronic tilt sensors   12
1.3 Principles of electronic distance measurement (EDM)   13
   1.3.1 Principles of distance measurement using phase measurements   13
   1.3.2 Errors influencing EDM measurements   17
   1.3.3 Methods of instrument calibration   20
   1.3.4 EDM instruments and applications   25
1.4 Electronic tacheometers   28
   1.4.1 Integrated design   28
   1.4.2 Modular design   28
   1.4.3 Data recorders   31
   1.4.4 Feature coding   34
1.5 Electronic coordinate determination systems (ECDS)   35
   1.5.1 Hardware   35
   1.5.2 Principles of coordinate determination using ECMS   38
   1.5.3 Applications   43
1.6 Conclusions   45
   References   45

2 Laser-based surveying instrumentation and methods   48
   G. PETRIE
2.1 Introduction   48
   2.1.1 Characteristics of lasers   48
   2.1.2 Laser construction   48
   2.1.3 Laser action   49
   2.1.4 Properties of laser radiation   50
   2.1.5 Classification of lasers   51
   2.1.6 The application of lasers to surveying   54
2.2 Distance measurement using lasers   54
   2.2.1 Interferometric methods   55
   2.2.2 Beam modulation/phase difference method   55
   2.2.3 Pulse echo method   58
2.3 Alignment and deflection measurements using lasers   65
   2.3.1 Alignment lasers mounted on optical instruments   65

		2.3.2 Specially designed and constructed alignment lasers	65
		2.3.3 Vertical alignment lasers	68
		2.3.4 Measurements of deflections using lasers	70

2.4 Laser levels 71
    2.4.1 Manually-levelled laser levels 72
    2.4.2 Laser levels employing optical compensators 72
    2.4.3 Laser levels employing electronically controlled self-levelling devices 75
    2.4.4 Laser detectors and control systems 78
    2.4.5 The application of laser levels in surveying and civil engineering 81
2.5 Lasers and safety 82
2.6 Conclusion 82
    References 82

3   North-seeking instruments and inertial systems   84
D.A. TAIT

3.1 Introduction 84
3.2 Compass devices 84
    3.2.1 Principle of compass measurement 85
    3.2.2 Compass devices used in surveying 85
    3.2.3 Applications 87
3.3 Gyroscope devices 87
    3.3.1 Principles of gyroscopic methods 87
    3.3.2 Gyroscopes applied to surveying instruments 90
    3.3.3 Methods of observation 91
    3.3.4 Applications 94
3.4 Inertial survey systems 95
    3.4.1 Principles of inertial measurement 96
    3.4.2 Construction of inertial survey systems 101
    3.4.3 Operation of inertial survey systems 101
    3.4.4 Post-processing of ISS data 105
    3.4.5 Inertial systems for surveying 105
    3.4.6 Applications 105
3.5 Conclusions 108
    References and Bibliography 108

4   Satellite position-fixing systems for land and offshore engineering surveying   111
P.A. CROSS

4.1 Introduction 111
    4.1.1 Review of satellite positioning observables 112
    4.1.2 Geodetic satellites 113
    4.1.3 Satellite positioning methods 114
    4.1.4 The role of the Earth's gravity field 115
4.2 Coordinate systems for satellite positioning 118
    4.2.1 Rotation of the Earth 118
    4.2.2 Satellite orbits 119
    4.2.3 Relationship between inertial and terrestrial coordinates 122
    4.2.4 Geodetic coordinate systems 124
4.3 Engineering satellite positioning systems 125
    4.3.1 The TRANSIT system 125
    4.3.2 The Global Positioning System (GPS) 130
    4.3.3 Models for geodetic relative positioning 137
    4.3.4 Applications of TRANSIT and GPS positioning 139
4.4 Related space geodesy methods 140
    4.4.1 Satellite laser ranging 140

	4.4.2 Satellite altimetry	141
	4.4.3 Very long baseline interferometry (VLBI)	143
	References	144

5 Analysis of control networks and their application to deformation monitoring 146
A.H. DODSON

5.1	Introduction	146
5.2	Theory of errors	147
	5.2.1 Types of error	147
	5.2.2 The principle of least squares	149
5.3	Adjustment of survey observations by least-squares method	150
	5.3.1 Observation equations	151
	5.3.2 Least-squares solution	152
	5.3.3 Rank deficiency and constraints	154
5.4	Analysis of control networks	156
	5.4.1 Reliability and tests for outliers	157
	5.4.2 Estimation of the weight matrix	159
	5.4.3 Precision estimates	159
	5.4.4 Accuracy and systematic biases	163
5.5	Network design	164
5.6	Networks for deformation monitoring	165
	5.6.1 Design of a monitoring network	166
	5.6.2 Analysis of deformation monitoring networks	166
5.7	Determination of deformation parameters	169
5.8	Applications to deformation monitoring	171
	References	172

PART B: DEVELOPMENTS IN REMOTE SENSING AND PHOTOGRAMMETRIC TECHNOLOGY

6 Remote sensing for topographic and thematic mapping 181
T.J.M. KENNIE

6.1	Introduction	181
6.2	Basic principles of remote sensing	181
	6.2.1 Characteristics of EM radiation	181
	6.2.2 Characteristics of remote sensing systems	181
6.3	Data acquisition	184
	6.3.1 Frame imaging systems	184
	6.3.2 Optical/mechanical line scanners	192
	6.3.3 Optical/mechanical frame scanners	197
	6.3.4 Linear array (pushbroom) scanners	198
	6.3.5 Microwave sensors	206
6.4	Digital image processing (DIP)	214
	6.4.1 Hardware	214
	6.4.2 Software	214
6.5	Applications of remote sensing	220
	6.5.1 Small-scale topographic mapping from satellite sensors	220
	6.5.2 Thematic mapping for engineering projects	226
6.6	Conclusions and future trends	232
	Bibliography and references	234

7 Analogue, analytical and digital photogrammetric systems applied to aerial mapping 238
G. PETRIE

7.1	Introduction: the role of aerial photogrammetric methods in engineering surveys	238
7.2	Stereoplotting instruments	239
7.3	Analogue photogrammetric instrumentation	240
	7.3.1 Optical projection instruments	240
	7.3.2 Mechanical projection instruments	242
	7.3.3 Digital data acquisition using analogue stereoplotting instruments	243
	7.3.4 Interactive graphics workstations attached to analogue stereoplotting instruments	247
7.4	Analogue instrumentation for orthophotograph production	248
	7.4.1 Optical projection instruments	248
	7.4.2 Non-projection methods of orthophotograph production	250
	7.4.3 Heights and contours obtained during the production of orthophotographs	255
7.5	Analytical photogrammetric instrumentation	255
	7.5.1 Algorithms for use with analytical instrumentation	258
	7.5.2 Comparators, monoplotters and image space plotters	259
	7.5.3 Analytical stereoplotting instruments	266
	7.5.4 Comparison of analogue and analytical plotters	271
	7.5.5 Analytically controlled orthophotoprinters	272
	7.5.6 Analytical plotters equipped with correlators	273
7.6	Digital photogrammetric instrumentation	279
	7.6.1 Digital image acquisition and processing	279
	7.6.2 Digital analytical instruments	280
	7.6.3 Digital production of orthophotographs	282
7.7	Applications of aerial photogrammetric instruments to engineering surveys	283
	7.7.1 Mapping from aerial photographs	283
	7.7.2 Terrain elevation data from aerial photographs	285
7.8	Conclusion	287
	References	287

8 Close-range photogrammetry 289
D. STIRLING

8.1	Introduction	289
8.2	Principles of photogrammetry	289
	8.2.1 Coordinate systems	290
	8.2.2 Rotation systems	291
8.3	Mathematical models used in photogrammetry	292
	8.3.1 Inner orientation	293
	8.3.2 Relative orientation	293
	8.3.3 Absolute orientation	293
	8.3.4 Collinearity equations	293
	8.3.5 Use of the collinearity equations	294
8.4	Cameras	297
	8.4.1 Metric cameras	298
	8.4.2 Stereometric cameras	300
	8.4.3 Phototheodolites	300
	8.4.4 Non-metric cameras	301
	8.4.5 Photographic media	302
8.5	Equipment for the measurement of photography	303
	8.5.1 Analogue stereoplotting instruments	303
	8.5.2 Analytical methods	305
8.6	Applications of close-range photogrammetry in engineering	308

	8.6.1	Applications in civil engineering	309
	8.6.2	Applications in mechanical engineering	317
	8.6.3	Applications in aeronautical engineering	318
	8.6.4	Applications in automobile engineering	319
	8.6.5	Applications in marine engineering	320
	8.6.6	Applications in process engineering	320
8.7	Automation of the photogrammetric process	320
	8.7.1	Digital image correlation	320
	8.7.2	'Real-time' photogrammetry	321
8.8	Conclusions	322
	References and bibliography	322

PART C: DEVELOPMENTS IN COMPUTER BASED MAPPING, TERRAIN MODELLING AND LAND INFORMATION SYSTEMS

9 Digital mapping technology; procedures and applications   329
G. PETRIE

9.1	Introduction	329
	9.1.1 The justification for digital mapping	329
	9.1.2 Relationship of digital mapping to geographic and land information systems and to digital terrain modelling	331
9.2	Data acquisition for digital mapping	332
9.3	Cartographic digitizing technology and procedures	332
	9.3.1 Manual point and line-following digitizers	335
	9.3.2 Automatic and semi-automatic line-following digitizers	342
	9.3.3 Automatic raster scan-digitizers	347
9.4	Digital map data structures	352
9.5	Digital map data processing	353
9.6	Cartographic display and plotter technology and procedures	356
	9.6.1 Graphics display devices	356
	9.6.2 Hard-copy output devices	363
9.7	Applications of digital mapping	380
	9.7.1 Digital mapping based on field survey data	380
	9.7.2 Digital mapping based on photogrammetric data	382
	9.7.3 Digital mapping at the Ordnance Survey	385
9.8	Conclusion	389
	References	389

10 Digital terrain modelling   391
T.J.M. KENNIE and G. PETRIE

10.1	Introduction	391
10.2	Data acquisition	391
	10.2.1 Ground survey methods	392
	10.2.2 Photogrammetric methods	392
	10.2.3 Graphics digitizing methods	392
10.3	Measurement patterns	393
	10.3.1 Systematic sampling	393
	10.3.2 Progressive sampling	394
	10.3.3 Random sampling	395
	10.3.4 Composite sampling	395
	10.3.5 Measured contours	396

10.4	Modelling techniques	396
	10.4.1 Grid-based terrain modelling	396
	10.4.2 Triangle-based terrain modelling	403
	10.4.3 Hybrid approaches to terrain modelling	407
10.5	Software and applications of terrain modelling in topographic mapping and civil engineering	408
	10.5.1 Grid-based packages	408
	10.5.2 Triangular-based packages	411
	10.5.3 Hybrid packages	419
10.6	Conclusions	425
	References	425

11 Land information databases   427
D. PARKER

11.1	Introduction	427
11.2	Spatial databases	428
11.3	User's spatial data model	430
	11.3.1 Topological data model	431
	11.3.2 Representation and description of features	434
	11.3.3 Simplified data models	436
	11.3.4 Representation of position (vector or raster)	439
	11.3.5 Information retrieval and analysis	441
11.4	Computer data structures	442
	11.4.1 Computer records and files	442
	11.4.2 Database techniques	448
11.5	Vector data storage	454
11.6	Raster data storage	457
	11.6.1 Data compaction methods	458
	11.6.2 Quadtrees	459
11.7	Land and geographic information systems	460
	11.7.1 Slimpac	460
	11.7.2 Arc/Info	461
	11.7.3 System 9	463
11.8	Applications of land and geographic information systems	468
	11.8.1 British telecom line plant graphics (LPG)	468
	11.8.2 South Australia Land Information System	474
	11.8.3 Connecticut GIS project	474
11.9	Conclusion	476
	References	476

Glossary of abbreviations   478

Index   481

PART A

DEVELOPMENTS IN FIELD INSTRUMENTATION AND PROCESSING TECHNIQUES

Part A of the book aims firstly, to consider the main technological changes which have taken place in the design of field surveying equipment in recent years, and secondly, to examine the mathematical techniques used to process field data for control surveys and deformation analysis.

The growth of the electronics industry and the development of microprocessors have had an enormous impact on the design of traditional field surveying equipment. These changes are discussed in Chapter 1. Not only have these changes enabled the standard angle and distance measuring procedures to be automated, but they have also provided the capability for on-site storage and processing facilities. These developments have had their greatest influence in large-scale mapping operations. Automated methods of recording digital field survey data using field data loggers are now one of the most important sources of input data for the digital mapping and land information systems to be discussed later in Part C. Also, the ability to perform basic survey computations, such as distance resection or polar coordinate calculations, in the field using stored coordinate data, now enables setting out to be carried out more flexibly.

High-precision measurement systems based on the use of two or more electronic theodolites linked on-line to a microcomputer have also been developed, although in practice their use has been slow to develop. By contrast, as already mentioned, although automated *recording* of field data is now commonplace, little progress has been made towards the production of automated field *measurement* equipment using unmanned surveying systems. Clearly there is a need for such devices, since they would enable more emphasis to be placed on the structuring and classification of field measurements rather than on observational procedures.

The surveying technology based on the use of lasers (Chapter 2) is now highly developed, and can no longer be considered as 'a solution in search of a problem'. Lasers operating in the visible part of the spectrum are now considered as the standard method to use for alignment purposes in projects ranging in size from the Channel Tunnel to the laying of small-diameter pipelines on building sites. Lasers are also widely used for high-precision deflection measurement (for example of dams).

Electronic distance measurement using laser sources of radiation and phase difference methods is now also a well established technology. However, more recently, the 'timed pulse' approach to distance measurement has been developed, particularly for offshore applications. In particular, the ability of such laser-based instruments to measure long ranges to passive targets (non-reflective surfaces) has proved to be a significant benefit. At short range, this characteristic has also been exploited for tunnel profiling.

On the other hand, while the methods for distance, angle, alignment and deflection determination have undergone dramatic changes in the recent past, the technology for determining height differences has remained virtually unchanged, with the notable exception of laser levelling systems. Yet, while laser levels instruments are now in widespread use on construction sites, they have not been readily accepted as a method of establishing height control, or as an alternative to precise levelling over short distances. However, recent research indicates that such operations may be quite feasible using such equipment.

The measurement of orientation using North-seeking instruments (Chapter 3) is fundamental to many engineering activities, ranging from the orientation of shafts in mining engineering to the deviation of boreholes in offshore oil exploration. By contrast, inertial surveying systems (ISS) have yet to achieve a prominent position in engineering surveying, despite their independence from refraction, line of sight, weather conditions and time of measurement. The main limitations to their widespread use at present, are high cost, the bulky nature of the equipment, and for some applications the accuracy which is achievable. However, for military surveying applications, the main attraction of ISS is their independence from external influences (such as breaks in transmission, or electronic jamming). Nevertheless, at present, ISS can only be considered economical in a limited number of cases, for example, in reconnaissance surveys in remote locations, or in situations where considerable amounts of data, often at high speed, are required. For the future, it may well be that a breakthrough in the hardware design of ISS systems, such as ring lasers, may provide the impetus to the more widespread application of this technology.

Yet another method of determining absolute and relative 3D positions is by measurements to, and from, artificial satellites. Both the theoretical background to these 'black box' devices and a description of some of the more important operational systems such as TRANSIT and the Global Positioning System (GPS) are discussed in Chapter 4. It is considered by many that GPS in particular will revolutionize the traditional methods of providing planimetric and height control for detail surveying. No longer will control points be based on a hierarchical series of control stations, generally located on high ground for plan control and on lines of communication for height. By contrast, 3D control point positions will be produced independently by measurements from GPS satellite transmitters. Although such operations are currently being performed to centimetric accuracy, the high cost of the receivers and the limited 'windows' available for recording measurements are limiting the more general application of the system. GPS is in use for a wide range of positioning applications, ranging from the measurement of the subsidence of North Sea oil platforms to the provision of Ordnance Survey (OS) control points for urban mapping. However, the uncertainty as to the degree to which the system will continue to be 'open' to non-military users is a major concern for the future.

Thus the modern engineering surveyor has an extraordinarily wide range of instruments available for position determination, each with its own specific merits and limitations. The choice as to which instrument to use will depend on factors such as the accuracy requirements of the project, the distances over which measurements will be required and, inevitably, the financial implications of using each technique. Before discussing the various techniques in the chapters which follow it is useful to be aware of their comparative performance. Figure A1 below illustrates graphically, in broad terms, the variations in accuracy which are achievable with each system.

However, the full potential of all of these techniques can only be fully exploited if a rigorous mathematical model is used to determine the 3D position. Chapter 5 therefore examines the use of least-squares estimation in the process of position determination. Furthermore, the use of such techniques is no longer restricted to organizations which possess large, expensive computing facilities, since most networks for engineering projects can now be analysed on relatively inexpensive microcomputers. However, it is the added benefit of being able to assess the precision, accuracy and reliability of the data which is equally significant. These analytical techniques are also particularly important when attempting to detect displacement or deformation of an engineering structure. Indeed, as the range of measurement techniques becomes greater, the need for more sophisticated mathematical models which enable different types of data (often of varying precision) to be combined will continue to increase.

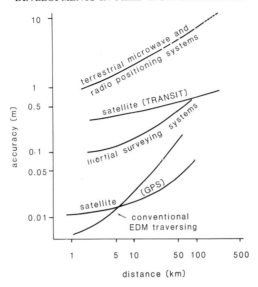

Figure A1 Comparative performance of terrestrial and satellite-based position-fixing systems (adapted from Stansell and Brunner, 1985).

Reference

Stansell, T.A. and Brunner, F.K. (1985) The first Wild-Magnavox GPS satellite surveying equipment: WM 101. *Proc. FIG Study Groups 5B/5C Meeting on Inertial, Doppler and GPS Measurements for Networks and Engineering Surveying*, Munich, 503–523.

1 Electronic angle and distance measurement

T.J.M. KENNIE

1.1 Introduction

The measurement of angles and distances is central to almost all land surveying activities, since from these observations it is possible to derive the relative positions and heights of points. Historically, angles have been measured using theodolites equipped with engraved metal or glass circles, and accuracies of a single second of arc have been achieved with such instruments. Recent developments in microelectronics, however, now enable these reading mechanisms to be replaced by electronic components. Also, the inclusion of microprocessors into the instrument design now allows a less rigorous observational procedure to be adopted without reducing the accuracy of the angular measurement. The basic principles and construction of these electronic theodolites are discussed in section 1.2.

Similarly, methods of distance measurement have changed dramatically in the recent past. Section 1.3 therefore examines the basic principles of measurement using electromagnetic waves. This section also examines the vitally important process of calibration. This has become of even greater significance with the birth of the surveying 'black box', an instrument which automatically outputs numerical values of distance and angle. Methods of independently checking the accuracy and integrity of such data are of crucial importance if systematic measurement errors are to be avoided.

The development of electronic surveying instruments which enable three-dimensional coordinates to be derived, either in the field or in the laboratory are also discussed. In the first case, the ability to electronically measure and store data has had a major impact on productivity levels in surveying. In the second case, the determination of coordinates to an accuracy of a fraction of a millimetre, over short ranges, has led to the greater use of surveying in the field of industrial measurement.

1.2 Principles of electronic angle measurement

The overall design of the traditional optical theodolite has remained virtually unchanged for the past 25–30 years. Although over this period improvements in materials and optical machining tolerances have enabled the instruments to become smaller and lighter and the optics more efficient, the general design has not altered significantly. A major change has, however, occurred in recent years with the introduction of electronic circle reading systems and electronic tilt sensing systems into the design of the theodolite. While electronic theodolites which incorporate such features have been available for the past 10–15 years, it is only in the 1980s that the cost of these instruments has become comparable with their optical counterparts, and consequently they have become more common in engineering surveying applications.

1.2.1 *Electronic circle measuring systems*

Most automatic circle measuring systems used in modern electronic theodolite designs are based on the use of a photolithographically coded circle and a photoelectric detector element or photodiode. Two measurement systems are used: absolute and incremental. Absolute systems provide unique angular values, normally by means of a coded circle; incremental designs, in contrast, normally involve measurement relative to a zero or index mark.

The concept of electronic angle measurement can be best illustrated by initially considering a glass circle on which a series of opaque sections has been photographically etched. If a light source is used to illuminate the glass circle, and a photodiode, with a width equivalent to one section, is used as an index mark, it is possible to measure the number of light and dark units which pass below the diode as the alidade is rotated. Hence the angle which has been turned can be determined. Unfortunately the degree of accuracy which can be achieved with such a single-track system is very limited. For example, assuming that the photodiode had an area of $0.5\,\text{mm}^2$ and the circle diameter was 80 mm, then the maximum number of transparent and opaque sections which can be accommodated is approximately 500, providing an angular resolution of about 40′ of arc.

An alternative to employing a single track of transparent and opaque sections is to use several concentric tracks. If each transparent section then represents an 'on' signal and is designated as the number 1, and each opaque section represents an 'off' signal and is represented by 0, then it is possible to represent circle readings as individual binary codes. Again, however, practical considerations restrict the use of this technique for high resolution. Firstly, it is essential that all the detector elements are precisely aligned, since a small error in the alignment of the detector (particularly when reading the innermost tracks) could have serious consequences on the accuracy of the circle reading. Secondly, the interval between successive transparent and opaque sections becomes extremely small, and hence difficult to manufacture, at high resolutions. The magnitude of this problem is illustrated by Table 1.1, which lists the number of tracks and the distance between sections on the innermost track for various angular resolutions. Even to achieve a direct angular resolution of 20″ of arc the distance between elements on the innermost track must be less than $2\mu\text{m}$. For higher resolutions, the problem becomes even more acute. Various designs have been proposed to overcome these problems (Gorham, 1976; Rawlinson, 1976). One of the most common techniques to be used for high-accuracy interpolation involves the use of moiré fringe patterns. Two typical instruments which use this approach are the Kern E2 and Wild T-2000 electronic theodolites.

1.2.1.1 *Kern E2 system.*
The Kern E2 electronic theodolite (Fig. 1.1) was launched in 1981 at the Montreux Congress of FIG (Fédération International de Géomètres), and is a development of an earlier design, the Kern E1.

The horizontal circle measuring system is an incremental design based on resolving, dynamically, a coarse measurement to about 30″ of arc, and, statically, a fine

ELECTRONIC ANGLE AND DISTANCE MEASUREMENT

Table 1.1 Pure binary coded electronic circle: distance between elements at varying angular resolutions (assuming circle radius of 80 mm and a track width of 1 mm)

Resolution	1'	20"	1"	0.1"
No. of sectors	2160	64 800	1 296 000	12 960 000
No. of tracks	12	16	20	24
Distance between elements on the innermost track (m)	1×10^{-4}	2×10^{-6}	1×10^{-8}	1×10^{-8}

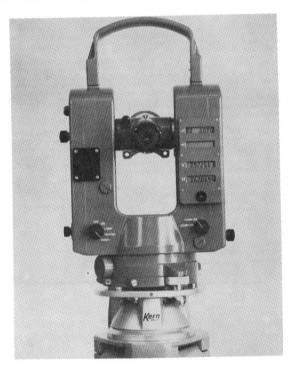

Figure 1.1 Kern E2 electronic theodolite. Courtesy Kern (Aarau) Ltd.

measurement to the nearest 0.3". Both measurements are performed by using a moiré fringe pattern. Interference patterns of this type can be formed when closely ruled parallel lines are superimposed at a slight angle to each other. In the case of the Kern E2, the pattern is formed by focusing a slightly magnified section of one side of the circle on to an identical section from the region of the circle which is diametrically opposite. Since the circle consists of some 20 000 opaque and transparent lines approximately 6μm wide, a moiré pattern is formed (Fig. 1.2). Four photodiodes monitor the sinusoidal brightness distribution of the moiré pattern, and the outputs are used to obtain a coarse and a fine measurement.

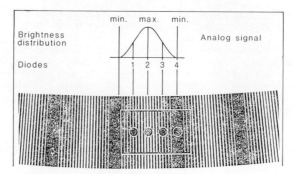

Figure 1.2 Kern E2 electronic angle measurement: formation and measurement of moiré fringe pattern.

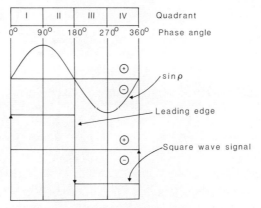

Figure 1.3 Conversion of sinusoidal output from photodiodes to a square wave and measurement of leading edge.

The coarse component of the circle reading is obtained by measuring the rotation of the array of photodiodes relative to the circle. This process is carried out by converting the dynamically measured sinusoidal output from the photodiodes into a square wave signal (Fig. 1.3) and then counting the number of leading edges (changes from negative to positive status) which occur during the transition from the initial to the final pointing. By this process it is possible to derive the angular change to approximately 30″ of arc.

The fine measurement is performed when the photodiodes are stationary in relation to the circle. The fine measurement aims to determine, by analysing the output from the four photodiodes, the location of the measuring point in relation to the leading edge of the last moiré period. The interpolation is carried out by measuring the phase angle of the moiré signal. This is achieved by processing the output from the photodiodes which are spaced at intervals of 0.25 of the moiré period (90°). Using the outputs from diodes 1 and 3, and from 2 and 4 and the process described by Munch (1984) it is possible to discriminate the position of the measuring mark to within 1% of the moiré period or to approximately 0.3″ of arc.

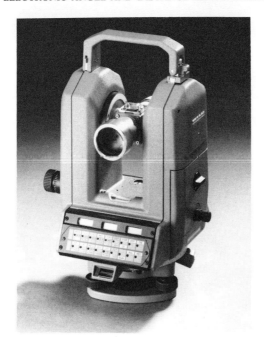

Figure 1.4 Wild T-2000 electronic theodolite. Courtesy Wild (Heerbrugg).

The vertical circle reading system is similar to that described above. However, in order to ensure that the vertical angles are measured with respect to a horizontal plane (and not relative to the arbitrary position of the telescope when the instrument is switched on) it is necessary to initially electronically orient the circle. This is achieved by an electronic zero pulse which is identified by the instrument when the telescope is rotated through the horizontal.

Accuracy testing, using the technique described by Maurer (1984) indicates that the extent of any circle graduation errors arising from the use of the electronic system described is less than 0.5″ of arc.

1.2.1.2 Wild T-2000 system. The Wild T-2000 electronic theodolite (Fig. 1.4) uses a dynamic, circle scanning technique to derive horizontal and vertical angles. In this case the circle is divided into 2096 equally spaced opaque and transparent segments (or 1048 phase units, ϕ_0) as illustrated by Fig. 1.5. A drive motor rotates the glass circle past two pairs of photodiodes which are positioned diametrically opposite each other (and thus enable circle eccentricity errors to be minimized). Figure 1.5 shows the case for one pair of photodiodes. One diode L_F is fixed, and is analogous to the zero mark of a conventional theodolite. The other, movable diode, L_M, records the direction of the telescope.

The determination of the angle ϕ between the two diodes is carried out by measuring the phase difference between the signals L_F and L_M. If, for example, the two signals are in phase, then the angle will consist of an unknown integer number (n) of phase units (ϕ_0). If, however, the signal from L_M is displaced relative to that from L_F, then the angle

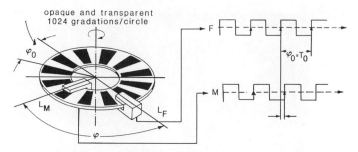

Figure 1.5 Wild T-2000 electronic circle measuring system.

will be defined as

$$\phi = n\phi_0 + \Delta\phi \tag{1.1}$$

where $\Delta\phi$ is the phase difference between the two signals. In order to resolve ϕ it is therefore necessary to derive a value for n, the coarse measurement, and a value for $\Delta\phi$, the fine measurement. Both measurements are carried out simultaneously after the circle has reached its specified rotation speed.

The coarse measurement is carried out by an electronic counter which sums the number of periods of the phase unit ϕ in the angle being measured. Since each phase unit is equivalent to a time interval (T_0) of about $300\,\mu s$, this measurement can be obtained directly and enables the angle to be resolved to the nearest 20 minutes of arc. The fine measurement is also carried out using an electronic counter in order to resolve $\Delta\phi$. Over 500 measurements (alternately from opposite sides of the circle) of $\Delta\phi$ are evaluated during a single rotation of the circle. The time taken for the entire process of obtaining a single measurement is 0.6 s, half of which is used to accelerate the motor to its design speed. A full description of the circle measuring system and of the laboratory tests to assess the accuracy of the instrument can be found in Katowski and Saltzmann (1983).

1.2.2 *Electronic tilt sensors*

A second unique feature of electronic theodolites compared with their optical counterparts is the inclusion of electronic tilt sensing devices. Components of this type can be used to automatically compensate horizontal and vertical angles for any residual inclinations of the vertical axis of the theodolite. Not only do these devices increase the accuracy of the instrument, but they also enable corrections to be applied directly to single face observations, so eliminating, for all but the most precise measurements, the need to record observations in the face left and face right positions of the telescope. Compensators may be referred to as single or dual axis devices. In the first case their function is normally only to correct vertical angles, whereas in the second case, corrections to horizontal angles may also be applied. The dual axis tilt sensor used by the Kern E2 is illustrated by Fig. 1.6(*a*) and (*b*).

The compensation system in this design is based on the reflection of a light spot from a liquid surface. The surface of the liquid will always remain horizontal irrespective of the inclination of the instrument, and can therefore be used as a reference surface. By redirecting the light source through various optical components and on to a photodiode, it is possible, by measuring the induced current, to determine the

ELECTRONIC ANGLE AND DISTANCE MEASUREMENT

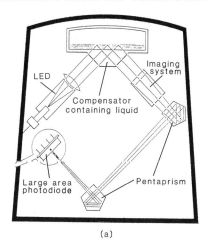

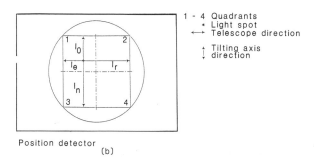

Figure 1.6 Kern E2: (*a*) dual axis electronic tilt sensor; (*b*) position detector (courtesy Kern Aarau).

position of the reflected light in relation to the zero point of the compensator. Values can then be derived for the inclination of the theodolite along the line of collimation and also along the trunnion axis of the instrument.

1.3 Principles of electronic distance measurement (EDM)

1.3.1 *Principles of distance measurement using phase measurements*
Although a wide variety of EDM instruments are currently available, most derive distances by measuring either the time delay between pulses of EM energy, normally referred to as the *pulse echo* technique, or by comparing transmitted and received signals and deriving the *phase difference* between them. The remainder of this section will deal specifically with instruments which utilize the latter technique. The former technique is normally restricted to laser-based instruments, and further details can be found in Chapter 2.

The fundamental stages involved in the determination of a distance electronically can be represented by a flow diagram (Fig. 1.7). The basic EM signal generated by the

14 ENGINEERING SURVEYING TECHNOLOGY

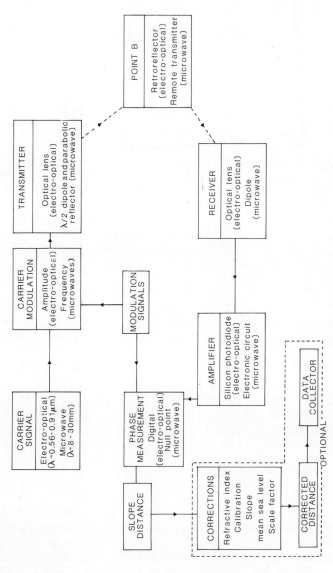

Figure 1.7 Flow diagram illustrating principle of operation of an EDM instrument.

ELECTRONIC ANGLE AND DISTANCE MEASUREMENT

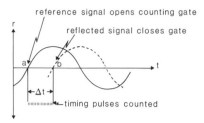

Figure 1.8 Phase measurement using digital pulse counting (Burnside, 1982).

instrument is termed the carrier signal, and may be of visible, infrared or microwave wavelengths. The first two cases may be termed *passive* EDM instruments, in which the EM signal is reflected from an unmanned reflector, whereas the latter forms the basis of *active* EDM instruments in which the return signal is generated by a second manned instrument at the terminal point.

Since the wavelength of the carrier signal is very short (0.5 μm–30 mm), in order to ensure that the phase difference can be measured precisely, a measurement or modulation signal is superimposed on to the carrier signal. This process is termed 'carrier modulation'. The transmitted signal is then either reflected or retransmitted along an identical path back to the instrument. With electro-optical devices, reflection of the signal is achieved by means of precisely manufactured retroreflective prisms, also referred to as corner cube reflectors. For short-range operations, acrylic retroflectors may also be used (Kennie, 1983, 1984). Reflectors of this type return the incident signal along an identical path back to the instrument. Generally speaking, alignment of the reflector is not a critical problem, although precise alignment of the measuring instrument is essential. In the case of microwave instruments, the signal is received by a second 'remote' instrument and retransmitted back to the master instrument.

The weak reflected signal is then amplified and compared in phase with the original transmitted beam. The difference in phase between the two signals is derived using either a digital pulse counting approach (electro-optical instruments) (Fig. 1.8) or a null point approach (microwave instruments) in which the phase difference between the two signals is brought to zero by introducing an additional phase shift (the required phase shift being indicative of the original phase difference).

The measurement of the phase difference between the two signals is representative of the fractional part of the total distance less than the integer value of the modulation wavelength. Figure 1.9 graphically represents the measurement of the distance AB. It can be seen that the distance AB is given by

$$D = M\lambda + \Delta\lambda \tag{1.2}$$

where M is an integer (2 in this case), and $\Delta\lambda$ is a fraction of a wavelength. Both M and $\Delta\lambda$ are unknowns. As mentioned previously, the signal is normally reflected back to the instrument, resulting in

$$2D = N\lambda + \Delta\lambda \tag{1.3}$$

where N is an integer (4 in this case).

This is equivalent to N revolutions of the vector OA (Fig. 1.9) plus the fractional excess equal to angle ϕ. Angle ϕ is a measure of the phase difference between the

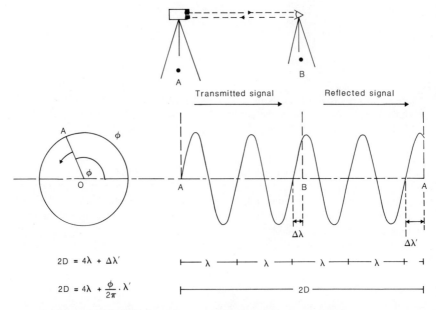

Figure 1.9 EDM distance determination: double path measurement.

outgoing and reflected signals and can be used to derive a value for

$$\Delta\lambda = \frac{\phi}{2\pi}\lambda \tag{1.4}$$

thus

$$D = N\lambda/2 + \frac{\phi}{2\pi}\frac{\lambda}{2}. \tag{1.5}$$

This is a rather idealized model, and equation (1.6) describes the situation more precisely.

$$D = N\frac{C}{2fn_a} + \left[\frac{\phi}{2\pi} \times \frac{C}{2fn_a}\right] + k_2 + k_3 \tag{1.6}$$

where C is the velocity of the EM signal in a vacuum
f is the modulation frequency (assumed error-free)
n_a is the refractive index of the atmosphere (see section 1.3.2.2)
k_2 is the zero error of the instrument (see section 1.3.2.3)
k_3 is the cyclic error of the instrument (see section 1.3.2.3).

The integer N in equations (1.3), (1.5) and (1.6) cannot be derived directly by a single measurement, and is normally obtained by introducing successive changes to λ, which lead to a series of equations of the general form

$$D = \overbrace{N_1}^{\text{unknown}}\overbrace{\frac{\lambda_1}{2}}^{\text{known}} + \overbrace{\left(\frac{\phi_1}{2\pi} \cdot \frac{\lambda_1}{2}\right)}^{\text{measured}} \tag{1.7}$$

$$D = N_2 \frac{\lambda_2}{2} + \left(\frac{\phi_2}{2\pi} \cdot \frac{\lambda_2}{2}\right) \tag{1.8}$$

$$D = N_3 \frac{\lambda_3}{2} + \left(\frac{\phi_3}{2\pi} \cdot \frac{\lambda_3}{2}\right) \tag{1.9}$$

In the case of microwave instruments, the solution to these equations is obtained by manually measuring $\Delta\lambda$ using five values of λ which increase progressively by a factor of 10. Dodson (1978) provides the following example to illustrate this principle:

λ(m)	$\lambda/2$(m)	$\Delta\lambda = \frac{\phi}{2\pi} \cdot \frac{\lambda}{2}$ (m)
2	1	.1243
20	10	6.124
200	100	76.12
2 000	1 000	376.1
20 000	10 000	2376.0

Distance = 2376.1243 m

With modern electro-optical instruments it is not necessary to manually alter the value of λ. It is changed automatically in a continuous manner, and phase difference values are measured several hundred times on both an internal calibration path and over the external light path.

For very high-precision distance measurement, the principle of measurement differs in some respects from that described so far. The Kern Mekometer ME 3000 is a typical example of this category of instrument.

The Mekometer was developed in the early 1960s at the United Kingdom National Physical Laboratory (NPL), was subsequently manufactured commercially by Kern of Switzerland and became available as a production model in the early 1970s. Two models are currently in use, the ME3000 and the ME5000, the latter being a recent development which uses a He-Ne laser.

The Kern Mekometer ME 3000 is one of the few instruments which uses a xenon flash tube as the source of the carrier signal. Also, it uses polarization modulation, with an extremely high modulation frequency and short modulation wavelength (30 cm). Furthermore, to reduce the effect of atmospheric variations on the measurements, the modulation wavelength is fixed by the resonance of a small microwave resonator containing air under pressure. It is a well-established instrument and has been used throughout the world for deformation monitoring projects. For example, Murname (1982) and Keller (1978) have used the instrument for monitoring reservoirs and arch dams respectively. A full account of the principle of measurement using EDM instruments can be found in Burnside (1982) and Rüeger (1988).

1.3.2 *Errors influencing EDM measurements*

Many different sources of systematic error can affect EDM measurements, including those caused by the instrument operator, the atmosphere and instrument maladjustment.

1.3.2.1 *Instrument operator errors.* Clearly it is important that, for reliable, precise results, due care and attention should be paid to the operation of an EDM instrument.

In particular, care should be taken with the following operations:

(i) Precise centring—regular checking of the optical plummet of a theodolite (if the EDM is theodolite-mounted), or of an integral optical plummet, is important; inaccurate centring is often one of the most common reasons which account for distance variations occurring between points measured using different EDM instruments

(ii) Pointing—careful pointing to the correct position on a reflector target is particularly important when using an EDM which has the facility for automatic slope reduction, since any pointing error may not become evident, particularly if the results of the slope reduction are not independently checked using, for example, the observed vertical angle

(iii) Setting automatic correction values—for instruments which have the facility to store atmospheric and other correction factors, it is essential to ensure that the correct values for the prevailing conditions are being used.

1.3.2.2 *Atmospheric errors.* The effect of the prevailing atmospheric conditions (temperature, pressure and relative humidity) on EDM measurements is analogous to the influence of temperature when measuring distances with a steel tape. Variations in temperature cause an expansion or contraction of the tape from its standardized length, which, if not corrected, will lead to a systematic error in the measured distances. Similarly, variations in atmospheric conditions from those which are assumed by the instrument will cause an increase or reduction in the measuring wavelength and lead to a systematic error if a correction is not applied.

Normally the effect of the atmosphere is defined by the change in the refractive index (n), where n is defined as

$$n = \frac{c}{v} \tag{1.10}$$

where c = velocity of electromagnetic waves in a vacuum (299 792.5 km/s)
v = the velocity in the medium (atmosphere).

The value of n is normally close to unity and is assumed by many EDM instruments to be 1.000320. In the visible and infrared parts of the EM spectrum, n is calculated using the Barrell and Sears formula:

$$(n_t - 1) = (n_s - 1)\frac{273}{273 + t} \cdot \frac{P}{760} - \frac{15.02E}{273 + t} \times 10^{-6} \tag{1.11}$$

where

$$(n_s - 1) \times 10^{-6} = 287.604 + \frac{1.6288}{\lambda_0^2} + \frac{0.0136}{\lambda_0^4} \tag{1.12}$$

and n_t = prevailing refractive index
n_s = refractive index of air at 0°C and 760 mm Hg pressure with a 0.03% CO_2 content
t = temperature during observation (°C)
P = pressure during observation (mm Hg)
E = water vapour pressure (mm Hg)
λ_0 = wavelength of signal in a vacuum.

Table 1.2 Precision required in measurements of t, P and E to obtain 1 ppm precision in n (after Burnside, 1982)

	Precision in measurement		
Carrier signal	$t(°C)$	$P(mm\,Hg)$	Humidity $(mm\,Hg)$
Microwave	0.8	±2.9	±0.17
Visible/infrared	1.0	±2.7	±20

Equation (1.11) is normally used to produce a nomogram or slide rule by the manufacturers which illustrates the variation in n (in parts per million of the distance, ppm) for various values of t and P. The variation can then be entered into the instrument as a scale correction.

Although equation (1.11) physically models the variations in n to an accuracy of 1×10^6, it is nevertheless based on observed values of t, P and E (normally at the terminal points of the line being measured). It is therefore essential that care is taken to accurately measure these quantities. Burnside (1982) has shown that to obtain a precision of 1ppm in the determination of n, the three quantities must be measured to the degrees of precision shown in Table 1.2.

Table 1.2 clearly indicates (i) that temperature and pressure need to be determined to about 1°C and 3 mm Hg, and (ii) that microwave instruments are very susceptible to changes in relative humidity and that this quantity should be measured to a much higher precision than for electro-optical instruments.

1.3.2.3 *Instrument errors.* EDM devices, like other surveying instruments, require careful use and regular calibration if they are to provide reliable and accurate results. Although it is rare for a modern EDM instrument to display results which are grossly incorrect (most instruments normally do not display any results if severely out of adjustment), most instruments are subject to relatively small but vitally important systematic errors caused by instrumental maladjustment.

Scale errors (k_1) occur if the modulation frequency of the EDM instrument does not correspond exactly with the design frequency value for the instrument. The error is proportional to the distance measured and may often be expressed in parts per million (ppm) of the distance. Ideally, it should be negligible (<1 ppm); however it may be as high as 20 to 30 ppm (20–30 mm/km) in extreme cases.

Zero error (k_2) (also referred to as additive constant error, index error or reflector/prism offset) represents the difference between the EDM measured distance between two points (corrected for scale, cyclic and atmospheric errors) and the known distance between the two points. It is caused if the internal measurement centre of the instrument (and reflector) does not coincide with the physical centre of the instrument/reflector which is plumbed over the measuring mark. It also accounts for the variation in the path length through the retroflector caused by refractive index variations at the glass–air interface. It is a systematic error of constant magnitude and therefore *not* proportional to the distance measured.

The value of zero error obtained during calibration refers to a particular EDM instrument and retroflector. If either the instrument or reflector are changed, then a new zero constant will exist.

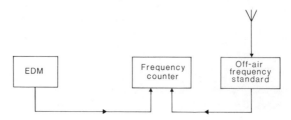

Figure 1.10 Laboratory determination of scale error using standard frequency.

Cyclic errors (k_3) (also termed periodic errors, resolver errors or non-linearity errors) are analogous to circle graduation errors in theodolites in the sense that they are periodic and can have a systematic effect, particularly at short distances. They are generally the smallest of the three sources of instrument error and are caused by internal electronic contamination between the transmitter and receiver circuitry, the effect being cyclic over the modulation wavelength. The effect of this source of error is reduced by the instrument manufacturers by electrical isolation and shielding of components within the instrument, although the effect becomes more apparent with a weak return signal (Bevin, 1975).

1.3.3 *Methods of instrument calibration*

In general terms the distinction can be made between the need for quick and simple calibration tests which could, for example, be carried out on site, and more sophisticated calibrations (normally involving the use of a laboratory or permanently sited baseline) for more accurate determinations of the state of health of an EDM instrument.

Three distinct approaches to instrument calibration can be identified: laboratory methods, field methods using known baseline lengths, and field methods using unknown baseline lengths.

1.3.3.1 *Laboratory methods.* Laboratory tests can be carried out to assess both the scale and cyclic error components. Their main advantages over field tests are, firstly, that they are quick and convenient to apply in practice, and secondly, they enable the surveyor to precisely control the measurement conditions. For practical reasons, however, they are limited to short-range calibration measurements.

Scale error is determined in the laboratory by comparing the observed modulation frequency of the EDM signal with a true reference frequency (Fig. 1.10). The reference frequency may be obtained either from off-air radio transmissions, or from a crystal generated laboratory standard.

Cyclic errors can be determined in the laboratory by recording distance measurements to a series of reflector positions which extend over a distance equivalent to the halfwavelength of the instrument. The simplest and cheapest arrangement involves the use of a graduated metal bar (Hodges, 1982).

The calibration bar is constructed so that the prism can be moved in increments of 10 mm (with an accuracy better than 0.02 mm). The bar is placed approximately 40 to 50 m from the EDM instrument to minimize pointing errors. If the effects of scale error are presumed to be negligible over such short distances, then each measurement to the bar (L^M) will be subject only to zero and cyclic errors, so that

$$L^M = L^T + k_2 + k_3 \tag{1.13}$$

where L^M is measured distance
L^T is true distance (unknown)
k_2 is zero error
k_3 is cyclic error.

For any other point i,

$$L_i^M = L_i^T + k_2 + k_{3i}, \tag{1.14}$$

Subtracting equation (1.13) from (1.14) to eliminate constant zero error:

$$\Delta L^M = \Delta L^T + (k_3 - k_{3i}) \tag{1.15}$$

or

$$(\Delta L^M - \Delta L^T) = (k_3 - k_{3i}) \tag{1.16}$$

A graph may then be drawn illustrating the differences between the distances (ΔL) at various values of L. The resulting graph will then illustrate the amplitude and phase of any cyclic error.

In situations where space is limited, the arrangement described by Deeth et al. (1979) may be adopted. A measurement range of 50 m was obtained by bending the path of the EDM signal using surface-silvered mirrors. The calibrated bar in this case was an 11 m track on which a rectroreflector was mounted. A laser interferometer (Hewlett Packard 5526A) provided a precise output of the movement of the reflector which could then be compared with that obtained using the EDM instrument.

1.3.3.2 *Field methods using unknown baseline lengths.* Measurements on a calibration baseline can be used to determine all three error components (scale, zero and cyclic errors). Both unknown and known baseline lengths may be used, although the former tends to be the more popular technique (Kennie et al., 1988). The main *advantage* of the unknown-baseline-length approach is that, although stability of the markers or pillars indicating the baseline is important, the points are required to be stable only for the duration of the survey observations. If a known baseline length is used, long-term pillar stability and periodic monitoring of the interpillar distance are particularly important. The main *disadvantage* of the unknown baseline length method is that it is not possible to derive a value for scale error by field observation. Two techniques which are based on the use of an unknown-length baseline are the three-point and Schwendener methods.

The simplest procedure for a quick and approximate determination of the zero error involves the use of a three-point baseline (Dodson, 1978), illustrated in Fig. 1.11. If the observed distances (l) are each in error by the same zero error (k_2), and if the actual distances from 1 to 2 and 2 to 3 are d_1 and d_2 then, ignoring all other errors,

$$l_{12} = d_1 + k_2 \tag{1.17}$$

$$l_{23} = d_2 + k_2 \tag{1.18}$$

$$l_{13} = d_1 + d_2 + k_2 \tag{1.19}$$

therefore

$$k_2 = l_{12} + l_{23} - l_{13} \tag{1.20}$$

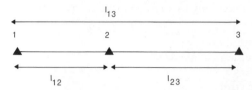

Figure 1.11 Zero error determination using three-point baseline method (Dodson, 1978).

If the baseline consists of more than two sections and each section l is measured, then if the total length is (L), a more precise value of k_2 can be derived using

$$k_2 = \frac{L - \sum l_i}{n - 1} \tag{1.21}$$

where L is the total length of baseline
l_i is the length of each baseline section
n is the number of baseline sections.

The standard error of the zero constant is given by

$$\sigma k_2 = \frac{(n+1)\sigma_d}{(n-1)^2} \tag{1.22}$$

where σ_d = standard error of the measured distances.

A further improvement in accuracy can be achieved by measuring not only the individual section distances, but all possible combinations of distances on a baseline, and using least squares methods to derive the most probable values for the zero error and (in some cases) cyclic error. One of the most common methods used is that devised by Schwendener (1972).

The Schwendener method improves the determination of k_2 by making use of a seven-point baseline (six sections), which, taking all combinations of distances, results in a total of 21 measurements. Since seven are required for a unique solution, the solution is overdetermined, with 14 redundant measurements.

The particular baseline design and interpillar distances suggested by Schwendener are illustrated in Fig. 1.12, and a typical pillar design and centring system in Fig. 1.13. The choice of interpillar distances is of vital importance and should, firstly, ensure that the distances to be measured are spread over the measuring range of the instrument, and secondly, that the fractional elements of the interpillar distances are evenly distributed over the half wavelength of the modulation signal (normally 10 m). Schwendener's basic observation equation is of the form

$$\left(\sum l_j\right)_i + k_2 = M_i + v_i \tag{1.23}$$

where l_j are the unknown interpillar distances $(l_{01} \ldots l_{06})$
k_2 is the zero error
M_i are the observed values (21 distances)
v_i are the residuals (21 values).

A tabular form of solution to determine the most probable values of l_{01}, l_{02}, l_{03}, l_{04}, l_{05}, l_{06}, and k_1 is presented in Schwendener (1972), together with equations to

ELECTRONIC ANGLE AND DISTANCE MEASUREMENT

```
   0   1    2      3        4          5            6
   o─o──o───o──────o────────o──────────o─────────o
   19.5 39.0 68.0  127.5    256.0      511.5        Total length 1021.5m
```

Distances measured

01	12	23	34	45	56
02	13	24	35	46	Total 21
03	14	25	36		
04	15	26			
05	16				
06					

Figure 1.12 Schwendener baseline design and inter-pillar distances (Schwendener, 1972).

Figure 1.13 (*a*) Typical pillar design for multi-point baseline; (*b*) Wild centring system. Courtesy Paisley College.

determine the standard error of a single measured distance and the zero error. An indication of any cyclic errors can be obtained by plotting the residuals (most probable values of the distances — observed values) as a function of the unit metres and fractions of a metre of the individual measured distances (Fig. 1.14). Although the method provides useful results, it should be noted that the cyclic error curve illustrates not only the effect of cyclic errors, but also the influence of random observational errors. Consequently, if the cyclic error is not significantly greater than the random variations, it may not be determined by this procedure.

1.3.3.3 *Field methods using known baseline lengths.* The use of known values for the length of the baseline and baseline sections is advantageous since it enables the value of the scale error to be derived in addition to zero and cyclic errors. A variation on the Schwendener method which includes measured distances and overcomes these limitations has been developed at the University of Nottingham (Ashkenazi and Dodson, 1975; Deeth *et al.*, 1979).

The Nottingham method uses a baseline consisting of seven pillars which span a total

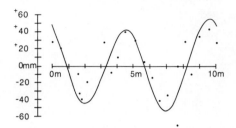

Figure 1.14 Graphical representation of cyclic error determination using Schwendener baseline (Schwendener, 1972).

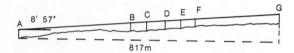

Profile of baseline

Pillar	Approximate interpillar distance (m)					
	A	B	C	D	E	F
B	380.5					
C	422.2	41.7				
D	473.0	92.5	50.8			
E	526.1	145.6	103.9	53.1		
F	577.5	197.0	155.3	104.5	51.4	
G	817.8	437.3	395.6	344.8	291.7	240.3

Figure 1.15 Nottingham University multipoint baseline (Ashkenazi and Dodson, 1975).

distance of 817 m (Fig. 1.15). By employing a Hewlett Packard 5526A laser interferometer, the four central bays (197 m) were measured to an accuracy of ± 0.1 mm. It has been shown (Ashkenazi and Dodson, 1975) that the basic Schwendener model can be varied to include the effects of scale and cyclic error to form the observation equation

$$\left(\sum l_j\right)_i + k_2 + k_1 \left(\sum l_j\right)_i = M_i - (k_3)_i + v_i \qquad (1.24)$$

where l_j are the known or observed distances
k_2 is the zero error
k_1 is the scale error
k_3 is the known cyclic error (from laboratory measurements)
M_i are the observed distances
v_i are the residuals.

ELECTRONIC ANGLE AND DISTANCE MEASUREMENT

Figure 1.16 Tellumat CMW 20. Courtesy Tellumat.

The inclusion of known distances into a Schwendener-type baseline has also been adopted by the Ordnance Survey (Toft, 1975) and the USA National Geodetic Survey (Spofford, 1982).

An alternative approach to the deduction of scale, zero and cyclic errors using known baseline lengths and least squares regression techniques has been developed by Sprent and Zwart (1978) and Sprent (1980) and is used widely in Australia.

1.3.4 *EDM instruments and applications*

Since their introduction, EDM instruments have had an enormous impact over the whole range of survey measurement processes. The following section highlights some of the features of the EDM instruments which may be used for these activities.

1.3.4.1 *Geodetic control networks.* Electronic distance measurements over long ranges (> 10 km) have been used since the original development of EDM instruments to measure baselines and to strengthen existing geodetic control measurements. The direct measurement of distances in excess of 15 to 20 km is almost always performed nowadays using active microwave measuring instruments such as the Tellumat CMW 20 (Fig. 1.16). Distance measurement using microwaves requires a second instrument at the end of the line to be measured, rather than a retro-reflector, to retransmit the microwave signal back to the instrument position. A further distinguishing feature of microwave instruments is the provision of a two-way speech link between the 'master'

Figure 1.17 Geodimeter 122 EDM instrument. Courtesy Geotronics.

and 'remote' instruments. 'Master' in this context refers to the instrument which is controlling the distance-measuring process.

Instruments in this category are also attractive because of their ability to measure in poor visibility conditions, for example fog, conditions in which electro-optical instruments would find it difficult to operate.

1.3.4.2 *Large-scale surveys.* Large-scale (< 1:1000 scale) route and site surveys form the central activity of most survey organizations. Although traditional techniques such as theodolite and tape traversing and stadia tacheometry are still used, they have been largely superseded in most large organizations by EDM traversing using instruments such as the Geodimeter 122 (Fig. 1.17) and polar detailing using electronic tacheometers and data recorders (section 1.4). The use of such EDM instruments not only decreases the observation time but also enables higher orders of precision to be achieved in comparison to more conventional techniques.

1.3.4.3 *Setting out.* The widespread adoption of EDM instruments in the construction industry has led to significant changes in the techniques used for setting out. Of particular importance has been the move towards polar methods of setting out using coordinates (either on a local or national grid) rather than the more traditional

ELECTRONIC ANGLE AND DISTANCE MEASUREMENT 27

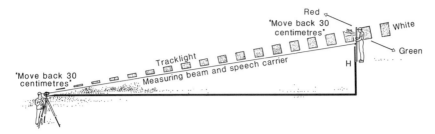

Figure 1.18 Instrument/prism reflector communication using UNICOM and Tracklight. Courtesy Geotronics.

techniques based on site grids and offsets. In order to improve the efficiency of such techniques in setting out, several additional facilities have been incorporated into many modern instruments.

Tracking: almost all modern electro-optical EDM instruments are capable of measuring to a moving target. Generally the level of accuracy is of a lower order of accuracy than for static measurements, typically 10–20 mm. Variations in slope or horizontal distance can thus be displayed and updated at intervals of a fraction of a second.

Integral speech communication: one-way communication between the instrument and target can be achieved by modulating the infrared carrier signal with speech information. The Geotronics UNICOM system uses a microphone built into the EDM instrument and a microphone mounted above the target prism to receive the transmitted instructions (Fig. 1.18).

Visible guidance system: a further facility offered by Geotronics is the tracklight system. In this case a flashing light source is built into the instrument. The correct alignment for the target occurs when the prism is located exactly along the direction being defined by the tracklight. When the target is in this position, a white pulsing light will be viewed by the survey assistant (Fig. 1.18). Any deviation off line will be indicated by a change in the colour of the pulsing signal, which turns red with movement to the right and green if the deviation is to the left of the correct position.

Remote receiver: an alternative approach to communication is adopted by the Kern (Aarau) company. In this case, rather than providing verbal instructions or visible clues to alignment, measured information is transmitted by the EDM instrument to a small receiver adjacent to the prism. A liquid crystal display (LCD) indicates up to five measured parameters, such as horizontal distance or northing and easting. An acoustic signal is also generated when the receiver is within the transmitted beam.

1.3.4.4 *Deformation monitoring*. The analysis of survey observations from different epochs to determine whether any structural deformation has occurred is the subject of Chapter 5. Several different measurement techniques can be used for such operations, including those based primarily on angle measurement, distance measurement or close-range photogrammetry. The determination of precise distances is of fundamental importance in all cases, and is generally performed using specialist instruments such as the Kern Mekometer (described previously) or the Comrad Geomensor. For very short-range distance measurements, laser interferometers may also be used (section 2.2).

1.4 Electronic tacheometers

The term electronic tacheometer refers to an instrument which is able to electronically measure both angles and distances and to perform some limited computational tasks using an integral microprocessor, such as reducing a slope distance to the horizontal or calculating coordinates from a bearing and distance. The instrument may also, but not necessarily, be able to electronically store data, either in an internal memory unit, or more commonly, on an external solid-state data recorder.

A further term which is in widespread use and is used to describe instruments in this category is 'total station'. While this latter term normally implies that the instrument is capable of electronic angle and distance measurement, it has also been applied to instruments which measure angles optically.

While there exist individual design features which are unique to a particular manufacturer, it is generally possible to differentiate between the integral and modular designs of electronic tacheometer. In view of the rate of development of microprocessor technology and the ever-increasing range of instruments on the market, it would, however, be inappropriate to attempt to describe the whole range of instruments in this field. Consequently, two examples which are representative of instruments in each group will be described in order to enable the general characteristics of each design to be more fully appreciated.

1.4.1 *Integrated design*

Many of the earlier examples of electronic tacheometers, such as the AGA Geodimeter 700 and the Zeiss (Oberkochen) Reg Elta, were of an integrated design. The primary feature of such a design is that the electronic theodolite and EDM instrument form a single integrated unit. One of the main advantages of such an approach is the need to transport only a single unit, and also to be able to dispense with auxiliary cables which are often required to link separate units. Furthermore, since the units are matched together during manufacture, and often use coaxial optics for telescope pointing and transmitting the EDM signal, instruments of this type rarely suffer from a lack of collimation of the telescope line of sight and the EDM signal. Two instruments which are typical of this design are the Sokkisha SET2/3 and the Geodimeter 440. The general characteristics of these instruments are summarized in Table 1.3.

The Sokkisha SET 2 (Fig. 1.19) and SET 3 are typical of the relatively low-cost electronic tacheometers which are being manufactured by Japanese electronics companies. Other recent examples include the Topcon ET-1 and the Nikon DTM-5. The system can also be linked to a data recorder such as the Sokkisha SDR2.

The Geodimeter 440 is a further example of an integrated instrument. The instrument has several unique features including an electronic level display, and the one way speech communication and visible guidance system discussed previously (see Fig. 1.18). It may also be linked to a data recorder.

1.4.2 *Modular design*

In this case the electronic theodolite and EDM instrument are separate units which can be operated independently. This arrangement is more flexible, since theodolites and EDM units with differing accuracy specifications can be combined. It may also be a more cost-effective solution, since the individual units may be replaced and upgraded

ELECTRONIC ANGLE AND DISTANCE MEASUREMENT

Table 1.3 Integrated electronic tacheometers

	Sokkisha SET 2	Sokkisha SET 3	Geodimeter 440
Angle measurement Standard deviation of mean pointing (face I and II)			
Horizontal	±2″	±5″	±1″
Vertical	±2″	±5″	±1″
Distance measurement Range (m) average conditions:			
1 prism	1300	1000	2300
3 prisms	2000	1600	3500
Maximum	2600	2200	5000
Standard deviation	±3 mm ± 2 ppm	±5 mm ± 3 ppm	±5 mm ± 5 ppm

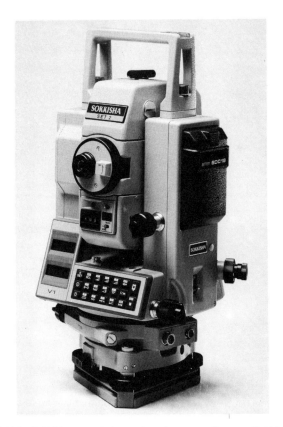

Figure 1.19 Sokkisha SET 2 integrated electronic tacheometer. Courtesy Sokkisha Instruments.

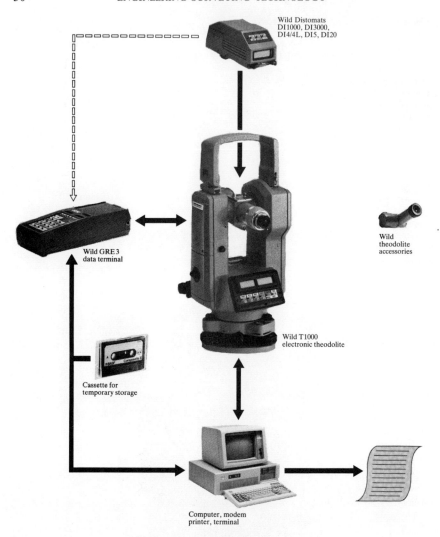

Figure 1.20 Elements of the Wild T-1000 modular electronic tacheometer. Courtesy Wild (Heerbrugg).

as developments occur. The Wild (Heerbrugg) and Kern (Aarau) designs are typical of instruments in this category.

Wild (Heerbrugg) offer two modular systems based around the Wild T-2000 (section 2.1.2) and T-1000 electronic theodolites (Fig. 1.20). Several combinations of theodolite and EDM are available, but the most popular arrangements are the Wild T-1000/DI-1000 and the Wild T-2000/DI-5. Both would normally be linked to a data recorder such as the Wild GRE3. The specifications of both systems are given in Table 1.4.

Kern (Aarau) also offer a range of components which can be combined to form a

Table 1.4 Modular electronic tacheometers

	Wild (Heerbrugg)		Kern (Aarau)	
	T-1000/ DI-1000	T-2000/ DI-5	E1/ DM 550	E2/ DM 503
Angle measurement Standard deviation of mean pointing (face I and II)				
Horizontal	±3″	±0.5″	±2″	±0.5″
Vertical	±3″	±0.5″	±2″	±0.5″
Distance measurement Range (m) average conditions:				
1 prism	500	2500	1800	2500
3 prisms	800	3500	3000	3500
Maximum	800	5000	3300	5000
Standard deviation	±5 mm ± 5 ppm	±3 mm ± 2 ppm	±5 mm ± 5 ppm	±3 mm ± 2 ppm

modular electronic tacheometer. Two of the most common arrangements are given in Table 1.4.

1.4.3 *Data recorders*

Traditionally the surveyor has recorded survey observations manually in a fieldbook. Standard fieldbook layouts have developed over the years, and these attempt to identify gross booking or reading errors. Some form of numbering and a field sketch are also normally required, particularly for recording survey detail. Conventional techniques such as these are quite adequate for many survey projects. However, with developments in digital mapping and terrain modelling in recent years, traditional methods of recording survey observations and subsequently transferring them via a computer keyboard have been found, in practice, to be slow and error-prone. Consequently the need arose for a device which could electronically store survey field observations. Such a device is referred to as a *data recorder*, or alternatively a data collector or data logger.

The data recorder and its associated peripherals for data transfer (Fig. 1.21) form a crucial link between the field observations and the final end product. In general, the distinction can be made between recorders dedicated solely to the logging and processing of survey observations, and the more general purpose hand-held computers which have been adapted to perform survey data collection but which can also operate as stand-alone computers.

1.4.3.1 *Dedicated surveying data recorders.* The design of data recorders dedicated solely to recording and processing surveying data has altered dramatically in the past 15 years. The first designs of data recorder, such as the Geodat 700, used punched paper tape as the recording medium. This approach suffered from many practical difficulties, including breakage of the tape (from mishandling or excessive humidity), a cumbersome and heavy recording mechanism and high power consumption. Although crude in design by present-day standards, this device was a significant development and

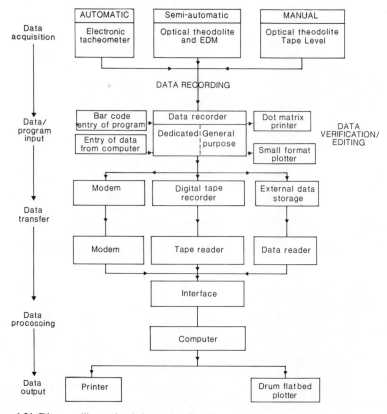

Figure 1.21 Diagram illustrating information flow from data acquisition to final output using data recorders.

provided the first effective method of automatically recording field observations.

The next development, which overcame many of the problems associated with paper tape recorders, involved the use of magnetic tape as the recording medium. The Wild TC-1 used this system. A digital tape recorder, protected against damp and dust by a sealed cover, enabled about 1200 points to be recorded on one digital tape. One of the main advantages cited by users of this technique was the relative ease of sending data recording from a remote location to a central processing facility for plotting. Although the magnetic tape units are no longer manufactured they are still used by several surveying organizations.

The most recent development has been the production of solid-state electronic data recorders. Most data recorders available at present are of this design. The distinction can be made between solid-state data storage modules which plug into an electronic tacheometer (such as the Wild GRM 10), and hand-held units with an integral keyboard. The latter are very much more common and range from devices based on electronic calculators (such as the Geodat 126), to more sophisticated devices such as the Kern Alphacord (Fig. 1.22). Kennie (1987) and Fort (1988) discuss the characteristics of these devices in more detail.

Figure 1.22 Kern Alphacord dedicated surveying data recorder. Courtesy Kern (Aarau).

Figure 1.23 Husky Hunter general-purpose data recorder.

In general terms, however, all dedicated designs suffer from two areas of weakness. First, users are often tied to a particular manufacturer's design of data recorder, since it can be difficult to interface recorders to different electronic tacheometers. Secondly, it may be difficult, if not impossible, to alter the software within the recorder.

1.4.3.2 *General purpose hand-held computers.* A more flexible approach which has become particularly common in recent years is the use of general purpose hand-held computers for data recording. Two different types of computer may be used: those which are fully weatherproof, of rugged construction and suitable for outdoor use (such as the Husky Hunter, Fig. 1.23), and those which are not designed for use outdoors (such as the Epson HX-20). This approach offers several advantages to the surveyor.

The first advantage is the ability to reprogram the form of data entry, for example, for specialist surveying operations or for other types of surveying equipment, which dedicated recorders would find difficult to accommodate. The second advantage is the flexibility to create specialist programs to process data in the field. Thirdly, in view of the general purpose nature of these devices it is also possible to use the computer for conventional surveying computational tasks and other office-based tasks such as word processing.

1.4.4 *Feature coding*

The method of storing data varies considerably, depending on the type of data recorder which is being used. It is possible, however, to distinguish between stored data which relates to the position of the object, and information which describes the object being surveyed.

Various options exist for recording field observations, ranging from recording raw observations (horizontal angle, slope distance and vertical angle), to storing three-dimensional coordinates. The former approach is preferred by many users, since it avoids the possibility of erroneously stored corrections (e.g. refractive index variations) inadvertently being applied to a series of measured distances.

Feature coding is the process of describing objects in the data recorder by means of some numeric or alphanumeric description. The coding process is normally carried out as an integral part of the field data acquisition phase, although instances of coding being performed after the survey has been performed have also been reported. While full alphanumeric coding may eliminate the need for a filed sketch over small sites, for large projects it is vitally important that a numbered sketch be prepared in the field. Coding systems vary considerably, and may reflect either the design of the data recorder being used or the in-house standards of a particular organization. Generally, however, they should enable the surveyor to describe certain characteristics of the object or how it should be represented on the final map. Thus a coding system could be used to define the type of annotation to be attached to a point feature. Similarly it may be used to describe whether points on a string (or series of linked points) should be joined by a straight or curved line.

1.5 Electronic coordinate determination systems (ECDS)

Electronic coordinate determination systems (ECDS), based on the use of two electronic theodolites and (normally) a portable computer, are becoming increasingly important for industrial metrology and deformation monitoring applications. They provide a portable, non-contact and real-time method of acquiring three-dimensional coordinates about objects ranging in size from vehicle sub-assemblies to oil rigs. While in theory objects smaller than about a metre in size can be measured by such systems, it is generally more appropriate in these circumstances to use direct methods such as coordinate measuring machines.

An alternative non-contact surveying technique for the measurement of objects such as those described above is close-range photogrammetry, and this approach is discussed in more detail in Chapter 8. Each method, however, offers complementary advantages to the surveyor and engineer. For example, photogrammetric techniques require only a relatively short time for data acquisition. This may be of value if it is important to reduce the interruption in the workplace to minimum during the survey. Photogrammetry also provides a permanent historical record of the raw data in the form of a pair of photographs. Thus, remeasurement can take place at a later stage if required. ECDS, in contrast, although requiring full access to the measurement surface during the survey, can provide real-time coordinate data. Real-time photogrammetric systems are also becoming available (Brown and Fraser, 1986) and section 8.7.2. However, they are at present generally research oriented, require substantial data processing facilities and are not widely available for routine measurements.

1.5.1 *Hardware*

The first commercial system developed for real-time coordinate determination was the Hewlett Packard CDS (Coordinate Determination System). This system, developed during the mid-1970s, was based on two HP 3820A electronic theodolites linked to a HP 9845 microcomputer (Johnson, 1980*b*). Although several systems were sold,

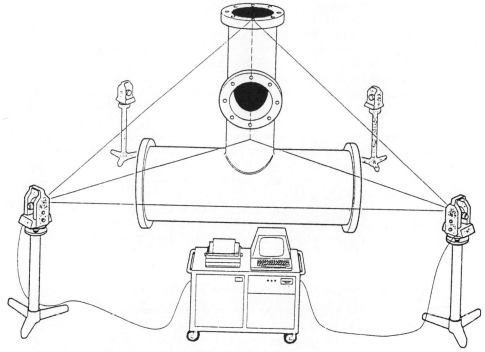

Figure 1.24 Wild RMS-2000 ECDS. Courtesy Wild (Heerbrugg).

primarily in the USA, development of the system ceased in 1980 following the decision by Hewlett Packard to discontinue its interest in surveying instruments. Two other systems which were also developed during this period were the Zeiss (Oberkochen) IMS (Industrial Measuring System) (Leitz, 1982), and the Keuffel and Esser AIMS (Analytical Industrial Measuring System) (Vyner and Hanold, 1982). Two systems which are currently particularly popular are the Wild Leitz RMS 2000 and the Kern ECDS 1.

1.5.1.1 *Wild Leitz Remote Measuring System (RMS)*. The Wild Leitz Remote Measuring System (Fig. 1.24) consists of between two and four Wild T-2000 or T-2000S (a high-magnification, short-focus model) electronic theodolites linked to a Wang 2200 microcomputer for on-line real-time operations, or to a Wild GRE3 data collector for off-line storage of angular measurements. (See Katowski, 1986.)

1.5.1.2 *The Kern ECDS 1*. The Kern ECDS 1 (Electronic Coordinate Determination System) (Lardelli, 1984; Grist, 1986) consists of two or more Kern E1 or E2 electronic theodolite linked to a DEC Micro PDP 11/23 or IBM PC/AT computer for data processing. For applications where it is difficult to provide mains power, it is also possible to use a Husky Hunter portable computer (Fig. 1.23) together with the Dimensional Control System (DCS) software package to perform most of the processing requirements.

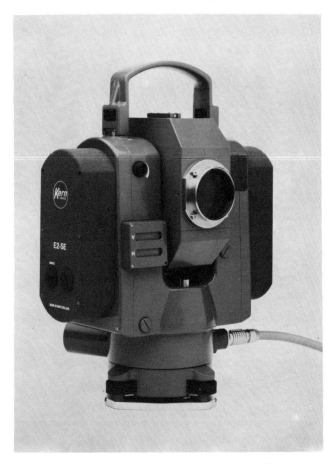

Figure 1.25 Kern E2-SE electronic theodolite. Courtesy Kern (Aarau).

1.5.1.3 *Kern SPACE system.* A development of the technique of real-time intersection (section 1.5.2) which attempts to automate the coordinate determination process is presented in Gottwald and Berner (1987) and Roberts and Moffitt (1987). Both papers discuss the development of a 'System for Positioning and Automated Coordinate Evaluation (SPACE)'.

The data acquisition element of SPACE is based on a modified version of the Kern E-2 electronic theodolite, while the processing phase is carried out using image processing techniques on an IBM PC/AT microcomputer. The theodolite, the Kern E2-SE (Fig. 1.25), incorporates several unique design features which aim to automate the target pointing and angle measuring process. First, servo-motors drive the horizontal and vertical axes of the instrument and also focus the telescope. The motors are designed to enable a positioning accuracy of up to $\pm 0.1''$ to be achieved, at speeds of up to $54°$ per second. Second, the theodolite includes within the telescope a Charge Coupled Device (CCD) sensor. The output from the CCD arrays are then fed to a video

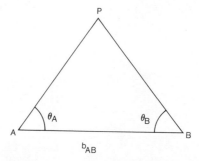

Figure 1.26 Two-dimensional intersection.

frame store in the computer. Image processing algorithms are then used to automatically drive the theodolite to either predetermined or targeted points (Zhou, 1986). A similar instrument has been designed by Wild (Heerbrugg) and is discussed in Katowski (1987).

1.5.2 *Principles of coordinate determination using ECDS*

1.5.2.1 *Real-time three-dimensional intersection.* All ECDS are based on the surveying technique known as intersection. In its simplest form this involves the simultaneous measurement, using two theodolites, of horizontal angles θ_A and θ_B from either end of a precisely measured baseline (Fig. 1.26). Standard surveying computational techniques can then be used to derive the two-dimensional coordinates of the unknown point P. Data acquired in this manner can also be more rigorously processed off-line. This approach is discussed subsequently in more detail.

The solution for the real-time three-dimensional coordinates of remote points is central to most ECDS applications. Two different approaches are used to solve this problem; the first uses trigonometric functions, and the second is based on the use of vector algebra. Both approaches can utilize least-squares estimation techniques if redundant observations have been recorded.

A set of equations can be derived by trigonometric methods for the general case in which the three-dimensional coordinates of the point P are required. Johnson (1980a) provides the following solution based on the measurement of the horizontal and vertical angles shown in Fig. 1.27.

$$X_P = \frac{b}{2}\sin(\phi_T)\left[1 + \frac{\sin(\theta_B - \theta_A)}{\sin(\theta_B + \theta_A)}\right]$$

$$Y_P = b\sin(\phi_T)\frac{\sin\theta_B \sin\theta_A}{\sin(\theta_B + \theta_A)} \quad (1.25)$$

$$Z_P = b\sin(\phi_T)\frac{\sin\theta_B \cot\phi_A}{\sin(\theta_B + \theta_A)}$$

The solution of these equations provides unique values in X and Y for the position of P.

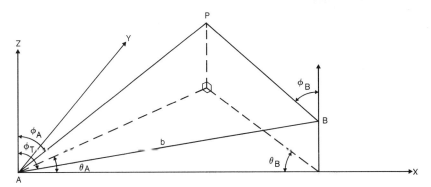

Figure 1.27 Three-dimensional intersection. At A, record horizontal angle BAP and zenith angles to P and B. At B, record horizontal angle ABP and zenith angle to P only.

It is possible, however, to derive two values for Z_P using the zenith angles ϕ_A and ϕ_B. A mean value for Z_P may then be derived. The difference between the two values, dZ, is used by several systems to check for gross errors in the pointing of the theodolites in the Z direction.

The solution of the intersection problem using vectors has been examined by Allan (1987) and Cooper (1986). Cooper derives the solution using the following argument. Consider two unit vectors p and q with scalar components δ and μ as illustrated in Fig. 1.28. The rays at the unknown point P will not in general intersect, since p and q are skew as a consequence of random observational errors. The position of P will, however, lie at some point along the vector e joining δp and μq, where e is given by

$$e = -\delta p + b + \mu q \tag{1.26}$$

Geometrically, the most probable position of P is at the midpoint of the vector of minimum length. The values of δ and μ which arise when e is a minimum are given by

$$\begin{bmatrix} \delta_0 \\ \mu_0 \end{bmatrix} = \frac{1}{\Delta} \begin{bmatrix} 1 & p \cdot q \\ p \cdot q & 1 \end{bmatrix} \begin{bmatrix} b \cdot p \\ -b \cdot q \end{bmatrix} \tag{1.27}$$

If the coordinate system and angles measured in Fig. 1.27 are used, then

$$b = \begin{bmatrix} X_b \\ 0 \\ Z_b \end{bmatrix} \quad p = \begin{bmatrix} \sin\phi_A \cos\theta_A \\ \cos\phi_A \sin\theta_A \\ \cos\phi_A \quad 0 \end{bmatrix} \quad q = \begin{bmatrix} -\sin\phi_B \cos\theta_B \\ \sin\phi_B \sin\theta_B \\ \cos\phi_B \quad 0 \end{bmatrix} \tag{1.28}$$

The solution of equation (1.28) for δ and μ can then be used to derive a value for the position vector of P relative to A from the following expression:

$$r = \delta_0 p + \frac{e}{2} = \tfrac{1}{2}(\delta_0 p + b + \mu_0 q) \tag{1.29}$$

The values of b, p and q can also be substituted in equation (1.26) to determine the components of the residual vector e.

The coordinates of point P can also be determined more precisely, using either of the approaches outlined above, by least-squares methods. Estimates of the unknown

40   ENGINEERING SURVEYING TECHNOLOGY

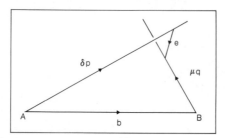

Figure 1.28 Analytical intersection using vectors (Cooper, 1986).

coordinates of P in the X, Y plane are initially determined using the horizontal angles measured from either end of the baseline. Next, one of the vertical angles and the estimated horizontal distance to the target (based on the approximate coordinates obtained previously) is used to determine the approximate height. Observation equations of the form described in Chapter 5 are then produced for each of the observed quantities (two vertical and two horizontal angles). These four equations in three unknowns (the corrections to the provisional coordinates) can then be determined by forming normal equations. The corrections obtained from the solution of the normal equations are then applied to the provisional coordinates, the process being iterated until the corrections become negligible. Although in theory it is possible to determine various statistical parameters to describe the precision of the coordinates (see Chapter 5), in practice these are not normally computed, since they would be of little practical value because statistically inconclusive.

The solution of equations (1.25) and (1.29) require a value for the length of the baseline b. This may be derived either by direct measurement or by the introduction into the object space of a scaling mechanism. The most common technique involves the use of a bar of known length. Calibration of the scaling bar is essential if systematic errors are to be avoided. It is also important to ensure that the bar is manufactured from a material with a low coefficient of linear expansion, such as carbon fibre, invar or in some cases glass. Alternatively, distances may be measured by laser interferometry (Chapter 2), or known distances on the object may be used. The scaling 'bar' is used to indirectly establish the length of the baseline. The technique is analogous to scaling a pair of photographs during the absolute orientation phase of setting up a stereoplotter. The use of a scaling bar generally reduces the measuring time and, more importantly, enables the baseline to be determined to a high level of accuracy with relative ease.

The simplest case occurs when a single distance is measured in the object space, as illustrated by Fig. 1.29. In this case an initial estimate of the length of the baseline (b') is used in conjunction with the observed horizontal angles θ_A and θ_B to determine the X and Y coordinates of the targets placed at either end of the scaling bar. From these coordinates the distance between the targets can be computed. By comparing this computed length of the bar S' with the known calibrated length S, a corrected value for the base b can be determined from the following expression.

$$b = \frac{S b'}{S'} \tag{1.30}$$

This revised value of b can then be used in equations (1.25) and (1.29) to determine the coordinates of the unknown points.

ELECTRONIC ANGLE AND DISTANCE MEASUREMENT 41

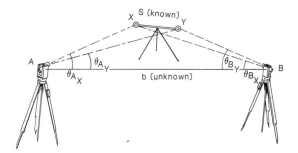

Figure 1.29 Indirect scaling using scaling bar.

An alternative approach to the processing of electronic theodolite data may be adopted in those situations where the simultaneous use of two theodolites providing real-time coordinates is not required. In such cases, the data acquisition may be accomplished at lower capital cost by using a single theodolite. The use of a single instrument is also advantageous since more flexibility in the positioning of the control stations is possible, and also there is no requirement to align the theodolite in relation to another instrument prior to obtaining measurements. More importantly, however, is the ability to rigorously process the data by least-squares techniques to fully exploit any redundant data. One particular approach which has been advocated involves the use of the bundle adjustment technique (Brown, 1985). The bundle technique is a computational method which is used in photogrammetry to simultaneously solve for the positions of unknown ground points and camera orientations using measured x and y photographic coordinates of known control points and unknown ground points. The method rigorously accommodates redundant data and provides a full error analysis. The application of the method to close-range photogrammetry is discussed in Chapter 8.

In order to utilize the bundle adjustment method in this context, it is currently necessary to transform angular measurements obtained from an electronic theodolite into equivalent photographic observations. These 'fictitious' photographs may then be processed using software packages such as STARS (Simultaneous Triangulation and Resection) (Brown, 1982), to provide ground coordinates and statistical information about the precision and accuracy of the position determinations.

Offline processing of data is also used in cases where it is necessary to transform coordinates from one coordinate reference system to another. For example, a common requirement is to transform from the 'theodolite system' to the 'object system' coordinates. In order to perform such a transformation it is necessary to know the coordinates, in both reference systems, of a minimum of three points. If more than three sets of coordinates are known, then a least-squares solution for each of the seven transformation parameters (three rotations, three translations and one scale change) can be carried out.

An alternative process which avoids the necessity of transforming coordinates from one system to another is to carry out all measurements in either the theodolite or the object reference frame. In the former case, if it is necessary to move the theodolites, their positions are re-established by recording angular measurements to a minimum of three points on the object and performing a resection calculation. In the latter situation, particularly if the object to be measured is relatively small, the object may be mounted on a rotary table and repositioned relative to two fixed theodolite stations. Offline processing of the data can then be performed to compute theodolite-based coordinates (Griffiths, 1985).

1.5.2.2 *Factors influencing the accuracy of real-time intersection.* The overall accuracy of an ECDS based on real-time intersection is a function of several interrelated factors. These factors are, firstly, the instruments being used for angular measurements, and secondly the geometry of the measurement problem, and thirdly the influence of the environment.

The use of high-precision electronic theodolites able to resolve to a fraction of a second is essential if high-accuracy measurements are to be obtained. Two instruments which are particularly well suited to this are the Wild T-2000 and Kern E-2 which have been discussed in section 1.1. The internal precision of such theodolites will, however, only be translated into high-accuracy measurements if, firstly, the effect of any instrumental errors is eliminated, and secondly, if the pointing of the telescope to targets on the structure is precise.

The first requirement is normally achieved by conventional observational techniques (double face readings), although in some cases where a reduced-accuracy specification is applicable, single face pointings electronically corrected for instrument maladjustments may be acceptable. The second requirement is equally important and is primarily dependent on the design of the targets. A wide range of possibilities exist for targeting an object, including

(i) Natural points of detail
(ii) Adhesive targets of known dimensions
(iii) Points of detail artificially created during manufacture
(iv) Artificially created targets formed by projecting a laser beam through the objective of one theodolite.

Whatever technique is used it is important that the design of the targets enables precise centring of the telescope crosshairs over a wide angular range (60–120°). It is also important that the software can compensate for any systematic errors which may be introduced by the target design, as, for example, in (ii) above.

The second group of interrelated factors are those concerned with the geometrical relationship between the theodolites and the points on the object to be measured. The decrease in accuracy caused by an increase in the length of the *baseline distance* between the two theodolites is illustrated in Fig. 1.30, in which a nominal angular error is assumed. In order to limit this effect for high-accuracy measurements, the length of the baseline is normally restricted to between 5 and 10 m. The *angle of intersection* on the object of rays from adjacent theodolites has a major impact on the accuracy of the coordinate determination. Figure 1.31 illustrates this in two dimensions for a particular case where the baseline is 10 m and the standard error of the angular measurements is $\pm 1.5''$. The intersection angle range suggested by Wild (Heerbrugg) for high-accuracy projects is between 78° and 142°.

The position and orientation of the scaling bar is also a significant consideration. It is important not only to locate the bar so that the angle of intersection at the target points on the bar is close to 90°, but also to ensure that the orientation of the bar enables a clear view of the targets to be achieved.

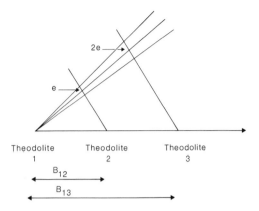

Figure 1.30 Influence of baseline length on coordinate determination.

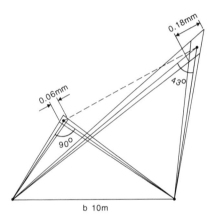

Figure 1.31 Influence of intersection angle on coordinate determination (adapted from Leitz, 1982).

Environmental effects such as vibration, wind, temperature fluctuations, atmospheric refraction and varying lighting conditions are some of the most influential factors which limit the accuracy of this technique. While several of the factors can be controlled, for example by ensuring that machinery adjacent to the object to be measured is not operated during measurement, ultimately uncontrolled environmental fluctuations which cause varying refraction conditions limit the accuracy achievable with such a measuring system.

1.5.3 *Applications*
As mentioned previously, ECDS are predominantly used in large-scale industrial metrology. The following two examples illustrate the use of these techniques in this field.

1.5.3.1 *The Wild Leitz RMS 2000 and the measurement of a parabolic antenna.* Katowski (1985) describes the use of the Wild Leitz RMS system to measure,

Figure 1.32 Use of Wild-Leitz RMS 2000 to monitor paraboloid reflector. Photograph courtesy European Space Research and Technology Centre (ESTEC).

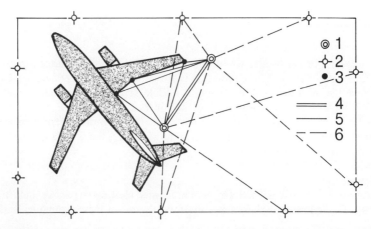

Figure 1.33 Kern ECDS-1: network configuration for aircraft dimensional control. Free choice of both base points (1), computation of its coordinates and the baseline data (4) by means of the directions (6) at the known fixed points (2), intersection (5) at the object points (3). Courtesy Kern (Aarau).

under varying environmental conditions, the shape of the paraboloid reflector and antenna for the Ulysses satellite by the European Space Research and Technology Centre (ESTEC) in the Netherlands. The reflector measured had a diameter of 1.65 m and measurements were recorded to 26 pre-targeted points using the measuring arrangement illustrated in Fig. 1.32. The measurements were subsequently processed to define the reflector focal point, the focal distance using the procedure outlined in Griffiths and Thomsen (1985). Several other applications of the Wild-Leitz system are discussed in Woodward (1987).

1.5.3.2 *The Kern ECDS and aircraft dimensional control.* In the assembly of large aircraft it is now becoming common practice to use systems such as the Kern ECDS to check the dimensional quality control of the various sub-assemblies which form the structure of the aircraft. Also, when the aircraft has been fully assembled, a full dimensional survey of the entire structure is carried out to ensure that the final dimensions are within tolerance and consequently unlikely to generate excessive stresses within the structure. Figure 1.33 illustrates how a network of control points (2) can be used as a basis for resecting theodolite stations (1) from which measurements can be taken onto the aircraft (3) to determine its dimensions. More recent applications of the ECDS have involved submarine hull mapping (Bethel and Sedej, 1986), and the alignment and positioning of the magnets for the Stanford Linear Collider (Oren *et al.*, 1987).

1.6 Conclusions

This chapter has discussed the current technology for the measurement and storage of angle and distance observations. The impact of these developments on surveying practice and productivity levels has been significant. Although such devices are now widely used in the land surveying community, it is only recently that they have become more common in general civil engineering and building projects. However, if the cost of these instruments continues to drop, this trend is likely to increase.

Finally, for the future it has been suggested (Ashkenazi, 1986), that the current hierarchical geodetic control frameworks based on angle and distance measurements may eventually be replaced by point measurements to satellite systems such as the Global Positioning System (GPS) (Chapter 4) or by inertial systems (Chapter 3). This seems unlikely to occur during the next 10–15 years, and in any case it has been argued that it would be inappropriate to become dependent on a system such as GPS which could, for political reasons, become inaccessible to non-military users. Consequently, the need for angle and distance measurements will continue, providing complementary data to those produced by satellite and inertial surveying systems.

References

Allan, A. (1987) The principles of theodolite intersection systems. *Proc. 2nd FIG/ISPRS Industrial and Engineering Survey Conf.*, September 2–4, London, 16–31.

Ashkenazi, V. (1986) Satellite geodesy—its impact on engineering surveying. *Proc. 18th FIG Congress, Toronto*, Paper 502/1, 70–85.

Ashkenazi, V. and Dodson, A. (1975) The Nottingham multi-pillar baseline. *Proc. 16th UGG General Assembly, Grenoble*, 22 pp (also reprinted in *Civil Engineering Surveyor*, June 1977, 1–8.

Bethel, J. and Sedej, J. (1986) Submarine hull mapping and fixture set out: an operation of the ECDS system. *46th ASP-ACSM Convention, Washington DC*, 92–96.

Bevin, A.J. (1978) Accuracy and calibration of short range EDM. *New Zealand Surveyor* **28** (1) 15–33.
Brown, D.C. (1982) STARS: a turnkey system for close range photogrammetry. *Int. Arch. Photogramm.* **24**, V/1, *Precision and Speed in Close Range Photogrammetry*, 68–89.
Brown, D.C. (1985) Adaptation of the bundle method for triangulation of observations made by digital theodolites. *Proc. Conf. South African Surveyors (CONSAS)*, Paper 43, 18 pp.
Brown, D.C. and Fraser, C.S. (1986) Industrial photogrammetry—new developments and recent applications. *Photogramm. Record* **12** (68) 197–217.
Burnside, C.D. (1982) *Electromagnetic Distance Measurement*. 2nd edn., Granada, London, 240 pp.
Cooper, M.A.R. (1986) Spatial positioning, vector notation and coordinates. *Land and Minerals Surveying* **4** (12) 623–625.
Deeth, C.P., Dodson, A.H. and Ashkenazi, V. (1978) EDM accuracy and calibration. *Proc. Symp. on EDM*, Polytechnic of the South Bank, London, 24 pp.
Dodson, A. (1978) EDM systematic errors and instrument calibration. *ASCE Journal* (now *Civil Engineering Surveyor*) **3** (4) 20–22.
Fort, M. (1988) Data recorders and field computers. *Civil Engg Surveyor*, July/August, 8 pp.
Gorham, B.J. (1976) Electronic angle measurement using coded elements with special reference to the Zeiss Reg Elta 14. *Survey Review* **23** (180) 271–279.
Gottwald, R. and Berner, W. (1987) The new Kern system for positioning and automated coordinate determination—advanced technology for automated 3-D coordinate determination. *47th ASP-ACSM Spring Convention, Baltimore*, 7 pp.
Griffiths, J. (1985) Measuring an object with the aid of a rotary table. Wild RMS-2000 Supplement No. 8, 11 pp.
Griffiths, J. and Thomsen, P. (1985) Measuring a parabolic antenna using a 3D surface element. Wild RMS-2000 Supplement No. 2, 9 pp.
Grist, M.W. (1986) Real time spatial coordinate measuring systems. *Land and Minerals Surveying* **4** (Sept.) 458–471.
Hodges, D.J. (1982) Electro-optical distance measurement. *University of Nottingham Mining Magazine* **34**, 19 pp.
Johnson, D.R. (1980a) An approach to large scale non-contact coordinate measurements. *Hewlett Packard Journal*, September, 16–20.
Johnson, D.R. (1980b) An angle on accuracy. *Quality*, November, 4 pp.
Katowski, O. (1986) RMS-2000, a system for non-contact measurement of large objects. Wild Heerbrugg, 15 pp.
Katowski, O. (1987) Automation of electronic angle measuring instruments. *Proc. 2nd FIG/ISPRS Industrial and Engineering Survey Conf., September 2–4, London*, 230–240.
Katowski, O. and Salzmann, W. (1983) The angle measurement system in the Wild Theomat T-2000. Wild Heerbrugg, 10 pp.
Keller, W. (1978) Deformation measurements on large dams. Kern (Aarua) Publication, 15 pp.
Kennie, T.J.M. (1983) Some tests of retroreflective materials for electro-optical distance measurement. *Survey Review* **27** (207) 3–12.
Kennie, T.J.M. (1984) The use of acrylic retro-reflectors for monitoring the deformation of a bridge abutment. *Civil Engineering Surveyor*, July/August, 10–13.
Kennie, T.J.M. (1987) Field data collection for terrain modelling. *Proc. Short Course on Terrain Modelling in Surveying and Civil Engineering*, University of Surrey, 12 pp.
Kennie, T.J.M., Brunton, S., Penfold, A. and Williams, D. (1988) Electro-optical distance measurement: calibration methods and facilities in the United Kingdom. *Land and Minerals Surveying*, January, 12–16.
Lardelli, A. (1984) ECDS-1: an electronic coordinate determination system for industrial applications. Kern Aarau, 11 pp.
Leitz, H. (1982) Three dimensional coordinate measurement with electronic precision theodolites. *Proc. NELEX'82, 14–16 September*, Paper 4/1, 10 pp.
Maurer, W. (1985) Kern E2: accuracy analysis of the graduated circle. *Kern Bulletin* **37**, 3–6.
Munch, K.H. (1984) The Kern E2 precision theodolite. *Proc. 17th FIG Congr. Sofia*, Commission 5, 22 pp.
Murname, A.B. (1982) The use of the Kern Mekometer ME3000 in the Melbourne and Metropolitan Board of Works. *Proc. 3rd Int. Symp. on Deformation Measurement by Geodetic Methods, Budapest*.

Oren, W., Pushor, R. and Ruland, R. (1987) Incorporation of the Kern ECDS-PC software into a project oriented software environment. *47th ASPRS-ACSM Convention, Baltimore*, 7 pp.

Rawlinson, C. (1976) Automatic angle measurement in the AGA 700 Geodimeter—principles and accuracy. *Survey Review* **23** (180) 249–270.

Roberts, T. and Moffitt, N.M. (1987) Kern system for positioning and automated coordinate evaluation: a real time system for industrial measurement. *47th ASPRS-ACSM Convention, Baltimore*, 5 pp.

Rüeger, J.M. (1988) *Introduction to Electronic Distance Measurement*. 2nd edn, 128 pp.

Schwendener, H.R. (1972) Electronic distances for short ranges: accuracy and calibration procedures. *Survey Review* **21** (64) 273–281.

Spofford, P.R. (1982) Establishment of calibration baselines. NOAA Technical Memo NOS/NGS/8, National Geodetic Information Centre, 21 pp.

Sprent, A. (1980) EDM calibration in Tasmania. *Australian Surveyor*, **30**(4) 213–227.

Sprent, A. and Zwart, P.R. (1978) EDM calibration—a scenario. *Australian Surveyor* **29** (3) 157–169.

Toft, D. (1975) Tests of short range EDM equipment in the Ordnance Survey. *Commonwealth Survey Officers Conference*, Paper C-1, 15 pp.

Vyner, N.A. and Hanold, J.H. (1982) AIMS- analytical industrial measuring system. *Int. Arch. Photogramm.* **24**, V/1, *Precision and Speed in Close Range Photogrammetry*, 524–532.

Woodward, C.A.W. (1987) Practical experiences with the Wild-Leitz RMS 2000 system. *Proc. 2nd FIG/ISPRS Industrial and Engineering Survey Conf., September 2–4, London*, 42–53.

Zhou, G. (1986) Accurate determination of ellipse centres in digital imagery. *ASPRS-ACSM Convention, Washington DC*, 256–264.

2 Laser-based surveying instrumentation and methods

G. PETRIE

2.1 Introduction

For a period after its original invention, the laser was often described as being a device or instrument in search of an application. This description was not one which could be supported or repeated by surveyors and engineers who very rapidly adopted lasers for alignment work. This successful application was followed soon after by the incorporation of lasers into instrumentation designed specifically for levelling and setting-out operations and the invention of various devices based on differing measuring principles which could be used to measure distance. The characteristics of these laser-based instruments are often decidedly different to those of traditional surveying instrumentation. In particular, they allow certain operations to be carried out at a larger range or with greater facility or efficiency than can be achieved with standard surveying instruments such as theodolites, levels and EDM devices.

2.1.1 *Characteristics of lasers*

A *laser* is a device which emits radiation which is more intense, monochromatic, coherent and directional than the light emitted by a conventional source such as an incandescent lamp. The latter emits radiation at many different wavelengths; it is much less intense in its radiation and the light is incoherent. In a laser, the intensity, coherence and chromatic purity all result from the fact that excited atoms are stimulated to radiate light cooperatively rather than to do so spontaneously and independently as they do in an incandescent lamp. All lasers consist of three basic elements:

(i) An *active material* which is the source of the radiation
(ii) A *pumping source* which starts up and continues the lasing actions
(iii) Two *mirrors*, one silvered, the other partly silvered, which are an integral component or feature of a laser.

2.1.2 *Laser construction*

Most lasers consist of a column of an active material which has a fully reflecting mirror at one end of the column and a semi- or partly-reflecting one at the other end. The best known active material is probably *ruby*, the material first used by Dr Theodore Maiman to generate laser light in July 1960 (Schawlow, 1961, 1968). The ruby is a crystal of aluminium oxide (Al_2O_3) in which are interspersed chromium atoms from the small amount (0.05%) of chromium oxide which forms the active material in the crystal.

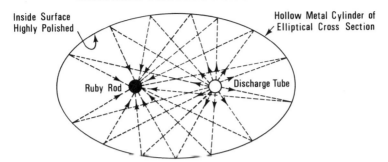

Figure 2.1 Construction of a laser.

For use in a laser, the crystal is cut into the shape of a cylinder or rod typically 5 cm in length and 0.5 cm thick. The end faces of the cylinder are made accurately parallel, with one of them fully silvered and the other partly silvered. The ruby rod is then placed with its axis along one of the foci (focal points) of a hollow metal cylinder of elliptical cross-section which is highly polished in its inside surface (Fig. 2.1). Along the other focal axis is a *discharge tube*, emitting high energy violet or green light which is delivered highly focused into the ruby crystal or rod.

2.1.3 *Laser action*

A flash from the discharge tube energizes a large number of the chromium atoms to an excited state. As their electrons cascade down from this high-energy state to their ground state, they pass through an intermediate level called the *metastable state* in which they spontaneously emit energy in the form of photons.

Figure 2.2 shows this sequence in more detail. In Fig. 2.2a, the active material is shown in its normal unexcited state, in which nearly all the atoms are at a low energy state, and Fig. 2.2b depicts the situation which arises when excitation of the atoms by the light from the discharge tube pumps a large number of the atoms to the high energy state, i.e. a population inversion takes place. Figure 2.2c shows how each atom falling spontaneously from the high level to the low level passes through the intermediate level or metastable state where it can emit a photon of light. Other atoms are also stimulated to emit photons with the same frequency and phase. Thus light amplification is produced by stimulating coherent emission from other excited atoms. Finally, Fig. 2.2d is a cross-section showing that light is emitted (via a photon) when an electron drops from a high-energy state to a lower one in an atom.

For a photon which travels parallel to the axis of the ruby cylinder, when it reaches the silvered end, it is reflected back along its path. In the course of its repeated backward and forward paths along the cylinder, it is joined by other photons emitted from other excited chromium atoms whose electrons have reached the metastable state. The emission of these photons is stimulated by the passage of the primary photon. Since the amplitude of the primary photons has been increased by the addition to them of exactly similar photons which are precisely in phase with the primary photons, so the term LASER was devised as an acronym from the description Light Amplification by Stimulated Emission of Radiation.

The overall effect may be depicted as comprising an in-phase tidal wave of photons sweeping up and down the axis of the ruby cylinder, increasing in amplitude as more of

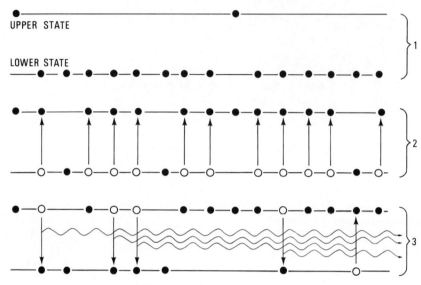

Figure 2.2 (a)–(d) Laser action.

the excited chromium atoms are stimulated into emission by the passing wave. All of this happens very quickly. The ruby cylinder or column (whose length is a whole number of wavelengths at the selected frequency) acts as a *cavity resonator*. The beam of monochromatic coherent light rapidly builds up in intensity till it emerges as a short, intense, and highly directional pulse of red light from the partially silvered face.

2.1.4 *Properties of laser radiation*
This light or radiation will have the following rather unusual properties.
  (i) It will be highly *monochromatic*. The light or radiation from a single spectral line of the best low-pressure gas lamp is spread over a band of frequencies that is typically 1000 megacycles wide. Radiation from a laser can be confined to 1 megacycle in width, i.e. to a single very well-defined wavelength.
 (ii) It is entirely *phase-coherent*, i.e. all the light waves have the same phase. This property helps to make the laser useful as a radiation source for distance measurement in surveying.
(iii) The beam is very nearly *parallel-sided*. This results from the loss from the curved surface of the sides of the crystal or rod of all the photons which do not hit the end faces normally or which have been emitted sideways in any case. This leads to the high degree of *directionality* of the beam which is characteristic of the laser. This makes it very useful for alignment and levelling in surveying.
(iv) The energy of the beam is very highly concentrated to give a high *intensity*, since each photon is added at exactly the right moment so that the added radiation will be in phase with the rest of the beam. The high-intensity beam will have a narrow cross-section which can be focused down to 0.1 mm in diameter. This can be exploited to give a much increased range in distance measurement in land surveying.

2.1.5 Classification of lasers

Lasers are usually classified according to the characteristics of the material used as the radiation source:

(i) *Gas lasers*: e.g. the helium-neon (He-Ne) and carbon dioxide (CO_2) lasers
(ii) *Solid-state lasers*: e.g. the ruby (Al_2O_3) laser
(iii) *Semiconductor lasers*: e.g. the gallium arsenide (GaAs) diode laser
(iv) *Liquid lasers*: e.g. the organic dye laser.

2.1.5.1 Gas lasers.
Ali Javan of the Bell Telephone Laboratories produced the first gas laser in February 1961, only a few months after the appearance of Haiman's ruby laser. He used a gas mixture of *helium* (90%) and *neon* (10%) held at low pressure in a glass tube (Fig. 2.3), with the neon acting as the active light-emitting element. Electrical pumping in the form of a continuous high-frequency discharge was used to pump the neon atoms to a certain energy level at which they were stimulated to emit photons, all at the same wavelength ($\lambda = 632.8$ nm) and in step. As with the ruby laser, the intensity of the beam was built up by passing it backwards and forwards between two mirrors. When the appropriate intensity level was reached, the coherent beam was projected through the semi-silvered mirror as a continuous beam rather than in pulses as with the ruby laser. So if high power and energy per pulse are characteristic of solid-state lasers, the features of high spectral purity and directionality and the continuous output of a beam of coherent light are some of the special characteristics of the helium-neon gas laser. Furthermore, it has a very low power output, is inexpensive and can be produced as a relatively sturdy and compact unit. The low power output means that it can be used safely in the field, and so it has found widespread use in surveying instruments—in alignment lasers, in laser-levels and in certain types of EDM equipment.

Other gases which exhibit laser action are the singly-ionized rare gases. The most useful is the *argon ion* laser which has been used in various industrial applications. It operates at wavelengths of 460–470 nm, but at a rather low efficiency, and so has not till now been used in surveying work. More efficient and powerful gas lasers have been devised utilizing gases comprising heavier molecules. The *carbon dioxide laser* (first produced in 1967) has an continuous output of several thousand watts and operates in the thermal infrared region at a wavelength (λ) of 10.6 μm. The unfocused beam from a CO_2 laser will cause a wooden board to catch fire immediately and eat through thin metal in a few seconds. However, these lasers have not found an application in

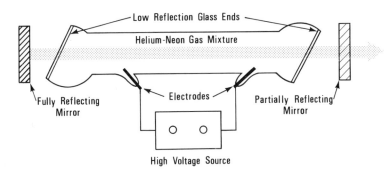

Figure 2.3 HeNe gas laser.

surveying, their high power and the invisible nature of their radiation making them impractical for use outside a laboratory where the necessary stringent precautions can be taken regarding safety.

2.1.5.2 *Solid-state lasers.* There are many solid-state lasers, the ruby laser having been the first laser of any kind. Its construction has already been described (Fig. 2.1). Quite high intensities result, but it is difficult to produce continuous operation of the ruby laser because of the large amount of energy required for this purpose, which can cause the ruby to overheat. So it is used almost invariably in a pulsed mode of operation, with the laser light emitted at a wavelength (λ) of 694.3 nm at the very end of the red part of the spectrum.

Besides the ruby, there are a number of other crystals (such as yttrium aluminium garnet (YAG), $Y_3Al_5O_{12}$; calcium tungstate, $CaWO_4$; and calcium fluoride, CaF_2) that can be used as the basis for a laser. Also there are a number of solid-state lasers which employ an active element held in a matrix of glass which is pumped by a light source such as a helical xenon flash lamp. Only certain glasses are capable of standing up to the special conditions (especially heat) existing during the operation of the laser. The glass used is often a silicate, such as SiO_2 or $(SiO_4)^2$, doped with a rare earth such as neodymium (especially), ytterbium (Yb) or erbium (Er). Neodymium (as Nd^{3+}) has intrinsic fluorescing properties which are very favourable for laser action, although performance depends greatly on the composition of the glass used. Very high energies can be reached with Nd-doped glass lasers, only rivalled by the ruby among crystalline materials, while the cost of the glass laser rods is much lower. Quite a number of laser-based instruments are designed to take either a ruby or a Nd-doped glass rod as the active source of the device.

The power of the ruby and other solid state lasers can be built up still further by a special technique called *Q-switching*, which employs a special type of shutter that delays the release of the energy stored in the pulse till it reaches a very high power level. The shutter is then opened to allow the laser emission to take place in the form of a single very intense pulse of coherent radiation. The most common Q-switch uses a Kerr cell equipped with switchable polarizing filters. When crossed, the transmission of light is attenuated. When the cell is pulsed to rotate the plane of polarization, the transmission attenuation disappears. In another version, the shutter takes the form of a light-absorbing dye (such as phthalocyanine) which is interposed before the partially silvered mirror where the laser beam will emerge. The energy builds up until the shutter is bleached, so releasing the pulse.

The main use of these high-energy solid-state pulsed lasers in surveying lies in the area of *distance measurement*. Much of the development has come from military requirements for rangefinders for use on tanks and with artillery, mortar and infantry units. So, whereas the helium-neon gas laser can emit perhaps 10 mW continuously, a Q-switched laser can emit 10 mW during a single pulse of 20 ns. The extremely bright, high-powered flash can be transmitted over very long distances to give an adequate return signal from a diffusely reflecting target several kilometres away even in daylight without the need for prisms or other reflective material. Quite a number of these devices have been utilized for distance measurement in military and civilian surveying applications. Ruby lasers, and neodymium-doped glass or YAG ($\lambda = 1.065\,\mu m$) lasers with Q-switching are used as the radiation sources in satellite laser ranging where

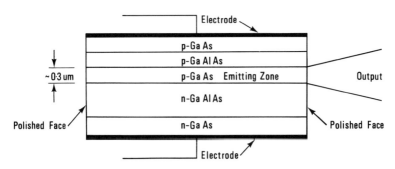

Figure 2.4 P–N junction.

corner cube reflectors are available to give an adequate return signal over distances of hundreds or even thousands of kilometers.

2.1.5.3 *Semiconductor lasers.* When the first ruby laser appeared in 1960 and the first gas laser a few months later in 1961, it was soon realized that there were possibilities for obtaining coherent emission from semiconductor materials, and that these could also form the basis for a laser. In September 1962, two American groups—Hall and co-workers at General Electric and Nathan and his team at IBM Research—successfully demonstrated semiconductor lasers within 10 days of one another.

While there are other (silicon- and germanium-based) semiconductor materials which can be used for laser action, the main material used is gallium arsenide (GaAs). At first, this was a dangerous material to work with—gallium is innocuous at room temperatures but highly reactive at high temperatures (e.g. when GaAs is at its melting point), when arsenide is also a poisonous explosive vapour. But the manufacturing difficulties were overcome, and now that the required chemical purity and crystalline perfection have been achieved, the material is used as very efficient non-laser (incoherent) light-emitting diode in many electro-optical EDM instruments.

But the same material can also be used as the basis for a laser, and under suitable conditions it can be made to emit an intense beam of radiation in the near infrared part of the spectrum. The active element is a p–n junction sandwiched between a p-type and an n-type semiconductor layer (Fig. 2.4). Electrons enter the junction from the n-side and holes from the p-side. The two mix, and a population inversion results. The electrons and holes move in the desired direction by the application of an electric voltage across the junction. This is produced by connecting a suitable electric power source to two electrodes affixed to the top and bottom surfaces of the material. Thus, in this case, the pumping action is electrical. As before, atoms within the material are excited to higher energy states and, in falling back to a lower state, they exhibit electroluminescence and emit photons. The photons travelling along the junction region stimulate extra photon emission and so produce more radiation. The actual reflective lasing action is produced by simply polishing the end faces of the material so that they act as mirrors with a reflectivity of about 30%. One end is then coated with a metal film of gold (a good infrared reflector) to increase its reflectivity to 100%, in which case, the emitted radiation emerges from the other end of the device. The wavelength of the HP3820 EDM device which uses a GaAs diode laser is 835 nm, in the near infrared part of the electromagnetic spectrum. Many other GaAs diode lasers, such as those used in the Vyner, Keuffel and Esser, and Fennel pulse-echo ranging devices, emit their radiation at 904 nm.

The semiconductor laser (often called an *injection laser*) is very efficient, since nearly half of all the electrons injected into the active material as electric current, produce a photon which is output as radiation. Besides this high efficiency, it is very easy to control the output power of the emitted radiation through small changes to the supply voltage. Also, the device is physically very small. The more negative points are that the output has a rather poor spectral purity and it has a rather wide divergence angle. If the semiconductor material is kept at low temperatures (77 K) using liquid nitrogen, the emitted radiation level is high and the power needed for its operation is low. But if operated at room temperatures, high current levels are needed to get comparable output power. It is possible by carefully controlling the operating temperature (using low temperatures) to achieve continuous operation of the semiconductor laser. Since the device is mostly operated at high current densities, this means that frequently the semiconductor laser will be operated in a pulsed mode to keep the power dissipation and heat in the device to an acceptable level.

2.1.5.4 *Liquid lasers.* Certain liquids, which basically are organic dyes, can be used as the basis for a laser. The use of such liquids as the active material in a laser brings the advantages of very high optical quality and of tunability which allows them to generate a wide range of colours. Furthermore, the overheating problems associated with solid-state lasers are readily overcome, since cooling is easily implemented by circulating the liquid. Although organic dye lasers have been used in applications such as spectroscopy, etc., they have not so far been used in surveying instruments.

2.1.6 *The application of lasers to surveying*

Having reviewed the characteristics and properties of the various types of lasers with specific reference to their suitability for surveying operations, the remainder of this chapter will cover the instruments developed specifically for such operations and give examples of their actual application to surveying and construction work. The three main areas which will be covered are distance measurement (section 2.2); alignment methods and deflection measurements (section 2.3); and laser levelling (section 2.4).

2.2 Distance measurement using lasers

Three main methods based on the use of lasers for the measurement of distance in surveying can be identified. These are:

(i) Interferometric methods
(ii) Beam modulation/phase comparison methods
(iii) Pulse echo or timed pulse methods.

2.2.1 Interferometric methods

In this method, the laser-based instrument acts essentially as an interferometer which is normally used to measure small changes in distance, though obviously the summation of a series of such changes will give rise to a measured distance. The optical arrangement of such a device is shown in Fig. 2.5. The coherent light from a continuous-wave laser, such as an He-Ne laser, is divided into two parts by a semi-reflecting mirror which acts as a beam splitter to produce two wavefronts of the same shape and intensity. The two plane reflecting mirrors M_1 and M_2 then reflect back these wavefronts, which are recombined at the beam splitter and pass on to the observation point where they are seen as an interference pattern. Each interference fringe will correspond to a half wavelength of the emitted radiation which, in the case of the He-Ne laser, is $\frac{1}{2} \times 644$ nm $= 0.3\,\mu$m.

If the mirror M_1 is attached to an object such as the component of a machine or instrument, which is moved in a direction parallel to that of the incident beam, a change in the interference pattern will result. The movement can be observed in a suitable telescope by a human observer, but in practice, the counting of the large number of fringes generated by the movement of the mirror will be carried out automatically using a photodetector and its associated electronics. Knowing the wavelength of the laser light source and the number of fringes detected, it is possible to determine the *distance* through which the mirror has been moved. Also it is possible to obtain information about the *motion* of the object by measuring the Doppler shift of the reflected laser radiation to give the acceleration and velocity of the object.

The *resolution* and *accuracy* to which the distance has been measured will be very high, in the micrometre (μm) class, but the length over which this can be carried out is normally some tens of metres. Thus the application of the laser interferometric technique is largely confined to laboratory or industrial environments: it is used more for calibration work and applications in metrology rather than in routine field surveying. Thus, for example, it has been used extensively in standards laboratories to produce a working standard for linear measurement with accurates of the order of $\pm 0.3\,\mu$m over a distance of 1 m in the case of gauge blocks and linear scales (Rowley and Stanley 1965; de Lang and Bouwhuis, 1969; Zipin, 1969). The method has also been used by the National Physical Laboratory (NPL), Teddington, UK, to calibrate invar tapes to an accuracy of $\pm 50\,\mu$m over a distance of 50 m (Bennett, 1973). A similar 70 m facility has been established in Australia at the National Measurement Laboratory (Ciddor *et al.*, 1987). However, the method has also been used in the field, employing the special procedure devised by Vaisala for the establishment of very high-accuracy baselines intended to be used for the checking and calibration of electronic distance-measuring instruments (Ashkenazi and Dodson, 1977; Ashkenazi *et al.*, 1979). Another application of a laser interferometer has been the measurement of distances of up to 60 m with an accuracy of 0.01 mm during the construction of high-energy particle accelerators at CERN, Switzerland (Gervaise and Wilson, 1986).

2.2.2 Beam modulation/phase comparison method

Basically this is the familiar method of distance measurement employed with electro-optical EDM instruments already described in Chapter 1. However, instead of a diode or a lamp being the radiation source, a laser is employed instead. The advantages of doing so are as follows:

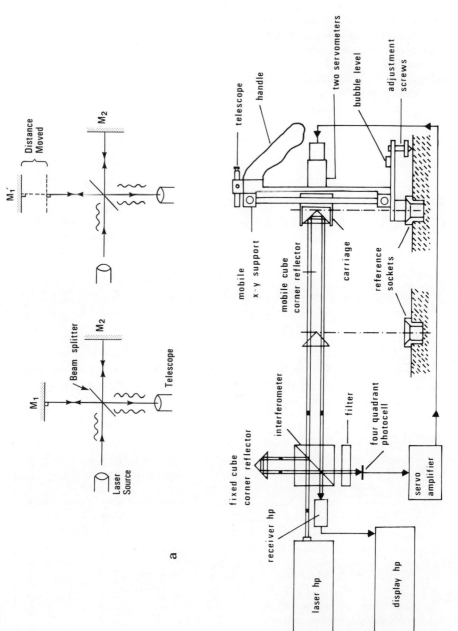

Figure 2.5 Laser interferometer.

(i) The higher power and efficiency of the laser permit a much greater *range* to be measured as compared with any other type of electro-optical EDM equipment
(ii) Its highly monochromatic characteristics result in a very narrow *bandwidth*, thus enabling a very high discrimination against stray light to be implemented, so assisting the operation of the measuring device even in conditions of bright sunlight
(iii) *Ground swing* is virtually eliminated because of the very small divergence of the beam.

A good example of the improvement in performance which can be achieved using a laser is the older AGA Geodimeter Model 6 EDM instrument. In its Model 6A and 6B forms, using either tungsten or mercury vapour lamps as radiation sources, the maximum ranges which were achievable were 2.5 km and 6 km respectively. However, in its Model 6BL form equipped with a helium-neon gas laser, the maximum range was increased to 20–25 km. In the still further developed Model 600 using a more powerful He-Ne laser, the maximum range was increased further, to 40 km. Other AGA Geodimeter instruments employing an He-Ne laser radiation source include the Models 700, 710 and 8. The latter in particular has a very great range—about 60 km in good conditions—which has resulted in its extensive use in traversing and trilateration for the establishment of surveying or geodetic control networks. The Keuffel and Esser Rangemaster has a similar range and application.

As noted in Chapter 1, the Kern ME 3000 Mekometer uses a xenon gas lamp as its radiation source. Again in the latest model, the ME 5000 Mekometer, an He-Ne gas laser has been substituted for the xenon lamp with a substantial (over threefold) increase in range—from 1.5 km to 5 km with a single reflector and from 2.5 km to 8 km with three reflectors—over the earlier model (Meier and Loser, 1986). The Kern Mekometer instruments have been used widely for the monitoring of deformation of dams, bridges, etc., and for the positioning, dimensioning and monitoring of industrial machinery, as well as for geodetic applications.

A further type of laser-based EDM instrument which has been developed for ultrahigh accuracy distance and deformation measurements is the Terrameter (Fig. 2.6) developed by Terra Technology Corporation, USA (Hernandez and Huggett, 1981). This utilizes two lasers—a helium-cadmium laser operating at $\lambda = 441.6$ nm in the blue part of the spectrum, and a helium-neon laser operating at $\lambda = 632.8$ nm. Since each beam will be refracted to a differing extent, a comparison of the two reflected signals will allow an accurate correction for atmospheric refraction to be established, so allowing distances to be measured to a very high resolution (0.1 mm). The application of this instrument to the establishment of the Large Electron Positron (LEP) Ring at CERN is described in Gervaise (1984). However, the very high cost of this type of instrument means that they are not widely available—indeed, the Terrameter has recently gone out of production.

The use of lasers in electro-optical EDM instruments has also spread to *electronic tacheometers* measuring over the much shorter distances encountered in detailed surveys for large-scale maps. Thus the Hewlett-Packard 3820A total station, manufactured in the early 1980s, used a solid-state GaAs diode laser (Gort, 1980). The optical design is shown in Fig. 2.7. The laser radiation is transmitted via a train of prisms on to the reflective surfaces of the objective lens and primary mirror of the instrument's main telescope and sent out to the remote station where retro-reflective prisms reflect the

Figure 2.6 Terrameter.

signal back to the instrument. The return signal retraces the path of the originally emitted beam, passing through a final beam splitter to the photo-diode which acts as a receiver. The phase change between the output and received signals is then measured to determine the distance to the remote station in the usual manner, as described previously in Chapter 1.

2.2.3 *Pulse echo or timed pulse method*

This method is an alternative to the phase-comparison technique described above and is associated almost exclusively with laser-based instrumentation. It involves the very accurate measurement of the time t required for a very short pulse of laser radiation to travel from the measuring instrument to the remote station or target and back. If the velocity v of the electromagnetic radiation along the path is known, then the distance d to the target can be derived from $d = v \cdot t/2$.

The method has been used as the basis for a variety of distance-measuring devices designed to operate over ranges from a few tens or hundreds of metres in the case of land or hydrographic survey applications, to thousands of kilometers in the case of satellite laser rangers. An unusual characteristic of certain of the devices is that they exploit the ultrahigh power achievable with certain types of laser to measure distances to objects which are not equipped with reflectors. Although this characteristic has been

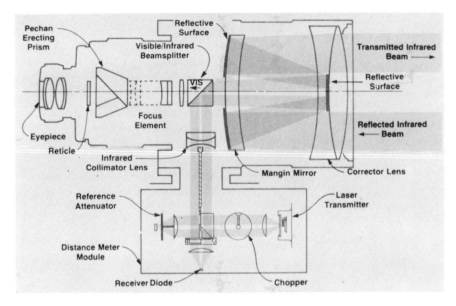

Figure 2.7 Optical arrangement of Hewlett Packard 3820A Total Station (courtesy Hewlett Packard).

exploited mainly for military purposes, it has also been used for special purposes in land and engineering surveying.

Since the basis for the distance measurement is a very short pulse of radiation, only those types of laser utilizing a *pulsed mode* of operation are suitable for inclusion in such instruments, in contrast to the continuous wave operation required for phase comparison. The resolution of a pulse echo device is governed by the length of the emitted pulse and the degree to which the centre of the return pulse (which may be quite distorted) can be determined. Since the speed of electromagnetic radiation is $300\,000$ km s^{-1}, a resolution of 1 m in a single measured range requires the timing of the pulse to be accurate to $1/300\,000\,000$ (3×10^{-9})s, while 1 cm resolution requires 3×10^{-11} s, and 1 mm resolution, 3×10^{-12} s (3 ps). A very accurate and stable quartz oscillator is required to control the measurement of the elapsed time which is the basis of the method.

When short distances are being measured, it is possible to make a large number of such measurements over a very short period of time, i.e. the pulse repetition rate can be high. With long distances, over which a longer time will elapse between the emission and reception of pulses, a corresponding slower rate, i.e. a longer interval between successive pulses, will be necessary. In the case where the object is stationary (such as a tripod-mounted survey reflector), the mean of a large number of ranges can be taken to increase the precision and accuracy of measurement. Where the object being ranged to is moving, such as a tank, a survey ship or a satellite, such an averaging procedure for the determination of a single distance is not possible.

The general principle of all laser pulse echo or timed pulse ranging devices is shown in Fig. 2.8. A high-energy laser has a lens or telescope positioned in front of it which

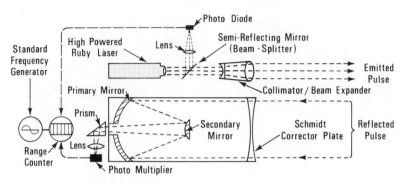

Figure 2.8 Principle of pulse echo or timed pulse method of distance measurement.

expands and collimates the pulse so that it has a small divergence. When the pulse is released by the laser, a tiny part of its energy is diverted by a beam-splitter on to a photodiode, whose signal starts up the timing device. On returning from the object whose range is being measured (with or without the assistance of retro-reflecting prisms), the pulse is picked up by the receiver optics, which may consist of either a simple lens or a more complex Cassegrain-type reflecting mirror-type optical system. These focus the returning pulse on to a photomultiplier, which delivers a stop pulse to the timing counter. In this way, the timing is determined to a resolution in the nanosecond to picosecond range (10^{-9} to 10^{-12} s).

2.2.3.1 *Semiconductor-based laser rangers.* A typical example of the simplest type of laser ranger using the pulse echo technique is the *RF2K Electronic Range Finder*, originally developed by D.J. Vyner in the late 1970s principally for marine surveying applications. This employed a pulsed diode laser emitting its radiation at an output power of 30 W at 904 nm wavelength (in the near infrared part of the spectrum). A co-axial lens was used, the inner component being the transmitting lens and the outer component acting as the receiver. On top of the main ranging unit is a separate telescope through which the surveyor or engineer could sight the object being ranged. The timing gave a resolution in range of 0.1 m, with an accuracy of $\pm$ 0.5 m over ranges up to 2 km if retro-reflecting prisms were used. A still more powerful version, the Model RF7K, could measure distances up to 7 km, again when used in conjunction with retro-reflecting prisms. However it was also possible to use the RF2K device for ranging to targets having no reflectors (brickwork, concrete, metal objects) at distances up to 100 m, and up to 250 m if car reflectors were attached to the object. The measured distances were shown directly on a bright display, the mean of 100 discrete readings being displayed every 0.2 s.

This particular type of laser ranging instrument was later produced by other manufacturers such as Keuffel and Esser (USA), with its Pulse Ranger (1982) with broadly the same specification as the RF2K instrument. The West German manufacturer Fennel has also followed the same general path with its FEN range of pulse rangers. The characteristics of these are as shown in Table 2.1.

Table 2.1 Fennel pulse rangers

Model	Measuring period (s)	Claimed accuracy	Range (m)						
			With prisms				With plastic reflectors		With reflective foil
			1	2	3	6	1	3	
2000	1	± 5 mm + 5 ppm	2000	2500			180	250	100
2000 Rapid	0.1	± 2 cm	2000	2500			180	250	100
2000 S	1		1500	1700	2000		150	200	80
4000	1	± 5 mm	2500	2800	3000	4500	180	250	100
4000 Rapid	0.1	± 2 cm	2500	3000	3500	4500	250	360	180
10 000		± 10 cm (20 m–4 km)							
Marine		± 20 cm (5 m–25 m)	2500	3000	3500	4500	250	360	180

More recently, a still more powerful model, the FEN 20 km (Fig. 2.9), has been introduced with a range of up to 20 km using 20 retro-reflective prisms. The claimed accuracy is ± 10 cm ± 5 ppm of the measured range. In addition, the Model FEN 101 has appeared, designed specifically to be used for reflectance off concrete and other comparable materials over ranges up to 50–250 m, depending on the characteristics of the surface used.

It is interesting to note that recently Wild has also introduced the DI 3000 Distomat instrument (Grimm *et al.*, 1986) which utilizes the same pulse echo measuring principle, based on laser diode technology, as that described above. In this respect, it therefore differs completely from the previous Wild DI Distomat series of electro-optical EDM instruments, which are all based on the phase comparison method. Like the Vyner, Keuffel and Esser, and Fennel instruments described above, the DI 3000 Distomat utilizes a GaAs laser diode (in this case, $\lambda = 865$ nm) which allows maximum ranges of 6 km (with one prism) to 14 km (with 11 prisms) in good conditions. In a standard (non-tracking) mode of measurement, the claimed accuracy (r.m.s.e.) is ± 5 mm ± 1 ppm. A variant of the instrument, the DIOR 3002 Distomat, has also been introduced to measure distances without reflectors up to 200 m.

All of these devices can be mounted on a yoke or, in the case of the Wild instruments, attached directly to a theodolite telescope, so permitting the pulse ranging to be carried out in conjunction with the theodolite which gives the horizontal and vertical angles required for reduction of measured distances to the horizontal and the determination of position and height. However, a further recent development from Fennel is the FET 2 Electronic Total Station, which, as its name suggests, is an electronic tacheometer, but based on the pulse echo technique instead of the phase comparison method described in Chapter 1. The resolution of the angle measurements is 1 mgon (3 s); the distance-measuring component of the FET 2 may be any one of the FEN ranging instruments described above. Digital data can be stored internally in an integral memory, or transferred via a serial RS-232C line to a portable microcomputer such as the Epson HX 20.

Figure 2.9 Fennel FEN 20 pulse ranger.

A similar type of device, designed specifically to measure vertical and horizontal profiles across rockfaces in quarries and open-cast pits, is the MDL Quarryman (Fig. 2.10). Again this comprises a semiconductor laser for the distance measurement with a resolution of 10 cm, and horizontal and vertical angle encoders with a resolution of 0.01° (36 s) of arc. The measured values of distance and angle are transferred to an electronic notebook where they are recorded and stored. On completion of the surveyed profiles, the measured data is transferred to a standard IBM-PC microcomputer, where a software package is used to compute each profile with output in the form of tables and graphical plots.

Still using the same technology, the *Atlas Polarfix* has been devised and produced by Krupp Atlas Electronik specifically to provide dynamic position fixing in a marine environment, for example during hydrographic surveys of coastal and inland waters and the position fixing and tracking of dredgers, tugs, rigs and other offshore vessels. On board the vessel, a ring of retro-reflecting prisms is set up in a convenient position to allow reception of the laser pulse from any direction. The position of the vessel is determined from the polar coordinates (distance and angle) measured by the Polarfix, the quoted accuracies being ± 0.5 mm ± 0.1 ppm per km of the measured range (up to 3 or 5 km with different models) and 0.001° (4 s) in angle. The device has an automatic tracking capability arising from the use of a sensing head and feedback loop to give signals to the drive motors. Thus after the initial set-up of the instrument and acquisition of the target by the operator, the tracking of the ship is carried out

Figure 2.10 MDL Quarryman.

Table 2.2 Characteristics of Simrad range of solid-state laser rangers

Model	Laser	Wavelength ($\lambda\phi$) (μm)	Min/max. range (m)	Resolution (m)	No. of measured ranges before recharging	Power supply
LP 3	Nd: glass	1.064	200/20 000	5	600	NiCad, 24 V
LP 7	Nd: YAG	1.064	150/10 000	5	600	NiCad, 12 V
LP 9	Nd: glass	1.064	200/20 000	5	600	NiCad, 24 V
LP 150	Nd: YAG	1.064	150/10 000	5	600	NiCad, 12 V

automatically. Where the positional information is required on board the vessel, a UHF telemetry link is used between the Polarfix and the vessel.

2.2.3.2 Solid-state-based laser rangers. A parallel development of laser-based pulse echo ranging devices for survey applications is derived mainly from military requirements and is based on the use of very high-powered solid-state lasers, especially Q-switched, ruby or neodymium-doped glass and YAG lasers. These are produced by manufacturers such as Simrad (Norway) and Carl Zeiss Oberkochen (West Germany). A large variety of models are produced, as may be seen from the Simrad range shown in Table 2.2.

With a measuring resolution of 5 m, obviously these instruments are designed primarily for use by artillery, mortar and infantry units. The LP7 is a hand-held unit, but the others are all tripod-mounted, with yokes and azimuth mounts on which metal circles provide measurements of horizontal angles for the determination of the positions of targets. However for topographic survey purposes, the LP9 pulse ranger can be attached to the Wild T20 theodolite, in much the same manner as the Wild Distomat EDM devices can be mounted on standard T1, T16 and T2 theodolites. Other possibilities are to attach those laser rangefinders to an automatic gyro compass, such as the Sperry or Wild units, to determine true north to give range, direction and positional coordinates.

At first sight, such developments would hardly seem to fit into an engineering survey context, but in fact the basic military laser ranging technology has been adapted for use in offshore surveying and engineering work. A well-known example, which has had a widespread use in offshore survey work, is the GOLF (Gyro-Oriented Laser Field-Ranger) system produced by Oilfield Hydrographic Systems (OHS), now Measurement Devices Ltd. (MDL), of Aberdeen, UK. This combines the Simrad LP7 hand-held laser ranger with a Robertson SKR82 gyro-compass (for north determination) and a rotary encoder (for the digital output of azimuth). Once the Golf device is pointed at the reference point on the target vessel's superstructure, the range and bearing are displayed and the coordinates computed if required.

As with the Polarfix, a telemetry link can be provided to pass the information to the vessel. Typically the system is used from a large stable platform or rig to fix positions of vessels such as a lay-barge, support ship, diving ship or survey ship, where accurate positioning (to metres) is required to ensure correct positioning over underwater objects such as well-heads and templates, or to ensure that anchors are not cast in a position where they will damage pipelines or other seabed installations.

The final version of this laser ranging technology based on the use of solid-state lasers is that associated with *satellite laser ranging*, where still more powerful lasers are utilized for distance measurements. While the basic principle is the same as that described above, the rangers use still higher-energy (usually Q-switched ruby) lasers; large mirror optics to receive and detect the very faint signals returned over ranges of hundreds or thousands of kilometres; and complex tracking mounts to allow continuous measurements to a fast-moving satellite. In view of their size, weight, complexity and cost, most of these ultra-powerful laser rangers are housed in fixed observatories and are used for the tracking of satellites such as LAGEOS, Starlette and GEOS, equipped with banks of retro-reflecting prisms. The main applications are for geodetic purposes such as determination of Earth rotational variations and polar motion; measurement of the Earth's gravity field via observations of orbital perturbations; or determination of the varying heights of the ocean surface in association with satellites such as GEOS-3, equipped with radar altimeters and similar instruments.

However, it is interesting to note a specific application to engineering survey in the USA, where the movement of the plates on either side of the San Andreas Fault is being monitored using satellite laser techniques, two stations being located on each side of the Fault some hundreds of kilometres apart. The observations are being repeated at fairly frequent intervals over a long period of time to detect any motion between the stations located on either side of the Fault.

2.3 Alignment and deflection measurements using lasers

The use of lasers for alignment work is well established in surveying and civil engineering. They are deployed extensively on building construction sites, in tunnels and mines, at sea to control dredging, bridge and pier construction, and in factories during the installation of machinery. The environmental conditions under which they are operated will often be very demanding—dust and extremes of temperature and humidity are often encountered—so the laser equipment used for most alignment work must be sufficiently rugged to withstand such conditions. Almost invariably, a continuous-beam He-Ne gas laser will be used.

2.3.1 *Alignment lasers mounted on optical instruments*
In the simplest application, a low-powered He-Ne laser is simply attached to the telescope of a standard level or theodolite by some type of clamping device. In this case, the beam is set parallel to the line of sight defined by the level or theodolite. However, it is more usual for the instrument to incorporate a special eyepiece including a beam splitter, which allows the laser beam to be injected into the instrument telescope (Fig. 2.11) to allow complete coincidence between the instrument's line of sight and the projected beam. The telescope then projects the beam towards the target where it is seen as a bright, sharply defined spot.

2.3.2 *Specially designed and constructed alignment lasers*
In other applications, for example where alignment is being carried out over long distances as in offshore dredging of trenches for tunnel or pipeline construction or along designated navigation channels, specially-built alignment laser instruments such as the *Spectra-Physics Transit-Lite* series have been used. These employ more powerful He-Ne lasers equipped with a collimating telescope and a sighting device, and are mounted on a specially designed base which can be levelled manually using footscrews (Fig. 2.12). Instruments of this type (called transit lasers in the United States) can project the laser beam to form a circular spot over distances up to 6 km in good conditions. It is also possible to add optical accessories which will transform the pencil-like beam into a well-defined narrow fan shape. In this way, either a vertical or a horizontal reference plane will be defined which will be seen as a line at the target.

For tasks which are less demanding in terms of range and accuracy, a number of manufacturers produce a much simpler type of alignment laser for use in trenches, sewers and drains, usually referred to as a *pipe-laying laser*. As would be expected for this type of application, the instrument is totally waterproof and the beam is projected on to a target which typically comprises a simple graticule of concentric rings engraved on a metal plate or a translucent plastic or fibreglass tile. Such a target is inserted at the end of the section of pipe to be laid, for example in a trench. The target is located at the centre of the pipe, and is kept in place by a template or spider. Next the pipe-laying laser, which may be positioned in a manhole or invert or in the trench itself, is switched on. The position and height of the pipe is then adjusted until the target coincides with the spot of the projected laser beam.

Since many pipes will require to be set to a pre-defined gradient, pipe-laying lasers are not only equipped with bubbles and footscrews to define a horizontal reference line but

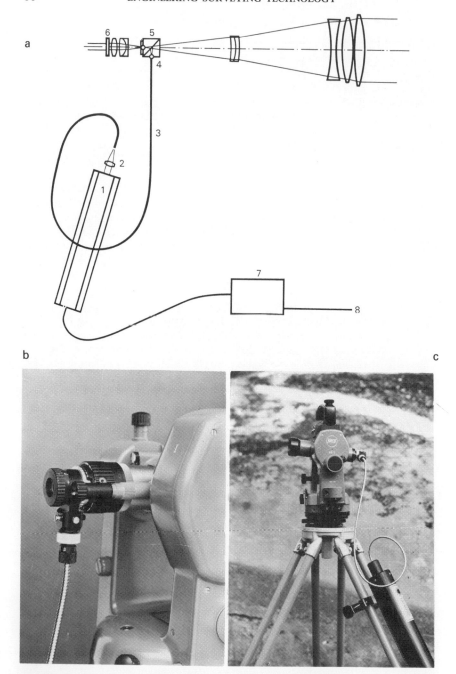

Figure 2.11 Beam splitter alignment laser device. (a) 1, laser light source; 2, objective; 3, light tube (glass fibre optics); 4, reticle; 5, light-splitting cube; 6, filter, 7, laser power converter; 8, power source. (b) and (c) Kern KO-S theodolite with laser attachment.

LASER-BASED SURVEYING INSTRUMENTATION 67

Figure 2.12 Spectra Physics Transite Lite (courtesy Spectra Physics).

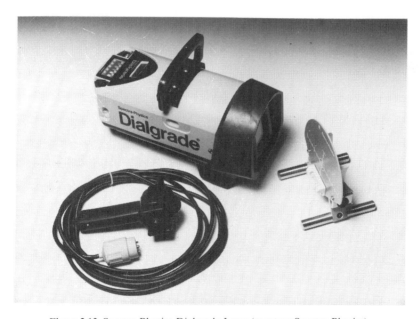

Figure 2.13 Spectra Physics Dialgrade Laser (courtesy Spectra Physics).

normally include a further rotational movement to allow the beam to be set to the required vertical angle or gradient. In certain types of pipe-laying laser instrument, such as the Spectra Physics Dialgrade (Fig. 2.13), an electronic self-levelling device is incorporated. Electrodes are positioned just above the liquid in the level bubble. If the bubble is not level, the electrode will touch the liquid, so completing a circuit which initiates the operation of a small stepping motor which rotates the bubble or the whole optical assembly (laser, bubble, etc.) until the electrode again comes clear of the liquid surface. The horizontal position having been reached or recovered, the motor is then switched off automatically.

Such an arrangement has a number of advantages over the purely manually-operated alignment laser. The unattended operation is an obvious point, but it is a simple extension of the self-levelling system to arrange that the laser beam can only be switched on when the instrument is level or at the required pre-set angle. In this way, if the instrument is disturbed or cannot reach the required direction or gradient, the laser beam will be shut off automatically.

For tunnelling work, a laser projector can be fixed to the sides of the tunnel at the appropriate height and in the desired direction, with the beam passing through two metal plates drilled with holes of an appropriate diameter to define and later to provide a check of the correct alignment (Fig. 2.14). The circular spot produced by the laser beam appears on a target plate which may be mounted on or attached to a tunnelling machine or 'mole'. This allows the operator to guide the machine so that the centre of the target is kept as near as possible to the projected spot. More detailed accounts of the use of lasers in tunnelling are given in Murray (1980) and Thompson (1980).

In a sophisticated extension of this technique, the positioning and direction of the tunnelling machine can be controlled automatically by using a four-quadrant photodetector as the target. If the detector is not positioned centrally on the projected laser beam, then each quadrant will receive and measure a different intensity of radiation falling on it. These differences are used to give signals to hydraulic actuators or electrically driven motors, which will re-set the position of the machine (and its detectors) back until an equal intensity of signal is recovered on each quadrant of the detectors. In this way, a feedback control loop is implemented to ensure constant alignment of the tunnelling machine.

2.3.3 *Vertical alignment lasers*

Vertical alignment is another very important application of laser techniques, widely used to ensure the verticality of buildings, towers, lift shafts, and other structures in the upward direction, and mine shafts, tunnels, etc. in the downward direction. The use of levelled horizontal alignment instruments to which a pentagonal prism is attached to turn the beam through a right-angle is a simple way of ensuring a vertically aligned beam. As will be seen in section 2.4, it is also possible to use certain types of laser level for vertical alignment purposes—those in which the laser is set in a vertical position and from which the rotating prism or mirror can be removed.

Specialized *auto-plumbing devices* have also been designed and constructed, by the UK National Physical Laboratory, for example, as described by Tolman (1976). The layout of the MKV auto-plumb instrument is shown in Fig. 2.15. The laser and its associated beam expander and collimating lenses are suspended as a unit in a gimbal mount to ensure verticality. The maximum operating range claimed is 125 m.

With all of these specialized laser-based vertical alignment devices, it is desirable that

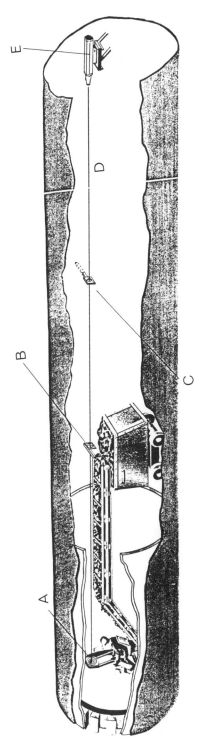

Figure 2.14. Laser for tunnelling operations. *A*, bullseye target. *B*, hole with the cross-hair in the plate mounted on back end of tunnelling machine can provide longitudinal alignment control of mole. *C*, aperture target: hole in board or plate drilled to the diameter of spot allows spot to pass through when correctly aligned. *D*, reference laser beam. *E*, laser mounted on side of tunnel, clear of surface movement and congestion. Laser spot is directed along the line and grade. Spot falls on bullseye target on mole. Operator centres spot.

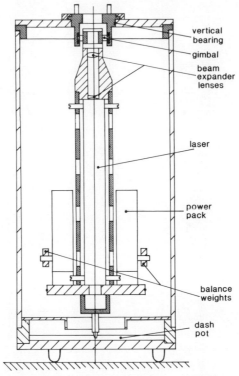

Figure 2.15 MKV Auto Plumb (courtesy NPL).

the instrument should be able to be rotated in azimuth around its vertical axis so that if the projected laser spot remains in a fixed position during such a rotation, it is a check that the beam is indeed set in the vertical direction.

2.3.4 *Measurement of deflections using lasers*

Closely associated with laser-alignment instrumentation and methods are the techniques used in civil engineering which measure deflections of dam walls, of floor shape in box girder bridges, of motorway paving, and of floors and walls under stress. A series of papers by Harrison (1973, 1978a, b) outlines the methods devised and implemented by the National Physical Laboratory to make such measurements.

The so-called *three-point alignment system* is used in each case (Fig. 2.16). This consists of three elements or components:

(i) A *laser radiation source*, almost invariably an He-Ne gas laser
(ii) An *imaging element*, which may take the form of a simple double convex lens, a Fresnel zone plate or a coarse grating
(iii) A *glass or plastic* screen on which a fine grid is marked and on which the projected laser spot is imaged.

The laser projects a spot which passes through the lens or Fresnel zone plate, producing a bright-well-defined spot on the screen. If a load is generated or imposed on the object or base (road surface, floor, wall, etc.) on which the imaging element rests, the result is a

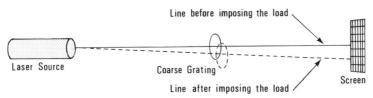

Figure 2.16 Three-point alignment system (courtesy NPL).

deflection of the lens or zone plate, which is seen and can be measured as a displacement of the projected spot on the screen. Accuracies of ±0.2 mm are claimed in the case of motorway paving deflections under high loading.

Harrison (1978b) also describes the measurement of dam deflections, using the same basic technique with a series of ten permanent stations set in the dam wall, the imaging element (a coarse grating) being located at each of the ten positions in turn. The claimed accuracy of determining the displacement as the dam wall comes under load from the impounded water is ±0.5 mm in both the horizontal and vertical directions.

2.4 Laser levels

A typical laser levelling system consists of the following components:

(i) A tripod-mounted laser which can be levelled and whose beam is projected by a rotating prism to form an illuminated horizontal plane
(ii) A detector unit which is attached to and can be moved up and down a levelling staff or rod, to allow the measurement of height differences relative to the laser-defined horizontal plane.

All laser levels use either a rotating pentagonal prism (or pentaprism) or an equivalent mirror system to project the horizontal plane defined by the laser radiation. The laser beam enters the prism in a vertical direction (upward or downward) and it is then reflected twice off the appropriate faces of the pentaprism to emerge in a horizontal direction (Fig. 2.17). The continuous rotation of this prism around its vertical axis ensures the projection of the horizontal plane. It is therefore of fundamental importance that the instrument be kept accurately and continuously level while it is operating in order to define the required horizontal plane.

To ensure that the instrument is set accurately in a horizontal plane, one of three possible methods can be used.

(i) The instrument can be levelled manually using simple tubular spirit levels (bubbles) in conjunction with footscrews in the same manner as a traditional surveyor's dumpy level.
(ii) Alternatively, the instrument may utilize an optical compensator device operating under the influence of gravity, which also ensures the definition and maintenance of the projected laser beam in a vertical direction before it enters the rotating pentaprism. Thus automatic levelling of the instrument takes place in a manner somewhat akin to that employed in a surveyor's automatic level.
(iii) A third possibility is for the instrument to incorporate some type of electronically-controlled self-levelling device, for example using the electro-levels and servo-motors described previously for the Dialgrade pipe-laying laser.

2.4.1 *Manually-levelled laser levels*

These devices are relatively simple in design, construction and operation. The He-Ne gas laser is mounted in a protective case sealed for all-weather use, with a base utilizing footscrews to ensure its deployment in a vertical position (Fig. 2.18). Two tubular level bubbles located at right-angles to each other are used to define the horizontal reference plane. The rotating pentaprism, whose rotation speed can be varied, then projects the laser beam to form a horizontal plane as described above.

The instrument is mounted on a tripod via its tribrach which has a vertical post to allow the instrument to be adjusted to the desired height at which the horizontal plane is to be defined. An alternative mount allows the instrument to be set or laid in a horizontal position to project a vertical beam or plane. A third level bubble is provided, attached to the side of the instrument to allow this vertical plane to be defined.

Obviously the accuracy achievable with an instrument of this type is wholly dependent on the inherent accuracy of the bubbles, their state of adjustment and the care with which the surveyor or engineer carries out the instrument levelling using the bubbles. Any lack of adjustment or error or lack of care in setting the bubbles will cause the resulting plane defined by the instrument to be inclined instead of horizontal or vertical.

These instruments can define the horizontal plane over ranges up to 200–300 m. A relatively flat site of this size is of course uncommon, so the laser level is well able to encompass the desired working area over which construction or surveying operations will normally be taking place. The detector unit used in conjunction with a levelling staff or rod can pick up the beam to give an accuracy in heighting of a few millimetres even at ranges of 200–300 m. Besides the obvious application of defining the elevation, depth or levelling of trenches, foundations, floors and walls, it is also possible to use the level for the control of tasks carried out indoors, such as the installation of ceilings, partitions, floors and walls, where the level can project a vertical as well as a horizontal reference plane. In such situations, the actual bright line projected by the laser on to a wall or ceiling will be used as a direct reference without the need to use a levelling staff or detector unit.

While most earlier models of laser levels were of the manually-levelled type, over the last five years or so the vast majority of laser levels produced and sold have employed one or other of the self-levelling systems, i.e. they utilize optical compensators or electro-levels and servo-motors.

2.4.2 *Laser levels employing optical compensators*

Devices of this type share with the manually-levelled type the basic principle of a vertically projected laser beam which passes into a rotating pentaprism to define a horizontal reference plane. However, in this case, a self-levelling capability is introduced by employing an *optical compensator* in which an optical component such as a lens is suspended in pendulum fashion above the output optics of the laser and operates under the influence of gravity to provide a vertical reference direction automatically. This vertical beam is then turned through a right-angle by the pentaprism, whose rotary movement gives the required horizontal reference plane.

The principle and construction of the Laser Beacon series of self-levelling laser levels manufactured by Laser Alignment Inc. is shown in Fig. 2.19. Two lenses are used in the system. The lower one is a plano-convex lens, which is mounted on a frame suspended freely from a bearing to allow its optical axis to take up a vertical direction. The

LASER-BASED SURVEYING INSTRUMENTATION 73

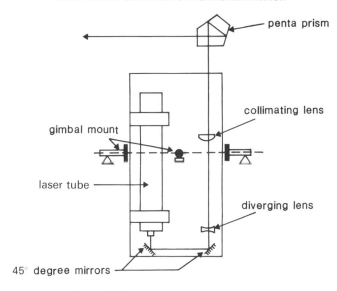

Figure 2.17 Rotating laser pentagonal prism.

Figure 2.18 Manually levelled laser levels

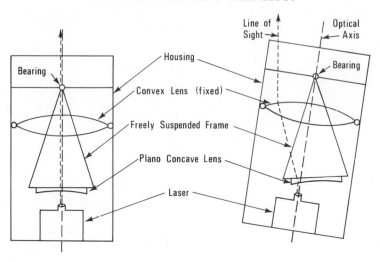

Figure 2.19 Laser Beacon.

suspended frame is fitted with a damping device to ensure that any movement of the frame comes quickly to rest. The upper lens is of the double convex type and lies in a fixed position within the main body of the instrument. The initial levelling of the instrument is carried out rapidly using a circular bull's-eye level bubble as reference.

When the instrument is truly vertical, the projected beam from the He-Ne laser will pass undeviated along the vertical path through the centres of both the suspended plano-convex and fixed double convex lenses. If, however, the instrument is not truly vertical, the laser beam strikes the suspended plano-convex lens, which refracts the beam, causing it to be deviated from its original (non-vertical) path. The beam then passes on to the fixed double convex lens which inputs a further refraction before it enters the pentaprism. The combined effect of all these optical components is to ensure the horizontality of the final output plane. The Laser Beacon instruments also incorporate a shut-off system which comes into operation whenever the range of the optical compensator device is exceeded.

An instrument such as the Laser Beacon can also be set horizontally on its side to project a vertical reference plane, as required in many building construction operations. However, the optical compensator cannot then be utilized, since the laser level is designed specifically to operate with the laser in a vertical position. In this case, the reference direction is defined by tubular spirit levels, whose bubbles are centred using footscrews as described above for manually-levelled laser levels.

The most recent developments in this area include the so-called electronic levels, such as the Spectra Physics EL-1, Laser Beacon LB-2, Pentax PLP-780, Sokkisha LPA 3, Topcon RL-10, and Wild LNA-3 instruments. These also use an optical compensator element to give a self-levelling capability, but the radiation source is a GaAs diode laser instead of the more usual He-Ne type of gas laser. The result is a considerable reduction in the size and weight of the instrument, and it is also possible to use rechargeable nickel-cadmium or alkaline batteries as the power source because of the very low power drain of this type of laser. The Spectra Physics EL-1, shown in Fig. 2.20, has a laser beam which is projected horizontally via continuously rotating mirrors located just above the

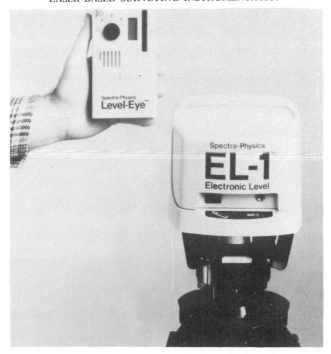

Figure 2.20 Spectra Physics EL-1 (courtesy Spectra Physics).

instrument's footscrews to give the required reference plane, rather than the more usual pentaprism.

Their small size and rugged construction are definite advantages of this type of instrument. The use of the GaAs diode laser means that the emitted radiation is in the near infrared part of the electromagnetic spectrum ($\lambda = 904$ nm), which is, of course, invisible to the naked eye. Thus it cannot be used to define a visible reference plane or line, as required in some of the applications discussed above. It can, however, be set and used in a vertical mode in conjunction with a suitable trivet and manually-controlled level bubbles, in the same manner as the Laser Beacon instruments.

2.4.3 Laser levels employing electronically-controlled self-levelling devices

The instruments which fall within this group have an automatic self-levelling capability once they have been set approximately in a level position, like the laser levels which use optical compensators described above. In this case, this capability is provided through the use of electro-levels and motors whose basic design, function and operation have already been described in connection with the Dialgrade alignment instrument. However, the complexity and sophistication of use of these devices is considerably increased when they are incorporated in an automatic self-levelling laser level.

A number of manufacturers, including Spectra-Physics, AGL (Gradomat) and CLS (Accusweep), offer such instruments based on the use of an He-Ne gas laser. The Spectra-Physics instrument will be used as an example. The entire instrument is contained within a waterproof, dustproof and shockproof casing in which the optical components of the instrument are set in two gimbal mounts which allow freedom of movement about two axes at right-angles to one another. Inside the main casing of the instrument, the He-Ne laser is set vertical, with a folded arrangement of the optical path using mirrors or prisms to ensure a compact configuration of the instrument. A more detailed diagram of the electro-levels and their associated motors and drives is given in Fig. 2.21. From this, it will be seen that two such electro-levels are used, each mounted at right-angles to the other to define the vertical reference direction. Each has its own motor, drive belt and pulley wheels, to allow the optical assembly to be rotated automatically around its own specific gimbal axis should the electro-level detect any lack of centring of its bubble. As in the Dialgrade instrument, if the electro-levels detect any lack of horizontality of the bubbles, the laser beam will be shut off, so that there is no possibility of the instrument being operated in a non-level position.

As with the laser levels using optical compensators, a number of the self-levelling instruments equipped with electro-levels and servo motors can be used to define or project a *vertical reference direction or plane*. In the case of the vertical direction used for the control of high structures, the removal of the rotating pentaprism means that the instrument is then essentially an auto-plumbing device. Where, however, a vertical plane needs to be defined, the instrument has to be laid on its side on a special trivet mount which provides a preliminary or rough levelling capability (Fig. 2.22). For the residual self-levelling, a third electro-level is provided in the direction perpendicular to the plane defined by the first pair of level bubbles which define the horizontal plane. The spinning pentaprism then projects the required vertical plane on to the wall, partition or building, where it can be seen by the construction personnel using the instrument.

The most recent developments in the area of self-levelling laser levels equipped with electro-bubbles parallel those which have taken place in the self-levelling instruments employing optical compensators, in particular, the use of GaAs diode (i.e. semiconductor) lasers instead of He-Ne gas lasers. Examples are the Realist David White ElectroBeam AEL-600, AMA DL-150 and CLS Mighty Mite instruments. Obviously these instruments are larger in size, heavier in weight, more expensive and require a greater power supply compared with their optical compensator equivalents, such as the EL-1 described in section 2.4.2. But they also have a much wider angular range (typically $\pm 4°$ to $6°$) over which they provide a self-levelling capability as compared with that of the optical compensator electronic levels (usually $\pm 10'$ to $12'$).

These new GaAs-based instruments using electro-levels are obviously more compact and lighter, and consume less power than their He-Ne-based equivalents. However, they can only be used in conjunction with special detectors operating in the near infrared part of the spectrum. This is no defect during laser levelling operations, where a detector unit has to be mounted on the staff or rod to intercept the laser-generated horizontal reference plane, no matter which type of laser level is being used. However, if a visual horizontal or vertical reference line is required for use by construction personnel working on the erection of floors, ceilings, walls or partitions, the bright red line produced by the He-Ne laser still offers advantages as compared with the invisible beam or plane of the GaAs laser diode which requires a detector to establish its position.

The accuracy of the levelling achieved by laser levels using electro-bubbles depends on several factors:

(i) The care with which the assembly and calibration of the optical components (lenses, mirrors, etc.) is carried out to ensure exact coincidence of the laser path with the optical axis of the whole optical assembly

LASER-BASED SURVEYING INSTRUMENTATION

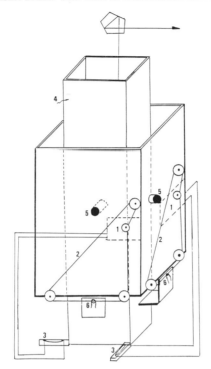

Figure 2.21 Laser levels employing electronically-controlled self-levelling device. 1, servo motor; 2, belt; 3, bubbles with electrodes; 4, optical assembly; 5, gimbal mount; 6, belt is fixedly attached to the optical assembly.

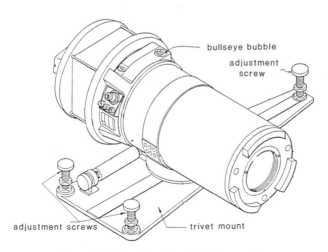

Figure 2.22 Provision of vertical reference line.

(ii) The sensitivity of the electrodes used in the spirit level tubes and the rapidity of their response to any small changes in the level of the liquid contained within the tube
(iii) The speed of response of the servo or stepping motors to the signals received from the electrodes, since any delay between the receipt of the electrical signals and the response of the motors will lead to an error in the defined reference plane
(iv) The electromechanical transmission and drive system, including gears, drive belts, etc., which are liable to wear over a prolonged period of use, so developing play, with a consequent effect on the overall accuracy of the laser level system.

As mentioned above, those laser levels using electro-bubbles can compensate for dislevelment over a very much wider range than those instruments using optical compensators. Thus it is possible to operate them with the instrument placed directly on the ground, which is usually not possible with the optically compensated type of level with its more restricted range of compensation, which normally demands a preliminary rough levelling using a quickset head, or footscrews requiring the use of a tripod. However, in practice, most types of laser level using electro-bubbles are also mounted on tripods, since this gives a much greater ground clearance, which in turn gives a wider range of operation over a specific site. Indeed, special tripods with extended legs are available, which allow the laser level to be set to heights between 3 and 4.5 m, specifically to allow such an increased operational range.

2.4.4 *Laser detectors and control systems*

From the discussion above, it will be seen that the directions, lines or planes defined by a laser level employing an He-Ne gas laser operating in the visible part of the spectrum can be seen against a wall, floor or target plate, and used as visual references by construction personnel on a building site or factory. However, if the laser level employs a GaAs diode laser operating in the infrared part of the spectrum, whose radiation is invisible to a human operator, then a suitable detector must be provided for the reference line or plane to be established. Furthermore, if the laser level is to be used for surveying work, for instance where a levelling staff is used to determine relative or absolute heights, or is to be used to control the operation of earth-moving equipment in an automatic or semi-automatic mode, detectors will need to be provided regardless of whether the emitted radiation is visible or non-visible.

2.4.4.1 *Laser detectors.* Such detectors normally employ photoelectric cells or photo-diodes, the strength of the output signals from these cells varying according to the amount of radiation falling upon them. Silicon or selenium-based cells are commonly used, the type used being matched to the wavelength of the radiation being emitted and projected by the specific laser level.

The simplest type of detector is a hand-held unit which can be used to make marks on walls, forms and other surfaces. If an elevation or a height difference has to be determined, the detector unit is mounted on a levelling staff or rod, and can be raised or lowered manually by the surveyor or engineer in a controlled fashion, a meter displaying the intensity of the received radiation (Fig. 2.23). The position where the intensity is highest is the point at which the centre of the projected beam intersects the staff. This reading can then be taken by the surveyor, using an index pointer mounted on the detector unit to give the height of the terrain or surface below that of the reference plane. In other types of detector units, such as the Spectra-Physics Laser Eye

LASER-BASED SURVEYING INSTRUMENTATION 79

Figure 2.23 Laser detector.

or the Laser Alignment Rod Eye units, a three-light indicator is controlled by an array of photodiodes which indicate whether the position of the central cell of the array is high, low or correct relative to the laser reference plane. Again, when the correct reference level is reached, the staff reading is read off by the person holding the staff or rod.

In a still further development, shown in the Spectra-Physics Laser Rod device (Fig. 2.24), the detector unit is driven by a small motor and drive shaft up and down a specially-built staff after the surveyor has actuated a start button to begin the measuring operation. As the detector unit enters the laser beam, an electronic control system progressively slows its rate of movement as the measured intensity increases. The unit then oscillates up and down the staff within the projected beam until the centre is defined exactly, at which point the movement stops. An audible bleep signals that this situation has been reached, and again the surveyor can observe the staff, reading the position of an index against its graduated scale.

2.4.4.2 *Laser control systems.* A further class of detector unit is designed specifically to control the use of earth-moving machinery, for example to control or set the height of the blade of a grader, bulldozer or trenching machine during excavation or filling operations. The unit consists of a series of silicon photodetectors which are positioned in a circle around a vertical post or mast to allow the interception of a beam from any direction. The detector unit and mast are mounted or located on the scraper blade of the machine. The distance from the reference plane defined by the laser level to the cutting edge of the blade can be set by extending the telescopic mast to the desired position and then locking it. In this way, the cutting edge of the blade cuts a surface parallel to and at a pre-determined distance from the laser-defined reference plane (Fig. 2.25).

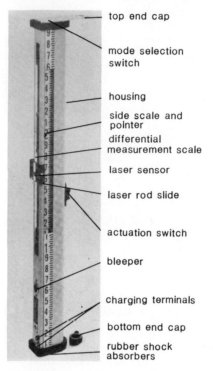

Figure 2.24 Spectra Physics laser rod detector device (courtesy Spectra Physics).

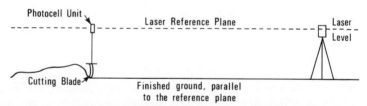

Figure 2.25 Control of earth-moving machinery.

It is even possible to use the blade and mast-mounted detector unit to carry out elevation surveys of an area, as in the Automatic Survey System devised by the Laserplane Corporation. In this case, the mast is not locked in position. The blade is allowed to follow the terrain surface, the detector unit being locked on to the laser-defined reference plane. The telescopic mast automatically lengthens or shortens as the blade height varies (Fig. 2.26). The variation in height so determined can be continuously recorded in a cab-mounted display/recording device over the whole length of each profile covered by the machine. The position of a specific elevation along a profile is determined by a special measuring wheel. A parallel series of such profiles can allow a map of spot height elevations and contours to be prepared.

Such control systems applied to the operation of earth-moving machinery are very expensive—$13 000 was quoted as the cost of a Laserplane control system in 1976. However this sum has to be evaluated in the context of the much greater cost of the earth-moving machinery itself and the potential savings involved in levelling or grading

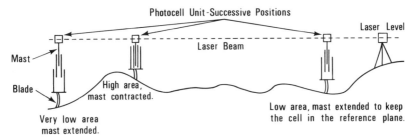

Figure 2.26 Laser Plane elevation survey

very large areas of land. Savings of 30% in the cost associated with such contracts have been quoted by Ross (1976).

2.4.5 *The application of laser levels in surveying and civil engineering*
Comparisons between conventional surveyor's levels and laser levels reveal that the latter have the following advantages:

(i) In conventional levelling, at least two people—one at the instrument and the second holding the level staff—are needed to carry out any type of level work. A laser level can operate unattended, so that only a single person is required to take readings at the staff.
(ii) A large number of engineers and surveyors located in quite different parts of a site can use the reference beam from a single laser level simultaneously for a variety of different purposes. Furthermore, this reference level is continuously available to all users, even at night.

Obviously the laser level finds its greatest applicability on construction sites which are flat or exhibit small height differences. In these circumstances, a large circular area of up to 300 m radius can be covered from a single instrument position. If large height differences are present, the area of operation of an individual laser level will be more restricted, but it can still be employed usefully on a wide range of surveying operations.

In the initial stages of work on a construction site, it can be used to carry out grid levelling over a large area from which a contour map can be derived using the spot heights determined by the laser levelling. Once construction has commenced, the laser level can be used to control site levelling, including excavation and filling operations, the height of forms for pouring concrete or the levelling of screeds and the marking of piles for cut-off. Once a uniform surface of stone, sand or other foundation material has been established, the laser level can be used to control the height and thickness of a layer of tarmac or concrete to the desired level. It can also be used to establish the depths of trenches, to set out kerbs and paving, to allow reference marks to be made on walls, partitions, columns and piles, and to set out grade stakes or sight rails.

When set on its side to define a vertical plane, the laser level can be used to mark out the positions of partitions or walls and to ensure that columns or other vertical objects or fittings are plumbed correctly. As mentioned previously, it is also possible to remove the pentaprism from the laser level to produce a single reference beam so that the instrument acts as a vertical alignment device. This is an especially useful feature of a self-levelling device, allowing it to be used as an auto-plumb controlling the verticality of walls, vertical slip forms and pre-cast panels, besides providing references on high-rise structures.

The accuracy with which a laser level can be used to determine or define heights has been established during an extensive series of tests carried out by Kattan (1981) and Mohammed (1988) on a large number of instruments from different manufacturers. The accuracies of ± 1 to 2 mm for ranges up to 150 m and ± 3 to 6 mm for ranges up to 300 m are quite sufficient for the vast majority of engineering surveying tasks and applications required to be carried out on construction sites. Certainly the accuracy is quite comparable, and the range much superior, to the values which can be achieved using conventional surveyor's levels. When one notes also the single-person operation, the possibility of multiple simultaneous use of an individual instrument, the possibility of defining inclined planes, and the easy night-time operation, it immediately becomes obvious why the laser level has displaced the traditional type of surveyor's level for many of the surveying activities carried out on construction sites.

2.5 Lasers and safety

The characteristic feature of the laser beam or pulse is that it stays highly collimated over long distances. The matter of safety, in particular with respect to the possibility of damage to the retina of the eye of the observer, must therefore always be borne in mind. It will be immediately obvious that the safety issues relating to a very low-powered alignment laser or a rotating beam He-Ne laser level operating in the visible part of the spectrum are very different to those associated with a Q-switched, neodymium-doped glass laser emitting very high-powered pulses in the non-visible infrared part of the spectrum for distance measurement, often over long ranges.

In the UK, all available laser products are classified by the British Standards Institution according to their hazard potential. Details of this classification are given in BS 4803, published in 1972. A further assessment and guide has been produced by the Royal Institution of Chartered Surveyors (1980), entitled *A Guide to the Safe Use of Lasers in Surveying and Construction*, which is aimed specifically at the land and engineering surveyor. In the United States, the use of lasers is regulated by the Occupational Safety and Health Authority. Further discussions of the issues and problems concerning lasers and safety are given by Price (1974) and Gorham (1980).

2.6 Conclusion

Laser-based instrumentation now forms a most valuable part of the surveyor's armoury which can be deployed with advantage on a large number of surveying tasks, ranging from levelling and alignment work to the measurement of distances from a few metres to several thousands of kilometres. For the future, it seems certain that the laser will be used increasingly in all surveying instruments which entail the optical projection of a beam, spot, fan or line towards a specific target, as required in levelling, alignment or deformation measurements. The development of laser rangers with very short pulse-lengths may also be very significant in the long term, and could even threaten the present dominance of EDM instruments based on the phase comparison technique.

So far, laser-based techniques have not been developed for angular measurements, and it will be interesting to see if such a development will take place over the next few years, for instance in measuring the direction of a specific point on a rotating reference plane using some type of encoder. Another possibility is the automation of levelling with digital read-out and recording of the staff reading, since at the present time, even with a laser level, the reading must be made by the surveyor or engineer.

References and bibliography

Ashkenazi, V. and Dodson, A.H. (1977) The Nottingham multi-pillar base line. *J. Assoc. Surv. Civil Eng.* (now *Civil Engineering Surveyor*), 1–8.
Ashkenazi, V., Dodson, A.M. and Deeth, C.P. (1979) The use of the laser interferometer for scale calibration of the Nottingham base line. *Paper presented to the XVII General Assembly of the IUGG*, Canberra.
Bennett, S.J. (1974) The NPL 50-metre laser interferometer for the verification of geodetic tapes. *Survey Review* **22**(172), 270–275.
BS 4803: (1972) *Guide on Protection of Personnel Against Hazards from Laser Radiation.* British Standards Institution, London.
Ciddor, P.E., Edensor, K.H., Loughry, K.J. and Stock, H.M.P. (1987) A 70-metre laser interferometer for the calibration of survey tapes and EDM equipment. *Australian Surveyor* **33**(6), 493–502.
de Lang, H. and Bouwhuis, G. (1969) Displacement measurement with a laser interferometer. *Philips Tech. Rev.* **30**(6/7), 160–165.
Gervaise, J. (1983) Applied geodesy for CERN accelerators. *Chartered Land Surveyor/Chartered Mineral Surveyor* **4**(4), 10–36.
Gervaise, J. (1984) Results of the geodetic measurements carried out with the Terrameter, a two-wavelength electronic distance measuring instrument. *Proc. FIG Commission 6 Symp., Washington, DC*, 23–32.
Gervaise, J. and Wilson, E.J.N. (1986) High precision geodesy applied to CERN accelerators. *Proc. Applied Geodesy for Particle Accelerators*, Publication 87-01, CERN, Geneva, 128–165.
Gorham, B.J. (1980) Lasers and Safety. *EDM & Lasers: Their Use and Misuse, Proc. Conf. Loughborough University of Technology*, 104–115.
Gort, A.F. (1980) A fully integrated, microprocessor-controlled total station. *Hewlett-Packard Journal* **31**(9), 3–11.
Grimm, K., Frank, P. and Giger, K. (1986) Timed-pulse distance measurement with geodetic accuracy. Wild Heerbrugg Technical Publication, 14 pp.
Harrison, P.W. (1973) A laser-based technique for alignment and deflection measurement. *Civil Engineering* **68**(800), 244–227.
Harrison, P.W. (1978a) 'Keeping to the straight'. *Surveying Technician*, April, 13.
Harrison, P.W. (1978b) Measurement of dam deflection by laser. *Water Power and Dam Construction* **30**(4), 52.
Hernandez, E.N. and Huggett, G.R. (1981) Two colour Terrameter—its application and accuracy. *Tech. Pap., ACSM Ann. Conv., Washington DC*.
Kattan, R.A. (1981) An investigation into the use of laser levels in engineering survey. M. Appl. Sci. Degree Dissertation, University of Glasgow, 135 pp.
Meier, D. and Loser, R. (1986) The ME 5000 Mekometer—a new precision distance meter. Kern (Aarau) Publication, transl. from *Allgemeine Vermessungsnachrichten*, 12 pp.
Mohammed, S.N. (1988) The design, construction, calibration and testing of electronic levels. M. Appl. Sci. Degree Dissertation, University of Glasgow, 168 pp.
Murray, G.A. (1980) The use of lasers at Dinorwic underground power station. *EDM and Lasers: Their Use and Misuse, Proc. Conf. Loughborough University of Technology*, 82–93.
Price, W.F. (1974) Lasers and laser safety. *Survey Review* **22**(173), 289–303.
Ross, R. (1976) Laser beam levelling. *Irrigation Age*, May/June.
Rowley, W.P.C. and Stanley, V.W. (1965) The laser applied to automatic scale measurement. *Machine Shop and Engineering Manufacture*, November, 430–432.
Royal Institution of Chartered Surveyors (1980) *A Guide to the Safe Use of Lasers in Surveying and Construction.* RICS, London.
Schawlow, A.L. (1961) Optical Masers. *Scientific American* **204**(6), 52–61.
Schawlow, A.L. (1968) Laser light. *Scientific American* **219**(3), 120–136.
Thompson, B.W. (1980) Lasers in tunnelling. *EDM and Lasers: Their Use and Misuse. Proc. Conf. Loughborough University of Technology*, 94–103.
Tolman, F.R. (1976) Automatic laser plumbs. *Proc. NELEX 1976 Metrology Conf.*, National Engineering Laboratories, East Kilbride, Glasgow.
Zipin, R.B. (1969) Laser applications in dimensional measurement and control. *Bendix Tech. J.* **2**(2), 58–66.

3 North-seeking instruments and inertial systems

D.A. TAIT

3.1 Introduction

One of the great recurring problems in field survey work has always been the determination of azimuth—the direction of north. This is true for the surveyor in the field, when wishing to start a traverse or radiation survey, but also in the office, where one finds that, for example, some resection computations initially require the computation of an azimuth, while others are considerably accelerated if at least the approximate azimuth of one line is known.

Survey procedures have been devised to overcome this problem to some extent. Traverses, almost by definition, start and end at points of known coordinate and from which other known points can be observed to obtain azimuth; stations for tacheometry and EDM radiation surveys are frequently traverse stations, the traverse legs giving the possibility of orientation.

The determination of azimuth at a point, without necessarily determining the coordinates of that point and without sight of another known point, has been a frequent requirement in all types of survey and, although it can be derived from by astronomical observations, these are lengthy, require considerable computation and are not always possible due to location (e.g. in mines or tunnels) or poor weather conditions.

Instruments have been developed which will give azimuth at an arbitrary point to a variety of accuracies, and will be discussed in this chapter. These fall into three groups, namely (i) compass devices, (ii) gyroscopic devices and (iii) inertial devices.

Although inertial devices do gave azimuth, and must therefore be regarded as north-seeking instruments, azimuth is by no means the only output from these systems. As will be shown in section 3.4, inertial devices can also operate as self-contained position fixing systems.

3.2 Compass devices

A compass consists of a magnetized needle or plate which is pivoted to rotate freely in a horizontal plane. The fact that such a device will come to rest in the magnetic meridian, which in most parts of the world approximates to a north–south line, has been known from early times and has provided navigators and surveyors with an instrumental method of azimuth determination.

NORTH-SEEKING INSTRUMENTS AND INERTIAL SYSTEMS

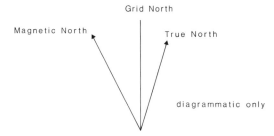

Figure 3.1 Typical relationship between magnetic, grid and true north, as shown on Ordnance Survey maps.

3.2.1 *Principle of compass measurement*

It must be realized at the outset that any magnetic compass device will come to rest in the magnetic meridian. In most cases, this is not coincident with the direction of north, nor will it coincide with grid north (when, for example, the Ordnance Survey National Grid is being used as a coordinate reference), as is shown in Fig. 3.1.

The difference between magnetic north and true north is known as magnetic variation or declination; this value will vary from place to place and, for a particular point, will also vary with time. Variations in the strength and direction of the magnetic field at any one place will produce changes in the magnetic declination of the following three types (Bannister and Raymond, 1984):

(i) Secular variation can be several degrees, and a full cycle can take up to 300 years for a given position, with annual irregular changes of several minutes of arc. As a consequence, when a magnetic bearing is quoted, the date of observation, the declination and the mean annual rate of change for the locality must also be given.
(ii) Diurnal variations are fairly regular and can be up to 10 minutes of arc in magnitude, with the maximum deflection from the mean position occurring about noon each day.
(iii) Annual variations are small and can be ignored for practical purposes.

In addition to these three effects, the compass will also be influenced by the close proximity of iron or steel objects, certain types of iron ore, and electric cables. These effects are known as local disturbance or attraction, and in some places are so unpredictable and irregular as to make the use of magnetic compasses impossible.

For these reasons, it is never possible to obtain a highly accurate value for magnetic variation, and therefore there is no need for compass devices themselves to be equipped with precision reading devices. A readout to a few minutes of arc is sufficient and an accuracy of ± 5 minutes in a bearing determined by compass equipment is about the limit possible.

3.2.2 *Compass devices used in surveying*

Compass devices used in surveying fall into two main groups: hand-held compasses and compass theodolites. The best known, and most widely used, hand-held compass encountered in field survey work is the military prismatic compass, such as the Compass Mk VII from Hall and Watts. These consist of a dial attached to a magnetic needle suspended on a steel pivot. The sighting device, a viewing slot in the prism block

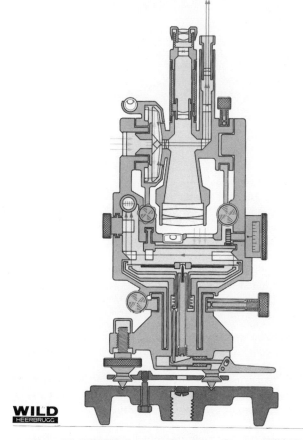

Figure 3.2 Wild TO Compass Theodolite. Courtesy Wild Heerbrugg.

and a vertical hairline in the lid of the box, allows observation of the target, with the hairline also providing the index mark for the simultaneous reading of the graduated circle scale through the prism. Readings can be taken to 10 minutes of arc.

Telescope compass devices, which are set on tripods and have a telescope for viewing, are also available from several manufacturers, and include the Wild B3 and the Stadia Compass BUMON from Breithaupt (Deumlich, 1982).

Compass theodolites are of three main designs. The simplest consists of a trough (box or tubular) compass which can be clamped to the frame of a normal theodolite. By rotating the theodolite around its vertical axis until the needle of the trough compass is over its zero mark, the theodolite telescope can be made to lie in the magnetic meridian. The current horizontal circle reading can then be used as an orientation correction to all future horizontal observations, or the bottom plate can be changed to read zero, in which case magnetic bearings are read directly. Kern provide both a tubular and a circular compass as accessories for their K1A Engineer's Theodolite, and simple tubular compass attachments are available for many other engineer's theodolites. These devices can normally be set to five minutes of arc.

More elaborate versions of the trough compass have been produced. Typical is the Datum Compass from Hilger and Watts, which was designed to fit the end of the trunnion axis of a modified Watts Microptic Transit Theodolite, from which all magnetic parts had been removed.

In the second type of compass theodolite, a compass needle is floated or suspended above a graduated circle. Usually the needle can be raised from its pivot and clamped during transit to prevent damage. This design was used by Kern and by Ertel on their BT I theodolite.

The third possibility is to use a magnetized circle, as in the case of the Wild TO theodolite (Fig. 3.2). When clamped, the circle can be read as in a normal theodolite, with a mean reading optical micrometer. When the circle is placed on pivot, the circle will orient itself to magnetic north, and when re-clamped, subsequent readings will be magnetic bearings.

The base plate of the theodolite is also the base of the metal carrying case. The telescope has a 20 × magnification and the horizontal and vertical circles can be read to one minute of arc.

3.2.3 *Applications*

These instruments can be used to full advantage when high accuracy in absolute bearing is not required, for example in field completion work, forestry surveys, and exploratory surveys of possible route lines, especially when low-order traversing is to be employed, when two interesting consequences result. First, only every second station need be occupied by the instrument, which could lead to considerable savings in time and give more flexibility in the way in which the survey is carried out, although a valuable check is lost when using this procedure. Secondly, when using a north-seeking device, every bearing will be of the same accuracy, no matter what its position in the traverse, and the unfavourable error propagation of a conventional theodolite traverse is avoided.

3.3 Gyroscope devices

A gyroscope is a device which tends to maintain its orientation in inertial space. The conventional form of a gyro is a spinning mass, usually a wheel or disc, turning about an axis, an idea familiar to anyone who has played with a child's top. However, its importance from a surveyor's point of view is that when external forces, such as the Earth's rotation, act on a gyro it responds in such a way that it can be employed as a north-seeking instrument.

In contrast to the magnetic devices discussed in 3.2, gyro devices seek true north rather than magnetic north. They are normally more accurate than compass devices and are free from the effects of magnetic anomalies. Gyro devices are especially useful in mines and tunnels, where they can be used to determine azimuth at the foot of deep shafts and to maintain bearings on long underground traverses. These devices have also been highly developed for military purposes, especially for artillery and missile use.

3.3.1 *Principles of gyroscopic methods*

About 100 years after Euler studied the behaviour of spinning rotors, the French physicist Foucault carried out experiments in the 1850s at the Paris Observatory. He discovered that a long, heavy pendulum set to oscillate in a north–south plane would

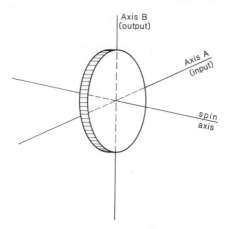

Figure 3.3 Schematic diagram of a gyroscope.

maintain this orientation fixed in space, although to an observer the plane of oscillation would appear to rotate with respect to the Earth. Foucault also realized that a rapidly rotating wheel mounted on gimbals would also maintain its axis of rotation fixed in space, but the demonstration of this had to wait some 50 years until the development of electrically driven motors.

The idea of Foucault was brought into use by Hermann Anschutz-Kaempfe (Strasser and Schwendener, 1964), who was proposing a submarine voyage under the north polar ice cap, where magnetic compasses would be of no use. Other navigation devices were to follow, much of the early work being carried out in Germany, and investigations into the possibilities of this technology for surveying were started. However, because of the bulk and weight of these early devices, they were of very little practical use of the field surveyor. In 1959, Professor O. Rellensmann, making use of the small gyro motors developed for use in aircraft inertial navigation systems, proposed a design for a gyro attachment which was small and light enough to be mounted on a standard theodolite (Caspary, 1987).

A mechanical gyroscope has three mutually perpendicular axes (Fig. 3.3), called the spin axis, the input axis and the output axis. The spin axis contains a disc, wheel or sphere which is rotated at speeds of up to 20 000–30 000 revolutions per minute about this axis.

If any attempt is made to tilt the spin axis once the gyro has reached working speed, by applying a rotation about the input axis, resistance will be felt and the gyro device will react by rotating or precessing about the output axis. The precession about the output axis will depend on the rotation applied to the input axis and the spin speed.

The angular rotation or velocity of the Earth can, for any arbitrary point, be resolved into two components, one around the local vertical, the other around the horizontal north/south line at the point. These components are shown in Fig. 3.4.

If a spinning gyro is suspended in such a way that the spin axis is horizontal, the input axis will pick up part of the horizontal component of Earth spin. In Fig. 3.5, the spin axis of the gyro makes an angle of θ with north. The horizontal component of Earth spin acting around the north–south line (meridian) can be further resolved into two components: $\omega \cos \phi \cos \theta$ about the spin axis, and $\omega \cos \phi \sin \theta$ about the input axis.

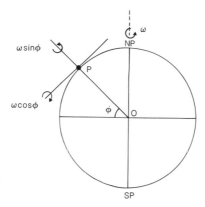

Figure 3.4 The resolution of Earth spin at point P.

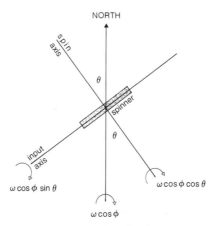

Figure 3.5 The influence of Earth spin on a gyroscope.

This second component, acting on the input axis, will cause a rotation about the output axis and if the gyroscope is free to rotate about the vertical axis, it will precess, and the couple causing this precession is proportional to $H\omega \cos\phi \sin\theta$ where H is the angular momentum of the spinner.

The component causing the rotation will be zero when the spin axis is in the meridian and the angle θ is zero. A gyro held with its spin axis horizontal is thus a north-seeking instrument.

Although the directional couple will disappear when the angle θ is zero, the mass inertia of the device causes the spin axis to overshoot the meridian and is then subject to a directional couple in the opposite direction. The gyro will tend to oscillate about the meridian, and these oscillations form the basis of the operational procedures.

Near the equator, when the latitude angle is small and its cosine is close to unity, the gyro will work most effectively, but the strength of the directional couple rapidly decreases at latitudes above 70°, where gyro work to survey standards should not be attempted.

Table 3.1 Gyro theodolites and attachments. Adapted from Deumlich (1982)

Manufacturer	Model	Measuring time (minutes)	St. dev. of one measurement	Weight (instrument)	Weight (complete)
English Electric	P.I.M.	40	30″		60 kg
MOM	Gi-B1	35	15″	11 kg	55 kg
Wild	ARK-1	10	20″	5 kg	27 kg
Wild	GAK-1	20	20″	2 kg	13 kg
Sokkisha	GP-1	20	20″	4 kg	15 kg
Fennell	TK-4	30	30″	4 kg	6 kg

3.3.2 *Gyroscopes applied to surveying instruments*

It is possible to make a distinction between gyro theodolites and gyro attachments. A gyro theodolite is a complete unit, containing the gyro and theodolite components, whereas a gyro attachment is designed to be mounted on a standard theodolite only when required, allowing the theodolite to be used conventionally at other times. A further distinction can be made between pendulous (or suspended) gyros, in which the unit is suspended and hangs under gravity, and the floating type, in which the gyro floats on a film of liquid, thus relieving the bearings of its weight. Suspended gyros are usually employed in surveying instruments.

Gyro theodolites and attachments have been produced by several manufacturers and are listed in Table 3.1, which has been modified from Deumlich, 1982. Figure 3.6(*a*) and (*b*) illustrates two typical examples.

A gyro attachment which has enjoyed widespread use is the Wild GAK 1 (see Fig. 3.7). The oscillating system consists of the mast and the gyro, which is held under gravity with its spin axis horizontal, both supported by a metal suspension tape. The spinner is rotated at a speed of 22 000 rpm. The supporting system consists of three columns with a chimney-like extension, at the bottom of which is found the frosted glass index with a graduated scale on which the gryo mark is projected (see Fig. 3.8). The whole unit weighs about 2 kg and can sit on a special bridge on the theodolite standards. A separate battery and electronics unit rests on the ground, with a cable leading to the gyro unit.

When using a gyro, observations are made using a combination of the horizontal circle of the theodolite, a stop-watch and the gyro scale.

3.3.3 *Methods of observation*

The detailed operating procedures for gyro devices differ from instrument to instrument and are therefore always given in the manufacturer's handbook. However, for pendulous gyros, the ones with greatest application in surveying, there are two main methods in current use. Both depend on the fact that a pendulous gyro will oscillate around the meridian, following a slightly dampened motion. A reasonably good initial orientation is required for these precise methods, and this can be achieved by using a good compass or by one of the pre-orientation procedures (Schwendener, 1966; Thomas, 1965, 1982; Popovic, 1984).

3.3.3.1 *Turning point (or reversal) method.* For the turning point method, the theodolite telescope must be oriented to within 1–2° of the meridian, before

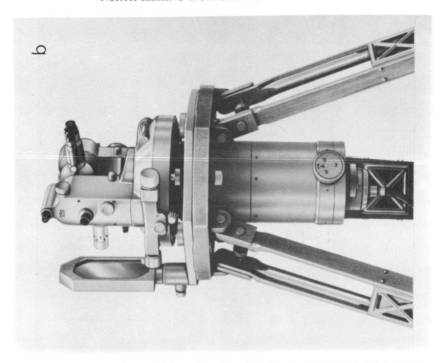

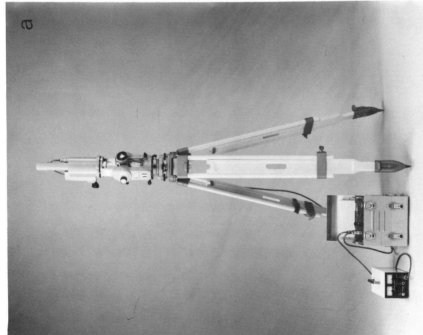

Figure 3.6 (*a*) Sokkisha GP1 Gyroscopic Theodolite. Courtesy Sokkisha Co. (*b*) MOM Gi-311 Gyrotheodolite. Courtesy MOM, Hungary.

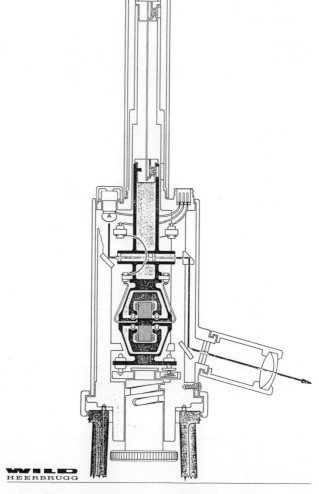

Figure 3.7 Wild GAK-1. Courtesy Wild Heerbrugg.

observations are started. When the gyro is released, the oscillation of the gyro mark against the reading scale can be observed and followed, using the tangent screw of the theodolite, so that the moving gyro mark is kept in the V-shaped notch in the middle of the reading scale (Fig. 3.8b). The gyro mark will move fastest at the centre of its oscillation and will appear to stop for a few seconds at each reversal or turning point, indicated by $U_1 - U_4$ in Fig. 3.9. At each turning point, the horizontal circle reading is noted and the gyro mark is followed again, but in the opposite direction. At least three readings are taken, and the Schuler Mean gives the horizontal circle reading of the meridian:

$$N = 1/2\left[\frac{U_1 + U_3}{2} + U_2\right]\text{etc.} \tag{3.1}$$

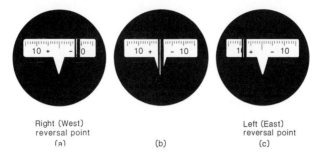

Right (West)
reversal point
(a)

(b)

Left (East)
reversal point
(c)

Figure 3.8 Three views of the reading scale of Wild GAK-1. Courtesy Wild Heerbrugg.

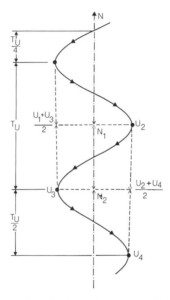

Figure 3.9 Principle of turning point method of observation.

3.3.3.2 *Transit method*. The preliminary orientation for the transit method must be good to 2' of true north, and therefore considerably better than for the turning point method. The theodolite remains clamped throughout the transit method observations, with no following of the gyro mark. Once the gyro has been released, the times at which the gyro mark makes three successive transits through the V-notch are noted on a stopwatch and the elongation values aw and ae are read on the gyro scale (Figs. 3.8a, c, 3.10).

If the telescope is already in the meridian, the east and west transit times would be equal, as the oscillation of the gyro around the true meridian would have been observed as the oscillation around the V-notch. However, if the telescope is not yet in the meridian, the east and west transit times will be different. The difference between the two transit times, Δt, is proportional to the deviation ΔN of the telescope direction from true north.

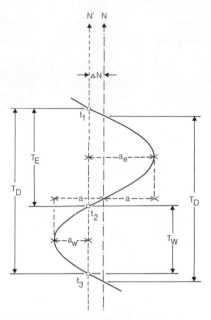

Figure 3.10 Principle of transit method of observation.

Table 3.2 Comparison of turning-point and transit methods of observation

Turning point method	Transit method
Continuous tracking necessary	No continuous tracking
Operator tracking affects precession	No disturbance of precession
Tiring for operator	Easier operating procedure
No accurate timing required	Accurate timing essential
Suspension tape torque eliminated	Suspension tape torque must be known
Specially extended tangent screw may be required	Normal tangent screw sufficient

The deviation from true north can be computed from:

$$\Delta N = c.a.\Delta t \tag{3.2}$$

where c is an instrument constant, a is the amplitude of oscillation, Δt is the difference in half oscillation times.

The relative merits of these two methods are listed in Table 3.2.

Modern gyro attachments give azimuths in an accuracy range of $\pm 5''$ to $\pm 40''$, using between 2 and 7 reversal points in times ranging from 15 to 60 minutes (Smith, 1977). By choosing a suitable instrument, procedure and number of observations, it is thus possible to obtain results suitable for a wide range of survey applications.

3.3.4 *Applications*

Gyroscopic devices have found application in a wide range of survey and engineering tasks but are particularly useful in underground surveys (Hodges, 1970, 1979).

The long wire methods of correlating surface and underground surveys are difficult to apply and limited in accuracy. Gyro theodolites offer an ideal and efficient alternative solution to this type of azimuth transfer, especially in very deep mining operations. Where mining or tunnelling is being carried out in the vicinity of dangerous deposits, when adjacent mine workings are to be connected and when attempting to connect tunnels driven simultaneously from each end, the gyro theodolite offers a better solution than conventional theodolite methods.

The gyro theodolite allows the mine surveyor to establish underground baselines away from the shaft areas in long, straight and uncluttered roadways, with a simple traverse connection to the shaft bottom for coordinate connection.

At the Royal School of Mines (RSM), work has been carried out in various tunnelling projects in Britain, Europe and the United States (Smith, 1983), including a project in Portugal which involved setting out the alignment for an inclined shaft on which it was intended to work from four headings at once. Gyroscopic methods have proved superior to conventional theodolite techniques in situations where tunnel boring is being carried out on multiple faces simultaneously and where precise alignment at break-through is essential.

Gyro attachments have played an important role in the construction of tunnels to house particle accelerators at CERN, Switzerland (Fischer *et al.*, 1987). The LEP tunnel is a remarkable underground construction, some 27 km in circumference and lying in a tilted plane. The basic control framework was laid out on the terrain surface and then transferred down vertical shafts to provide the basic underground framework from which tunnelling operations were controlled. Gyroscopic methods were used in this control transfer and in the surveying of the tunnels themselves. The CERN Applied Geodesy Group developed an automatic gyroscope, based on the Wild GAK-1, by replacing the graduated scale by a CCD line of 1024 photodiodes, and the manual adjustment knob by a motor encoder, to achieve computer-controlled release of the spinning motor.

Experimental work has also been carried out to determine azimuth on or between offshore oil platforms, espeically in areas where weather conditions make astronomical observations difficult or impossible. In spite of the problems caused by vibrations on these structures, gyro azimuths have been obtained, although the repeatability is poorer than on land baselines.

Gyro theodolites can also be used to bring orientation to an independent survey network, where connection to a national network is either impossible or unnecessary. They have also been used for the alignment of radar antennae, directional aerials and relay stations, and for the calibration or layout of landing guidance systems for aircraft.

Gyro theodolites have been used in dockyards for transferring headings throughout a ship undergoing refit, rather than using conventional traversing techniques starting from a known baseline within the dockyard. A plane parallel to the fore–aft axis of the ship, defined by three steel points welded into position, is established near the bow and serves as a reference or master plane for all other planes situated in the ship. This method is independent of the angle of tilt of the vessel in the dry dock and has been accepted as a standard method by one of HM Dockyards (Smith, 1983).

3.4 Inertial survey systems

The use of gyroscopes to give a self-contained, independent system for azimuth determination can be extended to give a navigation or positioning system which is also self-contained and automatic. These systems, known as inertial survey systems (ISS) or inertial navigation systems (INS) are free from external interference, do not need communication with cooperative ground stations and are independent of weather conditions, all factors making them particularly useful for military purposes, their original field of application.

From the early military systems, several inertial devices have been produced for survey purposes. These systems were seen to represent one of the biggest technological advances in survey instrumentation at the time of their introduction and were expected to revolutionize the whole art of position-fixing on the Earth's surface (Cross and Webb, 1980). These early hopes have had to be modified in the light of experience. For the accuracies demanded by the land surveyor, inertial surveying devices used as stand-alone systems have been disappointing. However, when used in conjunction with a conventional survey system (such as a triangulation network) or a satellite positioning system (such as GPS) (Chapter 4) the full potential of ISS can be exploited (Cross, 1986).

The basic concept of inertial surveying is very simple; the necessary mechanical, electrical and electronic components and the associated software required for a successful implementation of this concept are rather more complex.

3.4.1 *Principles of inertial measurement*

An ISS is basically a dead-reckoning system, that is, starting from a known position, measurements are taken to compute a new position. In conventional dead-reckoning methods, direct measurements are taken of heading and velocity, and these measurements require information from external sources, such as air speed indicator, log reading, Doppler counts or electronic position fixing (EPF) signals. In ISS, direct measurements are taken of the vehicle's acceleration, and this requires no external information. These systems are thus based on Newton's second law, which states that the acceleration of a body is proportional to the sum of the forces acting upon it.

With velocity defined as the rate of change of position and acceleration as the rate of change of velocity, a double integration is necessary to compute position from a measured acceleration.

An ISS system therefore measures accelerations and times, and for one coordinate axis, the basic principle can be illustrated in a simple diagram (Fig. 3.11).

For a full system, accelerations are measured by a gyroscopically controlled or monitored orthogonal triad of three accelerometers, mounted on a stable platform in order that position can be determined in three-dimensional space.

Although Fig. 3.11 is a very basic representation of an ISS system, it does illustrate some of the main problem areas in developing this principle to a working system. These are:

(i) Acceleration measurement
(ii) The definition and maintenance of the inertial coordinate system and reference frame
(iii) The need for efficient integration
(iv) The need for initial velocities and coordinate position
(v) The form in which the final output should be presented.

3.4.4.1 *Acceleration measurement.* Two simple instruments for measuring acceleration are a pendulum and a spring-balanced mass. However, the output from each of these devices is not linearly proportional to the acceleration applied. A 1g acceleration would

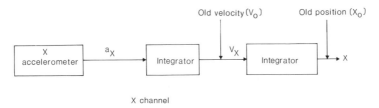

Figure 3.11 Basic outline of an ISS.

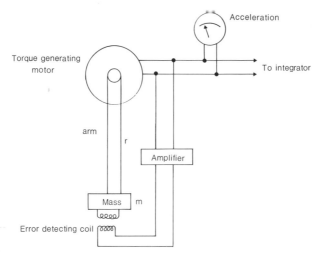

Figure 3.12 Pendulum-type accelerometer.

cause considerably more displacement around the rest position than at other points in the range.

In order for the accelerometer to give an output linearly proportional to the acceleration sensed, a more complex arrangement is required. One commonly used inertial accelerometer is a pendulous mass equipped with a sensor-torquer system (Fig. 3.12). A mass m is suspended on an arm of length r from a spindle containing a motor. If the assembly undergoes an acceleration in a direction perpendicular to r, then the mass m will experience a force $F = ma$. The torque T_i about the pivot will be

$$T_i = rF = rma \qquad (3.3)$$

When this torque is applied, the pendulum will tend to swing off vertical. This movement is detected by the error-detecting coil, and a voltage (V) proportional to the angular misalignment is generated and applied to the torque-generating motor, designed to give an output of $T_0 = KV$, where K is a constant.

When the torque T_i induced by the acceleration is balanced by the torque T_0 of the motor, the system is in equilibrium. At this point $T_i = T_0$ and

$$rma = KV$$

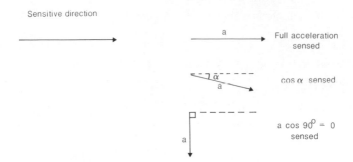

Figure 3.13 Sensitive direction of accelerometer.

or
$$V = \frac{rma}{K} \qquad (3.4)$$

With r and m known from manufacture, the voltage is linearly proportional to the acceleration, if the constant characteristic K of the motor is indeed constant.

As the mass of the pendulum is suspended from an axis, the accelerometer will be sensitive in one direction only. It will measure the full acceleration in this sensitive direction, but only the cosine projection of accelerations in other directions (Fig. 3.13). To measure in three-dimensional space, it is necessary to have three accelerometers mounted in an orthogonal triad. However, the gravitational force of the Earth is also an acceleration, and any accelerometer whose direction of sensitivity is not exactly perpendicular to the direction of gravity will always pick up a component of the gravitational pull of the Earth. In order to separate the acceleration due to the platform from that of Earth gravity, careful consideration must be given to the coordinate reference frames.

3.4.1.2 *Coordinate reference frames.* There are three frames of reference which must be defined and considered.

First, there is the inertial frame. The main properties of inertial reference frames are that they are non-rotating and have unaccelerated origins. For geodetic purposes, such a frame is well approximated by one which has its origin at the Earth's centre of mass and which is non-rotating with respect to the stars (Mueller, 1981). This is the frame of reference in which inertia occurs and in which the accelerations are assumed to take place.

Secondly, there is the accelerometer coordinate system. This is the three-axis system which contains the three accelerometers. The accelerations are therefore measured in this coordinate system. If accelerations occurring in the inertial coordinate system are to be measured by these accelerometers, then the accelerometer reference frame must be parallel to the inertial frame, or the relationship between the two frames must be known.

The third frame of reference is the geodetic system, the system in which the output is normally expressed.

An inertial device can only be of use in surveying if the relationship between the accelerometer frame and the geodetic frame is known and if from the accelerations which are measured the accelerations of the platform can be determined. The directional stability of gyroscopes is used to establish and maintain these relationships, and this can be accomplished in three ways.

If the accelerometers are held by gyroscopes in the inertial frame of reference, double integration will give coordinates in the inertial frame which then need to be numerically transformed into the geodetic system. This solution is called a 'space-stable' or 'space oriented' system, and is used in the Honeywell Geo-Spin system.

If the accelerometers are rigidly attached to the vehicle, and the vehicle rotations with respect to inertial space are measured, the system is called a 'strap-down' system. While this is the simplest from the measurement point of view, very complex computations are required. It is an ideal system for use with laser and fibre optic gyros, which have fewer mechanical components.

If the accelerometer frame is continually rotated to remain in the geodetic reference frame, the system is called 'local-level' or 'local vertical/local north'. This is a complex mechanical solution but requires simpler computations. It is the system used in most survey applications and by the Ferranti and Litton companies.

In each of these configurations, gyroscopes are used because of their ability to maintain orientation in inertial space. Some gyroscopes are similar to those used in theodolite gyro attachments. The Ferranti system runs on precision roll bearings, containing their own permanent lubricant. Others use air bearings (Litton) or electrostatic suspension (Honeywell).

Laser gyros, which have no moving parts, do not maintain their position in inertial space as do mechanical gyros, but measure angular rotation. These consist of a small solid-state laser source which sends its signal around a triangular light path (Fig 3.14). Any rotation of the block causes a phase difference between the two signals travelling in opposite directions and gives a measure of the angular rotation. Laser gyros are currently used for strapdown platforms in aircraft navigation systems, and may in the future, along with fibre optic gyros, replace some of the conventional suspended gyroscopes in survey systems.

However great the inertia of a gyro, it is not infinite, and with time there will be a drift from the initial orientation, causing a loss in the relationships between the three coordinate systems. Exterior references can correct for this drift, and this will be discussed in 3.4.4.

3.4.1.3 *Integration.* Precise integration is fundamental to the success of all inertial survey systems and should be carried out continuously if position is to be constantly displayed in real time on the output screen. To illustrate the problem to be solved, consider that the acceleration a, the velocity v and position x are known for station i. The position for station $(i + 1)$ can be determined if the acceleration at station $(i + 1)$ is measured.

$$v_{i+1} = v_i + \frac{(a_i + a_{i+1})}{2} \cdot t \qquad (3.5)$$

$$x_{i+1} = x_i + \frac{(v_i + v_{i+1})}{2} \cdot t \qquad (3.6)$$

This double integration will give only approximate results unless the time interval t is very short. For most ISS systems, acceleration is continuously sensed and fed to a capacitor which discharges when full, thus giving a quasi-continuous velocity

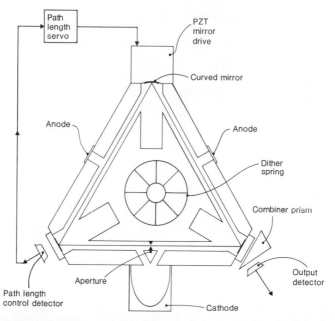

Figure 3.14 Schematic diagram of a laser gyro. Adapted from Street and Wilkinson (1982).

measurement which is sampled every 16–20 milliseconds. As this process has to be performed for each of the three coordinate axes, the operation is fairly demanding of the onboard computer.

3.4.1.4 *Initial conditions.* At the start of an inertial survey, the three components of the velocity (V_x, V_y, V_z) and the three initial position components (X_0, Y_0, Z_0) must be determined. Once the relationship between the inertial coordinate system and the geodetic coordinate system has been established, the change in this relationship is only a function of time due to the very uniform rotation of the Earth in the fixed inertial frame. The initial velocity in the inertial system might have to include the linear velocity due to the rotation of the Earth, even if the platform is at rest with respect to the geodetic coordinate system.

The initial positions (and therefore velocities) are usually determined from a precise knowledge of the astronomical coordinates of the point of departure of the platform. As with most conventional surveys, an inertial survey will normally start and finish on a point of known position.

3.4.1.5 *Transformation.* The output from an ISS is usually required in some geodetic coordinate system, whereas the measured accelerations may be referred to inertial systems not coincident with the system for output. The relationship between the coordinate systems described in 3.4.1.2 must be known so that the necessary transformations can be carried out by the system's computer. The computations necessary will depend on the way in which the accelerometers are aligned and some systems are more complex from the computational point of view than others (see section 3.4.1.2).

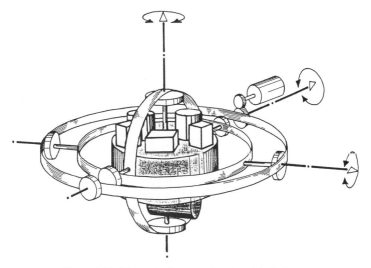

Figure 3.15 Schematic diagram of an inertial platform.

3.4.2 *Construction of inertial survey systems*

At the heart of an ISS is the inertial platform. This is a stable element on which three mutually orthogonal accelerometers and three single-degree-of-freedom (or two twin-degree-of-freedom) gyroscopes are mounted. The platform is isolated from its case and from the vehicle in which it is travelling by a series of gimbals (Fig. 3.15).

The accelerometers measure acceleration in their three mutually perpendicular directions; the gyroscopes control the attitude of the platform and therefore the directions in which the accelerometers lie. The information provided by the gyroscopes allows the transformation of the accelerations sensed in the inertial frame to coordinates in the geodetic frame, or, in the case of a local vertical/local north system (see section 3.4.3), allows the accelerometers to be held in the geodetic frame.

If, when the gyros reach their working speed, no disturbing torque is applied, the spin axes will maintain their orientation in inertial space. However, if a torque is sensed, caused either by a component of Earth spin or by a torquer motor acting on the outer gimbal axis, a precession will occur. The gyroscopes can then be used to stabilize the platform or to change its attitude in any desired manner.

The mechanical components of an ISS are therefore rather complex, and require engineering of the highest quality in their manufacture and thorough testing in their assembly.

The platform is mounted in a case with the on-board computer, a control panel and associated electronics. A typical survey system weighs about 50 kg and has outer dimensions of 540 by 470 by 300 mm (Fig. 3.16).

3.4.3 *Operation of inertial survey systems*

From the point of view of a user, there are few differences in the operational procedures for the different types of inertial system (see 3.4.1), but the internal workings and computations vary considerably. In order to relate the functions of the platform to the actions of the operator, further discussion will be restricted to the 'local vertical/local

Figure 3.16 Ferranti FILS-3 system.

north' system, the one most commonly met in land surveying.

Before measurement can commence with any ISS, the system must be aligned to a geodetic reference frame. This is a two-stage process taking between 60 and 90 minutes, and is carried out automatically under computer control. By driving the horizontal accelerometers until no acceleration is sensed, the platform will be levelled; using a gyrocompassing technique, the platform can be aligned with the local geodetic coordinate system. The accelerometer frame is thus parallel to the geodetic frame and the gyroscopes assume this to be the inertial frame at that instant.

However, this alignment will be disturbed immediately, even if the platform is at rest, because of the rotation of the Earth, which moves the geodetic frame but not the inertial frame. As the inertial frame in a local vertical/local north system is assumed to be coincident with the geodetic system, the platform containing the accelerometers must be torqued constantly. The amount of torque depends on the latitude, the measured elapsed time and the known uniform rotation of the Earth (Fig. 3.17).

The initial position, elevation and other survey data are fed in during this alignment procedure, so that the appropriate torques can be computed.

If the platform is moved over the surface of the Earth, the accelerometer frame and the geodetic frame will again move out of coincidence unless further torques are applied. Three torques are required for an arbitrary vehicle motion, two for a longitude change and one for a latitude change, and in each case the torque can only be applied if the change in position is known. The output from the accelerometers determines the degree of rotation to be applied to the accelerometers—this is the classic loop of inertial surveying (Fig. 3.18).

When the alignment procedures are complete and the coordinates of the starting point of the survey have been keyed in, the instrument can be switched to 'navigate'

NORTH-SEEKING INSTRUMENTS AND INERTIAL SYSTEMS

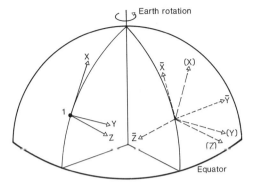

Figure 3.17 Torques required for Earth rotation. XYZ, inertial coordinates (= geodetic coordinates) at t_0; $(x)(y)(z)$, inertial coordinates at time $t_0 + dt$ ($\neq$ geodetic coordinates); $\bar{X}\bar{Y}\bar{Z}$, geodetic coordinates at time $t_0 + dt$. Inertial system must be constantly torqued into geodetic system, even when platform is at rest.

Torques are: $-w\,dt\sin\phi$ about Z axis

$+w\,dt\cos\phi$ about X axis.

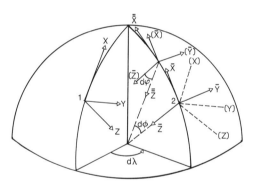

Figure 3.18 Torques required for vehicle movement. Longitude change $d\lambda\cos\phi$ about X axis; $d\sin\phi$ about Z axis. Latitude change $d\phi$ about Y axis. Torques can only be applied if movement $(d\phi, d\lambda)$ can be determined.

mode and the survey can commence. As the instrument is moved, the accelerations are measured, displacements computed and constantly applied to the coordinates to give a real-time output of position. Unfortunately, these raw coordinates will be in error for a number of reasons:

(i) Small outward centrifugal forces will also be sensed
(ii) The strength and direction of the Earth's gravity field varies, and therefore it is impossible to distinguish in this simple mode of operation between the vertical accelerations of the vehicle and gravity
(iii) There are deflections in the direction of the vertical, resulting in the predicted inertial frame vertical direction not being in coincidence with the direction of gravity
(iv) The gyros will drift from their initial alignments over time

(v) Errors will occur in the measurement of acceleration and in integration.

A (relatively) error-free output can be expected only if the accelerations actually measured or used for position determination are those of the vehicle alone in the desired coordinate frame. From the above, it is clear that contained within the accelerations measured will be accelerations due to other factors, and the influence of these accelerations on the output must be eliminated or reduced to acceptable levels.

Much of the accumulated error can be eliminated or at least greatly reduced if, instead of moving the platform from its initial position to the first point for which coordinates are required, the platform is brought to rest from time to time along its journey. These stops, which are of 10–30 seconds duration and should occur every 3–5 minutes, are termed zero velocity update or ZUPT stops. With the platform at rest, one would expect no accelerations to be sensed. If accelerations are present, these must be due to gravity, sensed because the horizontal accelerometers are not perpendicular to the local direction of gravity. This misalignment can be noted (or, in the case of the Litton system, the platform can be relevelled) and the strength of gravity measured. If the platform is moving with moderate velocity, such that centrifugal accelerations are negligible compared to those of gravity, it can be shown that the tilt error of a platform will not increase indefinitely but will follow an oscillation known as the Schuler Oscillation. If a platform is stopped regularly and the two horizontal velocities measured, these values can be used to predict future velocity errors, thus making real-time corrections possible.

This prediction of future errors can be carried out using curve-fitting routines, or by Kalman filtering, which is a statistical procedure for obtaining the most likely estimates (and therefore corrections) in real time. It is a recursive process, taking into account all previous measurements in the computation of each new estimate, and is based on some *a priori* knowledge of the nature of the errors. If external evidence shows that changes have taken place in the assumed nature of the errors, this evidence can be incorporated into the filter and applied to subsequent measurements.

During an inertial survey, there would be many stops, some at the points whose coordinates were required and others at arbitrary intermediate points (ZUPT stops), decided on a time basis. As it is not practical to have the instrument reference point exactly over the survey station, bearing and distance measurements to the ground stations would normally be taken, often using a theodolite/EDM combination mounted on the ISS.

As with a normal traverse, an inertial survey would normally finish on another point of known coordinate. From the closing errors, an adjustment can then be performed on all previously recorded coordinates to give better values, but a further, more rigorous adjustment is normally performed off-line later in the office to give final values (post-processing).

It should be apparent that ISS is an extremely powerful survey system. It can measure simultaneously what could only be achieved by the combined efforts of a theodolite, an EDM, a level, a field astro and a gravity party. Its traverses can be run without towers, cut lines or even lines of sight and in any weather, and can produce results of quite outstanding quality. Early results by Gregerson (1977) show mean errors for intermediate points of 20 cm in latitude and longitude and 15 cm in height for a 60 km traverse carried out in a flat gravity field with the system mounted in a vehicle. Cross and Webb (1981) report mean errors after post-processing for a 30 km helicopter survey of 26 cm for planimetry and 14 cm for height. Their corresponding figures for a four-hour Land Rover traverse over 30–40 km of very rough terrain were 63 cm and 31 cm respectively. By 1985, average errors of about 20 cm in planimetry and 10 cm in height were found by Cross and Harrison (1985), and Reuber (1985) quotes accuracies of the order of centimetres over short distances.

3.4.4 Post-processing of ISS data

Inertial systems are commonly used for the navigation of missiles, aircraft, ships and submarines. For these applications, a real-time output is essential, but the highest accuracies are not required, and the results obtained during the mission, corrected by Kalman filtering or by curve-fitting techniques, are adequate for most navigation applications.

In contrast, for geodetic purposes a final output in real time is not a usual requirement, but the highest possible accuracies are normally desired. The data generated during the survey are therefore recorded on tape to be processed later off-line on a more powerful computer. These operations, known as post-processing, can improve the coordinate information considerably.

If post-processing is intended, the operator would prefer to have raw data recorded, that is data not treated by a Kalman filter. However, as raw accelerations are not available in the Litton and Ferranti systems, the computed velocities for each channel at some multiple of the integration period would be used, but even raw velocities are not available on some systems. More refined corrections for the disturbances discussed in 3.4.3 would then be applied, making more extensive use of ZUPT information to give final ISS survey coordinates for all survey points. The processing is completed by performing a survey adjustment on all ISS lines, using existing control data. Ideally, the ISS lines should form a network with the known control points situated at the junction points (Cross and Webb, 1980; Cross, 1986).

3.4.5 Inertial systems for surveying

There are currently two main manufacturers of inertial systems for use in surveying. In UK, Ferranti has produced a number of systems derived from their navigation systems developed for military use. These include the Position and Azimuth Determination System (PADS) for use in artillery survey, the almost identical Ferranti Inertial Land Surveyor (FILS) and the Ferranti Inertial Navigation Directional Surveyor (FINDS) for casing surveys.

In the US, the Litton Company have produced systems for many survey applications, including road and rail surveys and for the precise navigation of aircraft flying aerial photography for photogrammetric mapping.

Several other companies produce inertial survey systems, but their equipment has not been as widely adopted as that of the companies mentioned above, or is not available for civilian use.

3.4.6 Applications

With such a versatile and accurate survey system, there are hardly any survey tasks for which an ISS could not be used. The technical specifications for a job, the type of terrain, the weather, or time limits, would not usually cause difficulties for an ISS.

As these systems are expensive, ISS has only been used when economic, and there are many survey and engineering tasks where this has proved to be the case.

106   ENGINEERING SURVEYING TECHNOLOGY

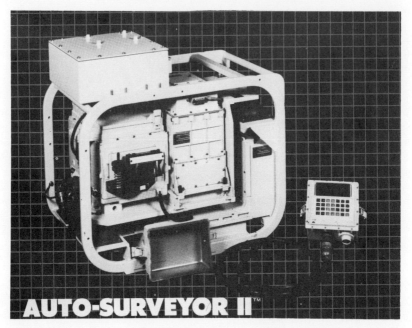

Figure 3.19 Litton Autosurveyor.

One of the first tasks in land surveying to be attempted using ISS was the densification of existing control networks in relatively remote areas for exploration and photogrammetric mapping. This requirement, expressed by Sheltech (Canada), encouraged Ferranti to develop the Ferranti Inertial Land Surveyor (FILS) from their PADS system. The Litton Autosurveyor, a development of the Litton LN-15 aircraft navigation system, was also used in Canada to provide points between Doppler stations (Fig. 3.19). Both of these projects showed that if a sparse but reliable control network of points already existed, the densification could be carried out economically using inertial survey methods (O'Brien, 1979).

Seismic surveys are often carried out in areas of sparse control, and inertial methods have been successful in determining the position of points in such surveys.

At the other end of the survey spectrum, inertial systems have been used in cadastral mapping. SPAN International report successful cadastral surveys in the USA, where monumented subdivision boundary survey stations were coordinated for whole counties. ISS has also been used to recover old corners of federally owned land in Alaska and the western United States (Harris, 1978).

ISS has been useful in surveys of road and railways, to determine the radii of curvature and the gradients. Ferranti have carried out such a survey for all trunk roads in Scotland, using their FIRS system, recording velocity data at 0.6 second intervals during transit, thus allowing position output every 10 m when travelling at 40 mph. A similar survey was carried out in Kent where, in addition to the gradient and curvature information, it was required to provide a locationally referenced inventory of the road-related infrastructure such as traffic signs, crossings and bridges, (Williams, 1982). A

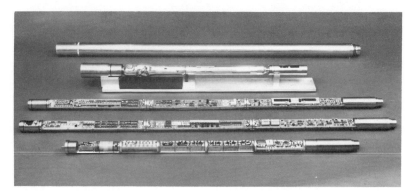

Figure 3.20 Ferranti Casing Surveyor (FINDS).

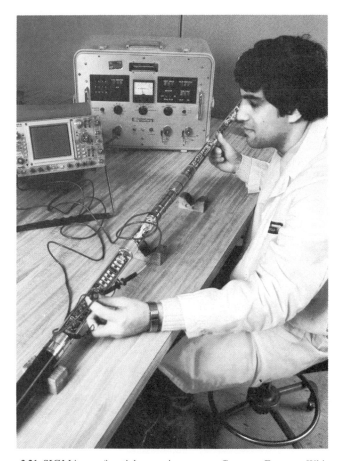

Figure 3.21 SIGMA gyro/inertial surveying system. Courtesy Eastman Whipstock.

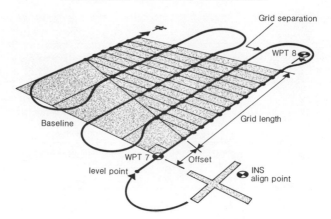

Figure 3.22 Guidance with Litton PICS system.

similar system has been developed for railroad surveys by Ferranti and by Span International.

The offshore oil industry has benefited from this technology. The Ferranti Inertial Navigation Direction Surveyor (FINDS) or Casing Surveyor has been designed to be lowered down within a well casing when the drill string has been removed (Fig. 3.20). Another system is being developed to fit inside a $2\frac{1}{2}$-inch diameter pipe, which will have wider applications with use on other types of borehole. A similar device, called the SIGMA-175 (Fig. 3.21), has been produced by Eastman Whipstock.

A navigation system for use on submersibles (HASINS), to be used alone or in conjunction with some acoustic navigation device, was also developed in the late 1970s, but experienced serious operational problems (Napier and Parker, 1982).

Inertial systems have also been used in photogrammetry and remote sensing to navigate the survey aircraft, especially when acquiring radar imagery which can be obtained through mist and cloud when normal visual navigation is impossible (Fig. 3.22). The Photogrammetric Integrated Control System (PICS) from Litton (DeCamp and Bruland, 1978) allows more regular flight lines to be flown, even in turbulent conditions, and also supplies the position of the camera at each exposure, leading to a reduction in the ground control required in aerial triangulation. ISS can also be used to record or correct the tilt of a side-looking airborne radar (SLAR) antenna, thus eliminating or reducing the distortion usually present due to the variable look angle.

3.5 Conclusions

The hand-held compass and the full inertial surveying system occupy positions at each end of a spectrum of survey equipment which could be used for the determination of the direction of north, and illustrate that in modern engineering surveying there is always a choice to be made in the selection of equipment for a particular task. Every year, this choice is made more difficult as more devices become available. The sensible surveyor should not forget the past with its simpler methods and equipment, nor should he fail to become acquainted with the modern sophisticated systems, but should choose the most appropriate solution for his particular survey or engineering problem from the range of options available.

The use of inertial survey systems for position determination would appear to be most beneficial in an integrated survey system where conventional triangulation, trilateration or satellite techniques are used to provide a control network which can then be broken down and densified using inertial methods (Mueller, 1981). Satellite and inertial methods should be seen as complementary, just as triangulation and EDM traversing complemented each other in classic surveying.

References and bibliography

Anon. (1975) *Fundamentals of Inertial Navigation*. Training Support Services, Litton Aero Products. 105 pp.

Anon. (1977) *An Introduction to Inertial Navigation*. Training Support Services, Litton Aero Products, 45 pp.

Bannister, A. and Raymond, S. (1984) *Surveying*. 5th edn., Pitman, 510 pp.

Caspary, W.F. (1987) Gyroscope technology, status and trends. *Proc. CAS Applied Geodesy for Particle Accelerators*, CERN, Geneva, 166–182.

Cross, P.A. and Harrison, P. (1985) Test network for inertial surveying: implications for lower order control in Great Britain. *Land and Minerals Surveying* **3**(11) 570–577.

Cross, P.A. and Webb, J.P. (1980) Instrumentation and methods for inertial surveying. *Chartered Land Surveyor*, Autumn, 4–27.

Cross, P.A. and Webb, J.P. (1981) Post processing of inertial surveying data. Paper presented to Association of British Geodesists, 14 May 1981, 22 pp.

DeCamp, S.T. and Bruland, R.V. (1978) *The Role of Inertial Navigation in Aerial Survey and Photogrammetry*. Litton Aero Products, No. 809005, November 1978, 27 pp.

Deumlich, F. (1982) *Surveying Instruments* (Translated by W. Faig). Walter de Gruyter, Berlin, 316 pp.

Fischer, J., Hayotte, M., Mayoud, M., Trouche, G. (1987) Underground geodesy. *Proc. CAS Applied Geodesy for Particle Accelerators*, CERN, Geneva, 183–203.

Gregerson, L.F. (1977) A description of inertial technology applied for geodesy. *70th Ann. Conv. Canadian Institute of Surveying*, Ottawa, 17–20 May, 13 pp.

Hadfield, M.J. (1982) Field test results on the GEOSPIN ESG inertial system. *ASCE Spec. Conf. on Engineering Applications of Space-Age Technology*, Nashville, 16–19 June.

Hargleroad, J.S. (1982) Gyrocompass survey system providing a north reference to better than one-arc second accuracy in a portable, lightweight unit. *Proc. ACSM/ASPRS Denver Convention*, 11 pp.

Harris, W.E. (1978) The Spanmark inertial surveying systems. *5th Nat. Cong. of Photogrammetry. Photointerpretation and Geodesy*, Mexico City, 3–5 May, 15 pp.

Hodges, D.J. (1970) The introduction and application of electromagnetic distance measuring instruments and gyrotheodolites to mining surveying practice in the United Kingdom. *Fourth Nat. Survey Conf.*, Durban, South Africa, July.

Hodges, D.J. (1979) The introduction of gyrotheodolites into British colliery surveying practice with particular reference to the Wild GAK-1 gyro attachment. *Survey Review* **XXV** (194) October, 157–166.

Mueller, I.I. (1981) Inertial survey systems in the geodetic arsenal. *Bull. Géodésique*, **55**, 272–285.

Napier, M.E. and Parker, D. (1982) Performance analysis of inertial positioning for submersibles. *HYDRO '82 Symp.* Hydrographic Soc., Southampton, December.

O'Brien, L.J. (1979) The combination of inertial and Doppler systems for secondary surveys. *Conf. Commonwealth Surveyors*, Paper B1, 6 pp.

Popovic, Z.G. (1984) Determination of the position of the null-line of gyroscope oscillations using the transit method. *Survey Review* **XXVII** (213) July, 303–310.

Rueber, J.M. (1985) High accuracy in short ISS missions. *Proc. 3rd Int. Symp. on Inertial Technology for Surveying and Geodesy*, Banff, 16–20 September.

Schwendener, H.R. (1966) Methods and practical experience in the determination of true north with a theodolite gyro attachment. *Allgemeine Vermessungs Nachrichten*, April, No. 4. (English version published by Wild Heerbrugg).

Smith, R. (1983) Engineering and the suspended gyrotheodolite. *Lands and Minerals Supplement*, December.
Smith, R.C.H. (1977) A modified GAK-1 gyro attachment. *Survey Review* **XXIV** (183) January, 3–24.
Strasser, G.J. and Schwendener, H.R. (1964) A north seeking gyro attachment for the theodolite, as a new aid to the surveyor. Wild Heerbrugg, 16 pp.
Street, A.J. and Wilkinson, J.R. (1982) The laser gyro. Royal School of Mines Meeting, *Applications of Gyro Systems to Engineering and Surveying*, 1 July, 14 pp.
Thomas, T.L. (1965) Precision indication of the meridian. *Chartered Surveyor* **97** (March) 492–500.
Thomas, T.L. (1982) The six methods of finding north using a suspended gyroscope. *Survey Review*, **XXIV** (203) January, 225–235.
Williams, E.P.J. (1982) Inertial survey applications. Royal School of Mines Meeting, *Applications of Gyro Systems to Engineering and Surveying*, 1 July, 14 pp.

4 Satellite position-fixing systems for land and offshore engineering surveying

P.A. CROSS

4.1 Introduction

Satellite positioning is the term used to describe the determination of the absolute and relative coordinates of points on (or above) the Earth's land or sea surface by processing measurements to, and/or from, artificial Earth satellites. In this context, absolute coordinates refer to the position of a point in a specified coordinate system, whereas relative coordinates refer to the position of one point with respect to another (again in a specified coordinate system). Relative positions are generally more useful in surveying and can usually be more accurately determined.

The first applications of the technique were made in the early 1960s, but at that time the lengthy observing periods (weeks or months at a station) and rather low accuracy (standard errors of several metres) meant that it was only really useful for global geodesy. Important results were, however, obtained, especially in the connection of various national and continental terrestrial networks and in the determination of the overall position, scale and orientation of national coordinate systems. Nowadays relative positions can, in certain favourable circumstances, be determined from satellite measurements with standard errors of a few millimetres within a few minutes. Clearly this makes satellite positioning a powerful tool for a wide variety of engineering applications.

Moreover, satellite positioning has two very important advantages over its traditional terrestrial counterpart. Firstly, as in the case of inertial surveying, the derived positions are genuinely three-dimensional. This is in direct contrast to traditional techniques where plan and height control have invariably been treated separately, both from a point of view of station siting (plan control points are normally on hilltops whereas height control points are usually located along roads and railway lines) and from a point of view of measurement and computation. Secondly, and perhaps most importantly, the traditional requirements of intervisibility between survey stations are not relevant. All that is required is that the stations should have a line of sight (in the appropriate part of the electromagnetic spectrum) to the satellite(s) being observed. These two advantages mean that all-purpose control points can usually be placed directly where they are needed, and are the essential reasons for the revolutionary effect that satellite positioning is having on many branches of surveying.

This chapter gives the basic theoretical background to modern satellite positioning and describes the systems that are currently operational. Not all of those that have been

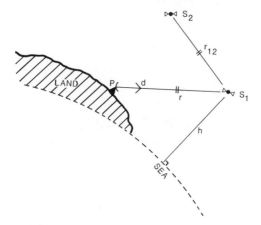

Figure 4.1 Satellite positioning observables.

included have direct engineering applications, but it is important to give an overall picture of the state of the science, because most of the systems directly used for engineering rely on other systems for vital related information such as calibration, Earth rotation parameters and coordinate system definitions. The chapter includes a general review of some of the land and offshore applications of satellite positioning and a look into its likely future.

4.1.1 *Review of satellite positioning observables*
Figure 4.1 shows an idealized geodetic satellite S_1 and ground station P. In principle, the following quantities are observable.

(i) The *direction d* from P to S_1 can be observed either by the use of a special tracking telescope equipped with encoders for angle recording (the theodolite version is called a kine-theodolite), or, more commonly, by means of a camera. In the latter case the result is a photograph of the satellite against a background of stars, and some rather straightforward photogrammetry leads to the computation of the direction cosines of the satellite from the known star directions. Computations with directions observed simultaneously from two ground stations lead to the direction between them, and a number of such interstation directions to 'satellite triangulation'. Obviously direction measurement can only take place at night, and the satellite must either have a light on board or must be lit by the sun (in the same way as the moon). Bomford (1980) gives a full description of this technique, but it is not considered further here because, although it does still have some rather specialist geodetic applications, it is completely obsolete as a positioning technique.

(ii) The *distance* or *range r* from P to S_1 can be observed in a number of ways (all based on the time of travel of an electromagnetic signal) and is probably the most useful observable in current satellite positioning. Three important distance measuring techniques are as follows.

   (a) Laser ranging: where a laser pulse is transmitted towards a satellite carrying a corner cube reflector and the return travel time measured. Laser ranging is

currently extremely important in geodesy, with a large number of applications (see 4.4.1). Also, it is the most accurate method for absolute positioning and for relative positioning over long distances.
  (b) *Pseudo-ranging*: where a satellite transmits a signal at a known time and the time at which it reaches a receiver at *P* is measured. GPS (Global Positioning System) navigation is based on pseudo-ranging (see 4.3.2.1).
  (c) *Phase measurement*: where the receiver measures the phase of a signal received from a satellite. This is not strictly a distance measurement but, rather as in an EDM instrument, when combined with a knowledge of the total number of complete cycles between the satellite and receiver, it can lead to a distance computation (although this is unlikely to be explicitly carried out). Currently precise relative positioning using GPS is based on measured phase (see 4.3.2.1).

(iii) The *radial velocity* $\partial r/\partial t$, of the satellite with respect to the receiver can be observed by measuring the Doppler shift of a signal emitted by the satellite. The measurement process is actually identical to the foregoing phase measurement but the rate of change of phase is computed rather than using its instantaneous value. This is the main observable in positioning using the TRANSIT satellites (see 4.3.1.1).

(iv) The *height h* of the satellite above the Earth's surface can be observed (by the satellite itself) by sending a radar pulse towards the ground and measuring the return travel time. The process is known as satellite altimetry (see 4.4.2), and currently only operates effectively over very flat surfaces (sea, lake, ice, etc.). Its main geodetic application is in the determination of the Earth's gravity field (which is indirectly of great importance to the engineering application of other satellite positioning techniques).

(v) The *satellite-to-satellite distance*, r_{12}, or *radial velocity*, $\partial r_{12}/\partial t$ between S_1 and another satellite S_2 is an observable that is of great theoretical interest to satellite positioning. Such systems are currently not operational (missions are planned for the mid-1990s) but would be of great value in the determination of the Earth's gravity field and in satellite orbit monitoring. These observables are not treated here, but Taylor *et al.* (1984) is recommended for further reading on this topic.

It is important to emphasize that Fig 4.1 represents an idealized situation. No current satellite or ground observing system is capable of all of these measurements. The foregoing has been included simply as an overview of what is possible.

4.1.2 *Geodetic satellites*
Details of particular geodetic satellites will be discussed when the relevant systems are treated. Here some general remarks are made with regard to the design of the types of satellites that are used for positioning. Generally such satellites are divided into two classes: passive and active satellites. Passive satellites have no power source, and all measurements are made on reflected energy. Examples are ECHO and PAGEOS satellites launched in the 1960s for satellite triangulation by photographic methods. These were large (up to 30 m diameter) spherical balloons that simply reflected sunlight. On a clear night a large number of passive satellites can be observed with the naked eye; they are usually parts of rockets that were released when launching special applications satellites and have simply remained in orbit. Similarly, the specialist laser ranging

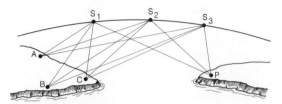

Figure 4.2 Geometrical satellite positioning.

satellites, such as LAGEOS and STARLETTE, are simply passive spheres covered in corner cube reflectors (see 4.4.1).

Most geodetic satellites are active and need a power source to receive, generate and transmit signals, to operate onboard clocks and computers, and possibly for orbit adjustment and attitude control. Such power is normally supplied by means of batteries which are charged via light energy collected by means of photoelectric cells (called solar cells). In order to collect sufficient energy, a large number of such cells are required, and because the surface area of the satellite is not normally sufficiently large to house them, they need to be mounted on panels attached to the main body of the satellite (see for instance Fig. 4.10). This results in satellites with rather complex shapes which in turn leads to problems in orbit prediction due to the subsequent difficulty in modelling some of the non-gravitational forces (see 4.2.2).

Most active satellites have directional receiving and transmitting antennae which must be constantly pointed towards the Earth, hence requiring some form of attitude control. This can be provided in a number of ways, the most common and simple of which is *gravity stabilization*, which is provided by a stabilization boom as shown, for example, in Fig 4.10. Furthermore, some satellites can also control their position in orbit by means of small jet engines, called thrusters. These can serve a number of purposes, including orbit alterations (for example to change ground tracks in satellite altimetry), and maintenance of drag-free orbits (as in DISCOS (DISturbance COmpensation System) whereby a small sphere is enclosed in a vacuum inside the spacecraft which is manoeuvred so that its overall motion is that of this inner sphere, i.e. motion in a vacuum.

4.1.3 *Satellite positioning methods*

Satellite positioning methods can be divided into three groups: *geometrical*, *dynamic* and *short-arc*. Geometrical methods solve the positioning problem by use of pure geometry as, for instance, in Fig. 4.2. Imagine a satellite S passing over four stations A, B, C and P, all of which can simultaneously observe, say, the distance to S. If the relative three-dimensional positions of A, B and C are known, and P is unknown, we can proceed as follows. Let all four stations simultaneously observe the distance to the satellite at points S_1, S_2 and S_3. Then distances AS_i, BS_i and CS_i can be used with the known ground station coordinates to compute the satellite coordinates S_i (for $i = 1, 3$) so enabling distances S_1P, S_2P and S_3P to be used to compute the coordinates of P (from the now known satellite positions). The process, which used to be called *trispheration*, is extremely simple, as no knowledge is required of the orbit (except to ensure the correct pointing of the distance measuring system), and it illustrates satellite positioning at its simplest level.

To apply the purely geometrical approach in practice leads to very serious

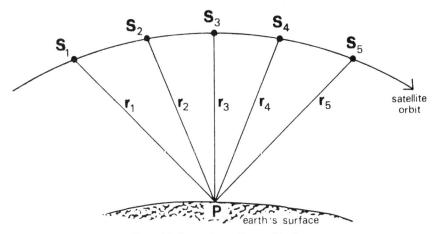

Figure 4.3 Dynamic satellite positioning.

difficulties, e.g. four simultaneous observations are virtually impossible to arrange, four stations must be occupied to obtain the coordinates of one and (unless the stations are hundreds of kilometres apart) the geometry will be extremely poor. The opposite scenario is represented by the dynamic positioning case shown in Fig. 4.3. Here the orbit is modelled (by the use of orbit dynamics) and the satellite position at the time of observation predicted. Hence the position of a single ground station P can be found from, for example, distance observations PS_1, PS_2 and PS_3, to a satellite in three positions or to three satellites. The major problem with this method is that any error in the orbit prediction is transferred directly into the station position. It is worth noting here, however, that if two receivers are placed near to each other, then dynamic positioning will lead to similar errors in each station position, and the relative position between the two stations may well be extremely well determined. This procedure is sometimes known as *translocation*.

The third positioning strategy, short-arc positioning is a combination of the two foregoing methods. A number of receivers are deployed at a combination of known and unknown sites and observations made approximately simultaneously. The relative positions of the satellite along a short arc of the orbit are predicted (much simpler models are needed than those for dynamic positioning which must model the satellite over many revolutions), and a computation made for the relative positions of the receivers and perhaps some general parameters describing the short arc of the orbit.

Most satellite positioning is nowadays carried out using the dynamic method. Short-arc methods are used when the highest accuracy is required, and geometric methods reserved for very special applications where multi-station campaigns can be organized so that large numbers of simultaneous observations are made.

4.1.4 *The role of the Earth's gravity field*

The Earth's gravity field plays a crucial role in satellite positioning. It is the dominant force acting on a satellite in orbit, and a knowledge of it can be critical (especially for low satellites) in orbit prediction for dynamic and short-arc positioning. In order to predict satellite positions, gravity field models must be constructed. These are usually based on computations with data from a variety of sources including terrestrial gravity

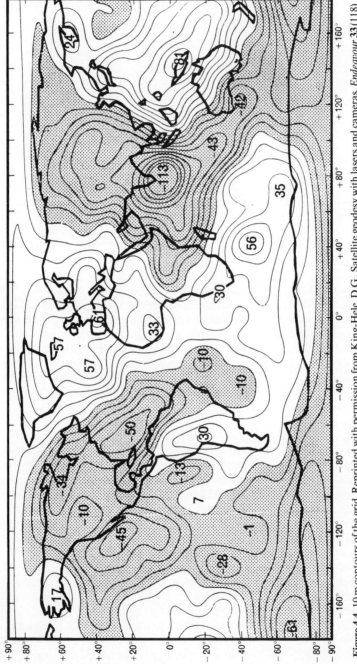

Figure 4.4 10 m contours of the grid. Reprinted with permission from King-Hele, D.G., Satellite geodesy with lasers and cameras, *Endeavour* **33**(118) 3–10, © 1974, Pergamon Journals Ltd.

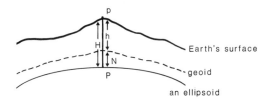

Figure 4.5 Section showing the Earth/geoid/ellipsoid relationship.

measurements, geodetic astronomy and satellite orbit perturbations. The topic is outside the scope of this book, but Reigber *et al.* (1983) is recommended for further reading.

Perhaps less obvious in this context is the role of the geoid, which is one of the equipotential surfaces of the Earth's gravity field. The geoid is actually the particular equipotential surface which corresponds, on average, to mean sea-level, and it (or an approximation to it) is the traditional height datum for most practical engineering surveying purposes. It can be determined in a number of ways, as detailed in, for instance, Bomford (1980), but nowadays it is normally computed either from a global gravity field model or directly from terrestrial and/or satellite observations, or perhaps from a combination of both. An approximate model is shown in Fig. 4.4.

Satellite positioning leads to a pure three-dimensional position (or difference in position) which can be transformed, using reverse forms of equations (4.10), (4.11), (4.12) to an *ellipsoidal height H* (or difference in height) with respect to some mathematically defined reference surface such as an ellipsoid. The situation is shown in Fig. 4.5. In order to compute heights above the geoid, known as *orthometric heights*, *h*, from (4.1), it is essential to know the *geoid-ellipsoid separation*, or *geoidal height*, *N*. Similarly, to compute differences in *h* a knowledge of the difference in *N* is required.

$$h = H - N \qquad (4.1)$$

This point is extremely important, because nowadays satellite positioning can, in many circumstances, yield differences in ellipsoidal heights more accurately than geodesists can predict differences in geoidal heights. This essentially means that our knowledge of the Earth's gravity field is the limiting factor in practically useful height determination from satellite positioning, and currently a tremendous effort is being made by most national surveying agencies to determine the geoid more accurately, both in a global sense and locally for areas in which satellite positioning is taking place. An example of this is the determination of the geoid in the region of the Channel Tunnel—see Best and Calvert (1987).

One increasingly important way to determine the geoid is to occupy established height bench marks (i.e. points with known orthometric height) with satellite receivers and to measure *H*. Then (4.1) can be rearranged in the form

$$N = H - h \qquad (4.2)$$

to yield the geoid. Sophisticated interpolation using models of the topography, and observed gravity if it is available, can then lead to contoured charts for *N*.

4.2 Coordinate systems for satellite positioning

A number of coordinate systems need to be considered when computing positions from observations to or from artificial satellites. Essentially these fall into two categories. Coordinate systems that are fixed in space, known as *inertial coordinate systems*, are normally used for orbit computations, whereas positions on the surface of the Earth are normally expressed in *Earth-fixed coordinate systems*. A knowledge of the relationships between these coordinate systems is crucial to satellite positioning, because without it the computation of the relative positions of points on the Earth's surface and a satellite cannot be undertaken.

Clearly these relative positions depend on both the motion of the Earth and that of the satellite, and these are discussed in 4.2.1 and 4.2.2 respectively. In 4.2.3 the relationship between these motions is considered, and the methods for determining the lengths and directions of vectors between points on the surface of the Earth and satellites are outlined. Finally, in 4.2.4 transformations between coordinate systems used for satellite positioning and the types of coordinate systems that have traditionally been used by engineering surveyors in practice are discussed.

4.2.1 *Rotation of the Earth*

The motion of the Earth in space, although extremely complex in detail, can be described rather simply: the Earth spins about its N–S axis once per day, orbits the Sun once per year and moves in an effectively secular fashion with respect to the Sun within our galaxy (which in turn translates with respect to other galaxies). All of these motions are important in varying degrees to satellite positioning. However, because an artificial satellite essentially moves in unison with the Earth through space, it is the *rotational* motion of the Earth about its own axis that is most important.

Earth rotation is usually considered in two parts: rate of rotation and axis of rotation. Both are variable and both directly affect the relative position of a satellite and a point on the Earth's surface. The rate of rotation of the Earth is often described by its first integral with respect to time: the length of the day (LOD). The LOD varies, according to Vanicek and Krakiwsky (1987), with the following characteristics.

(i) A secular lengthening of the LOD (slowing down of the Earth) of about 2 milliseconds per century due mainly to tidal friction
(ii) Periodic seasonal variation of up to several ms due to meteorological forces
(iii) Irregular variations of up to 10 ms, of unknown cause.

The axis about which the Earth rotates moves with respect to the solid body of the Earth. It does so in a fashion that leads to a motion of the instantaneous pole that can be described, as for instance in Bomford (1980), as follows:

(i) A twelve-month periodic motion with an amplitude of up to 3 m due to meteorological causes
(ii) A fourteen-month periodic motion (often called the Chandler wobble) with an amplitude of up to 5 m, due to the Earth's axis of inertia not being parallel to its spin axis
(iii) A small, apparently secular motion, of about 10 m per century, of unknown cause.

For positioning it is usual to adopt coordinate systems that are essentially fixed with respect to the Earth (otherwise the coordinates of fixed points on the Earth's surface

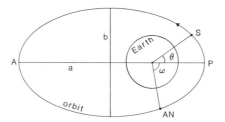

Figure 4.6 A normal orbit.

would be changing). Hence the rotation of the Earth must be constantly monitored and predicted so that surveyors carrying out precise satellite positioning can relate Earth-fixed coordinates to a satellite orbit. A number of organizations, the best known of which is the BIH, or Bureau International de l'Heure, have carried out this task in the past, but recently (due to start in 1988) a new International Earth Rotation Service (IERS) has been formed. For navigation satellite positioning systems (such as the TRANSIT and GPS systems in 4.3) satellite orbits that automatically take account of predicted variation of the rotation of the Earth are available, so the surveyor or engineer may not have to deal explicitly with this problem.

4.2.2 *Satellite orbits*

A satellite orbiting in a perfectly spherical gravity field, and sensing forces due only to that field, follows what is known as a *normal* orbit. Real Earth satellites follow *perturbed* orbits due to the non-sphericity of the Earth's gravity field and other forces (particularly due to the attraction of the Sun, Moon and planets, air drag, solar radiation pressure and tidal effects). First we consider a normal orbit.

Normal orbits obey Kepler's laws, viz.

  (i) The orbit is an ellipse with the centre of the Earth at one focus
 (ii) The velocity of the satellite is such that a line joining the centre of the satellite to the centre of the Earth (called the radius vector) sweeps out equal area in equal time
(iii) The orbital period T squared is proportional to the semi-major axis a cubed.

Figure 4.6 shows a normal orbit, and the following definitions refer to it. The *perigee* P and the *apogee* A are the points at which the satellite is closest to, and furthest from, the Earth respectively. The *ascending node AN* is the point on the orbit where the satellite crosses the equator when passing from south to north. The *perigee argument* ω and *true anomaly* θ are the angles between the perigee and the ascending node, and between the perigee and the satellite respectively.

Kepler's laws were deduced, in the early part of the seventeenth century, from observations of the planets in their motion around the Sun. Later in the same century they were proved mathematically (see Roy, 1982, for a modern version of this) by Newton using his universal (inverse square) law of gravitation. He also extended Kepler's third law by deriving the value for the constant of proportionality as

$$T^2/a^3 = 4\pi^2/GM \tag{4.3}$$

where G and M are the Newtonian gravitational constant and the mass of the attracting body respectively. For the Earth, the product GM is approximately $3.986 \times 10^{14} \, \text{m}^3\text{s}^{-2}$.

Table 4.1 Variation of satellite orbital period and velocity with height

Height	Orbital period	Velocity
0 km	84 minutes	7.9 km s^{-1}
200	88	7.8
2 000	127	6.9
36 000	24 h	3.1

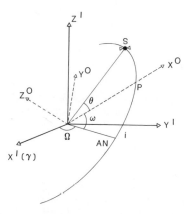

Figure 4.7 Satellite orbital and inertial coordinate systems. i = angle between orbit plane and $X^I Y^I$ plane.

Equation (4.3) allows an approximate computation of the basic characteristics of a circular normal orbit because it enables the orbital period (and hence velocity) to be computed for a satellite of a specified height (the semi-major axis in this case is simply this height plus the Earth's radius). Table 4.1 gives some typical values.

Note that a satellite with a height of approximately 36 000 km orbits the Earth with the same period as the Earth rotation. If it were in the equatorial plane it would remain above the same point on the Earth's surface. Satellites with a period of 24 h are termed *geostationary*, and are useful for communications. Up to now they have not been used for precise positioning, although some specialist navigation (rather than geodetic) systems have been based on them.

Satellite orbits are normally described and computed in inertial space. We define an inertial coordinate system as in Fig. 4.7 with its X^I–Y^I plane in the equator and the X^I direction towards the *vernal equinox* (or *first point of Aries*), γ. The vernal equinox is defined exactly in Bomford (1980); here we simply explain it by stating that it is the apparent direction of the Sun from the Earth as it crosses the equator when travelling northwards (i.e. on 21st March).

A normal orbit is usually described by six quantities known as *Keplerian elements* of orbit. These are as follows.

(i) a and e, the semi-major axis and eccentricity of the orbital ellipse, i.e. the size and shape of the orbit.
(ii) i, the inclination of the orbital plane to the equator, and Ω, the right ascension of

the ascending node (the angle between AN and γ), i.e. the orientation of the orbital plane with respect to inertial space.
(iii) the perigee argument, ω, i.e. the orientation of the orbital ellipse within its plane.
(iv) t_p, a time that the satellite passed (or will pass) through the perigee.

A knowledge of the Keplerian elements enables the computation of the satellite's position in the inertial coordinate system at any time t. The procedure is as follows.

Step 1 Compute the *mean anomaly* M from

$$M = (2\pi/T)(t - t_p) \tag{4.4}$$

where T is computed from (4.3), and the *eccentric anomaly* E from Kepler's equation

$$M = E - e \sin E \tag{4.5}$$

using an iterative scheme.

Step 2 Compute the *orbital coordinates* of the satellite, X_S^o and Y_S^o where orbital coordinates are defined as having their X axis towards the perigee, as in Fig. 4.7.

$$X_S^o = a(\cos E - e) \tag{4.6}$$

$$Y_S^o = a(1 - e^2)^{1/2} \sin E \tag{4.7}$$

Step 3 Rotate the orbital coordinates to equatorial coordinates using

$$\begin{bmatrix} X_S^I \\ Y_S^I \\ Z_S^I \end{bmatrix} = \begin{bmatrix} \cos\Omega & -\sin\Omega & 0 \\ \sin\Omega & \cos\Omega & 0 \\ 0 & 0 & 1 \end{bmatrix} \begin{bmatrix} 1 & 0 & 0 \\ 0 & \cos i & -\sin i \\ 0 & \sin i & \cos i \end{bmatrix}$$
$$\begin{bmatrix} \cos\omega & -\sin\omega & 0 \\ \sin\omega & \cos\omega & 0 \\ 0 & 1 & 1 \end{bmatrix} \begin{bmatrix} X_S^o \\ Y_S^o \\ 0 \end{bmatrix} \tag{4.8}$$

For perturbed orbits, the process of computing inertial coordinates of a satellite is rather more complex and beyond the scope of this text. Ashkenazi and Moore (1986) is recommended for further reading on this matter, but the following general remarks may be useful.

Perturbed orbits are computed by a numerical integration. The various forces acting on a satellite, such as the Earth's gravity field (usually modelled by spherical harmonics), air drag, solar radiation pressure, gravitational attraction of the Sun, Moon and planets, tidal forces, etc., must first be modelled. A starting state vector (position and velocity) is then assumed and the forces for that position summed. The acceleration is computed, and a standard numerical integration process, such as Runge-Kutta or Gauss-Jackson, used to compute a new state vector after a specified time interval. A new set of forces are then computed and the process repeated. Generally, satellite orbits behave in a highly non-linear fashion (i.e. they are very sensitive to the starting values adopted), so it is usually necessary to compare predicted satellite positions with observations and refine the starting values in an iterative scheme.

As an example of the differences between normal and perturbed orbits, Table 4.2 gives the effects of a variety of forces on a GPS satellite (height 20 000 km, inclination 55°) after five hours. Note that it is not possible to generalize from these figures to other

Table 4.2

Force	Effect on satellite position (m)		
	Radial	Cross-track	Along track
Non-spherical gravity field	3000	2600	4000
Gravity of Moon, Sun and planets	260	380	420
Solar radiation pressure	23	11	4
Earth tides	0.09	0.07	0.16
Ocean tides	0.05	0.06	0.07

satellites. Lower satellites would, however, exhibit larger perturbations due to the gravity field and tidal effects, but smaller ones due to solar radiation and lunar, solar and planetary forces.

For most of the satellites used in navigation and geodetic positioning, it is common for the satellites themselves to transmit their own orbital information. This is generally called a *satellite ephemeris* and typically takes the form of Keplerian elements and their rates of change, plus some additional parameters to enable the correction from a normal to a perturbed orbit. The computation of inertial coordinates then involves the use of the foregoing equations (4.3) to (4.8) plus the application of a three-dimensional 'correction', the form of which varies from system to system. It should be noted that broadcast orbital elements will themselves depend on past observations made by a number of tracking stations, followed by a prediction (orbit integration) based on an adopted force model. The resulting satellite positions (and any derived ground station positions) are consequently dependent on the tracking station coordinates and the force model used, and they must therefore be included in any proper description on the coordinate system. This point is important because it extends the classical concept of a coordinate system definition beyond the simple choice of position, direction and scale.

4.2.3 *Relationship between inertial and terrestrial coordinates*

In order to use satellite positions, either to compute ground station positions, or to predict vectors between ground stations and the satellite, it is necessary to transform them from an inertial coordinate system to an Earth-fixed coordinate system. Usually both systems are defined so that they use the same Z direction, hence requiring only a Z rotation to effect the transformation. Figure 4.8 shows an Earth-fixed coordinate system X^E, Y^E, Z^E with X^E towards a conventionally adopted zero of longitude (usually in the vicinity of Greenwich). The angle between X^E and X^I is known as the Greenwich Apparent Sidereal Time, denoted GAST, and the required transformation can be carried out using

$$\begin{bmatrix} X_S^E \\ Y_S^E \\ Z_S^E \end{bmatrix} = \begin{bmatrix} \cos \text{GAST} & \sin \text{GAST} & 0 \\ -\sin \text{GAST} & \cos \text{GAST} & 0 \\ 0 & 0 & 1 \end{bmatrix} \begin{bmatrix} X_S^I \\ Y_S^I \\ Z_S^I \end{bmatrix} \qquad (4.9)$$

The value of GAST at any particular time depends on a number of factors and is extremely complex to compute accurately. For a detailed discussion of the topic, and all the relevant formulae, the reader is referred to *The Astronomical Almanac*, published jointly each year by the USA and UK Nautical Almanac Offices (obtainable from HMSO). Here we simply mention that it depends on the motion of the Earth with

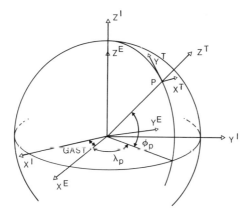

Figure 4.8 Inertial and terrestrial coordinates.

respect to the Sun (for the definition of the vernal equinox), and hence on the precession and nutation of the Earth, and on rotation of the Earth (both axis and rate). Information that enables its computation is usually broadcast by navigation and positioning satellites, along with their orbital information.

Satellite coordinates, as given by (4.9), are needed when carrying out satellite positioning by the dynamic or short-arc methods. It is, however, also often required to go one stage further and compute the distance and direction from a given point on the Earth's surface to the satellite at a specified time. This information is needed to predict the times when observations from a particular site to a particular satellite are possible and, in the case of laser ranging or direction measurement, the exact pointing of the tracking device. Such information is often referred to as a *satellite alert*. The first stage in this computation is to compute the Earth-fixed cartesian coordinates of the observing station P from the given latitude ϕ_P, longitude λ_P, and height H_P, using

$$X_P^E = (v + H_P)\cos \phi_P \cos \lambda_P \qquad (4.10)$$

$$Y_P^E = (v + H_P)\cos \phi_P \sin \lambda_P \qquad (4.11)$$

$$Z_P^E = (v(1 - e^2) + H_P)\sin \phi_P \qquad (4.12)$$

where

$$v = a/(1 - e^2 \sin^2 \phi_P)^{1/2} \qquad (4.13)$$

Then the satellite-to-ground distance r_{PS} is simply given by

$$r_{PS} = \{(X_P^E - X_S^E)^2 + (Y_P^E - Y_S^E)^2 + (Z_P^E - Z_S^E)^2\}^{1/2} \qquad (4.14)$$

To determine the direction to the satellite it is usual to transform the satellite position into a topocentric coordinate system (Earth-fixed but with a centre at P and with X^T, Y^T and Z^T in the east, north and up directions respectively). The satellite position in this coordinate system is given by

$$\begin{bmatrix} X_S^T \\ Y_S^T \\ Z_S^T \end{bmatrix} = \begin{bmatrix} 1 & 0 & 0 \\ 0 & \sin \phi_P & \cos \phi_P \\ 0 & -\cos \phi_P & \sin \phi_P \end{bmatrix} \begin{bmatrix} -\sin \lambda_P & \cos \lambda_P & 0 \\ -\cos \lambda_P & -\sin \lambda_P & 0 \\ 0 & 0 & 1 \end{bmatrix} \begin{bmatrix} X_S^E - X_P^E \\ Y_S^E - Y_P^E \\ Z_S^E - Z_P^E \end{bmatrix} \qquad (4.15)$$

and it can be shown that the altitude and azimuth of the satellite are given by

$$\text{altitude} = \sin^{-1}(Z_S^T/r_{PS}) \qquad (4.16)$$

and

$$\text{azimuth} = \tan^{-1}(X_S^T/Y_S^T) \qquad (4.17)$$

4.2.4 *Geodetic coordinate systems*

Except in the rather rare case of the purely geometric approach, satellite positioning will yield coordinates of ground stations in a coordinate system that is defined by the models (tracking station coordinate set, force models, etc.) used for the orbit prediction and those for the computation of GAST. For practical applications it is unlikely that such coordinates will be directly useful. They will normally have to be transformed into the local national coordinate system, or, for engineering work, perhaps into a special purpose system that may have been adopted for the particular project in hand.

Nominally the transformation between geodetic coordinate systems is expressed by a seven-parameter transformation, often referred to as a *Helmert transformation*, with the required geodetic coordinates (X^G, Y^G and Z^G) being related to those given by satellite positioning (X^S, Y^S and Z^S) by

$$\begin{bmatrix} X^G \\ Y^G \\ Z^G \end{bmatrix} = \begin{bmatrix} dX \\ dY \\ dZ \end{bmatrix} + (1+s) \begin{bmatrix} 1 & \theta_Z & -\theta_Y \\ -\theta_Z & 1 & \theta_X \\ \theta_Y & -\theta_X & 1 \end{bmatrix} \begin{bmatrix} X^S \\ Y^S \\ Z^S \end{bmatrix} \qquad (4.18)$$

where dX, dY and dZ are translation parameters, s a scale parameter, and θ_X, θ_Y and θ_Z are small rotation parameters. Values for these parameters between all of the major national coordinate systems and the main satellite positioning systems have been published, for example Ordnance Survey (1980). In areas where they are not known it is necessary to carry out satellite position fixing at stations with known local coordinates and to solve (4.18), usually by least squares, for the transformation parameters, treating X^G, Y^G and Z^G, and X^S, Y^S and Z^S, as observed quantities.

Note that it is rather rare in practice to describe the transformation using all seven parameters. This is for the following two reasons.

(i) Most geodetic organizations have attempted to use the same orientation for their reference systems; classically, the mean pole direction, known as the *Conventional International Origin* or CIO, has been used to define the Z axis, and the *BIH zero of longitude* has generally been used for the X axis. Had all organizations been successful in their aim of adopting common directions for these axes, and if all past geodetic networks were free of scale errors, all geodetic transformations would involve translation only. Unfortunately this has not generally been the case; in particular a lack of radio time signals in the earlier part of this century (when most modern reference systems were being established) has led to significant longitude rotations between most systems. Also, scale errors in existing networks mean that a scale parameter is usually necessary in the transformation. Hence it is common to adopt a five-parameter transformation.

(ii) When only a small part of the Earth's surface is being considered, the effect of small rotations can be equally described by translations. For example, for a point on the X axis (i.e. zero longitude) a small Z rotation is equivalent to a Y shift. Hence for small areas it is usually sufficient to use only translation and scale parameters.

Finally, it is important to mention here that many of the world's current terrestrial control systems have serious errors that cannot be described by a Helmert transformation. Hence a single set of transformation parameters is not sufficient to express the relationship between such a terrestrial system and a satellite system. The point is explored in more detail in Seppelin (1974). There are a number of possible solutions to this problem, the most common of which is to adopt different sets of transformation parameters for different parts of the existing network.

4.3 Engineering satellite positioning systems

A number of different techniques have been used for satellite positioning and, of these, two have been found to have extensive engineering applications: TRANSIT and the Global Positioning System (usually referred to simply as GPS). Both were initially intended for navigation and actually have rather low design specifications as far as positioning accuracy is concerned, but with specially developed receivers and software, it has been found that extremely accurate relative positions can be determined. In this section these two systems are described in detail.

4.3.1 *The TRANSIT system*

The TRANSIT system, alternatively named NAVSAT or NNSS (Navy Navigation Satellite System), is often referred to simply as 'satellite Doppler', since the position determining process is based on the Doppler shift principle. It was launched in 1963 as a military system aimed specifically at updating submarine inertial systems, the idea being that a submarine could predict when a satellite would pass overhead, and then rise to just below the sea surface (with the antenna just above it) to position itself. In 1967 it was released for civilian use and in the early 1970s began to be used extensively for positioning fixed points on the land and on platforms on the sea. The advent of TRANSIT essentially marked the birth of satellite positioning as an 'everyday' positioning tool.

Nominally the system consists of a 'birdcage' configuration of six satellites in polar orbits (inclination $90°$) at a height of approximately 1100 km, and with an orbital period of about 106 minutes, as in Fig. 4.9. In practice the orbit planes precess slowly (by up to several degrees in longitude per year) and, from time to time, actually cross each other, so destroying their even spacing. TRANSIT is maintained by the US Navy, and occasionally the number and type of satellites is altered. At present (1988) there is a full configuration of six satellites, of which four are of the original OSCAR type, shown in Fig. 4.10, and two are the new NOVA type. The NOVA satellite is in a drag-free orbit, as explained in 4.1.2.

When used in the navigation mode a single pass of a satellite (typically lasting about sixteen minutes) over a vehicle is sufficient for a position fix, but for precise static positioning a much larger number of passes (up to say fifty) are required. The number of available passes is primarily a function of latitude: at the poles there are about eighty passes per day, whereas on the equator a pass with reasonable geometry can only be expected, on average, every two hours.

Each TRANSIT satellite contains a highly stable oscillator which is used to generate two continuous signals, one nominally at 399.968 MHz and the other exactly 3/8 of this. Modulated on both of these signals is a binary coded message known as the *Broadcast Ephemeris* (BE). This contains, amongst other things, the predicted Keplerian elements

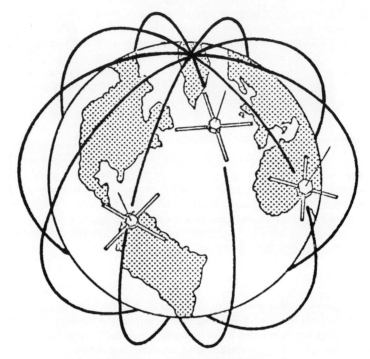

Figure 4.9 Birdcage pattern of TRANSIT satellites.

of the satellite orbit and other orbital information which enable the position of the satellite at any time to be computed, following the methods in 4.2.2, with a standard error of about 20 m. The binary sequence that forms the BE is repeated (with slight changes) every two minutes and is arranged into 25 'words' of information. These words are sometimes used by receivers to start and stop the 'Doppler counting'.

The satellites are tracked by an operations network of four US ground stations (known as OPNET), one of which computes, and uploads, a new BE for each OSCAR satellite every 12 h (every 24 h for NOVA satellites, as they have more stable orbits). They are also tracked by a larger number (about twenty) of globally distributed stations (called TRANET) who provide information for the US Defence Mapping Agency to produce, in arrears, a set of satellite positions and velocities known as the *Precise Ephemeris* (PE), which has a positional standard error of about 1 m. The PE is often used by national survey organizations for absolute positioning, but it will not be discussed further here as it is not normally available to commercial users, and even when it is the delay in obtaining it often makes its use not worthwhile.

4.3.1.1 *Principles of Doppler positioning.* The satellite in Fig. 4.11 is at points $S_1, S_2, \ldots$ etc. at times $t_1, t_2, \ldots$ etc. At time t_1 it emits a signal with (an assumed constant) frequency f_s that arrives at the receiver, such as shown in Fig. 4.12 at time τ_1. Then if Δt_1 is the travel time of the signal it follows that

$$\tau_1 = t_1 + \Delta t_1 \tag{4.19}$$

Figure 4.10 OSCAR transit satellite. From Stansell, T.A. (1978) *The TRANSIT Navigation Satellite System*, Magnavox, California, p. 3.

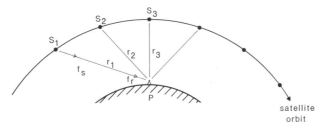

Figure 4.11 Doppler positioning.

Similarly for a signal leaving S_2 at t_2

$$\tau_2 = t_2 + \Delta t_2 \tag{4.20}$$

Due to the variable relative radial velocity between the satellite and receiver the signal arrives with a variable frequency f_R. This is mixed (using a process called *heterodyning*) with a signal of frequency f_G generated by the receiver, and cycles of the resulting 'beat frequency', $f_G - f_R$, are counted. The number of cycles of this beat frequency during a

Figure 4.12 Magnavox 1502 TRANSIT Doppler receiver with antenna and power supply. From Stansell, T.A. (1978) *The TRANSIT Navigation Satellite System*, Magnavox, California, p. 25.

known receiver time epoch, such as between τ_1 and τ_2, is denoted by N_{12} and is usually referred to as the *integrated Doppler count*. It is the fundamental observed quantity in satellite Doppler positioning. Note that some navigational receivers use the words of the BE to start and stop the cycle counting process, i.e. their integrated Doppler counting periods are based on the 'satellite clock' rather than the 'receiver clock'. The case of receiver timing is, however, more accurate and is used in most geodetic receivers. Also, it is more general, as it could be applied to any satellite emitting a radio signal and, for these reasons, it is the one considered here.

The number of cycles counted between τ_1 and τ_2 is then

$$N_{12} = \int_{\tau_1}^{\tau_2} \{f_G - f_R\} \, dt \qquad (4.21)$$

SATELLITE POSITION FIXING SYSTEMS

which can be rearranged using (4.19) and (4.21) as

$$N_{12} = \int_{t_1 + \Delta t_1}^{t_2 + \Delta t_2} \{(f_G - f_S) - (f_R - f_S)\} \, dt \qquad (4.22)$$

which, after integrating the first term and separating the second term becomes

$$N_{12} = (f_G - f_S)(\tau_2 - \tau_1) - \int_{t_1 + \Delta t_1}^{t_2 + \Delta t_2} f_R + \int_{t_1 + \Delta t_1}^{t_2 + \Delta t_2} f_S \, dt \qquad (4.23)$$

The second term of (4.23) can be written as

$$\int_{t_1 + \Delta t_1}^{t_2 + \Delta t_2} f_R \, dt = \int_{t_1}^{t_2} f_S \, dt \qquad (4.24)$$

because the number of cycles leaving the satellite between times t_1 and t_2 must be the same as the number arriving at the receiver between times $t_1 + \Delta t_1$ and $t_2 + \Delta t_2$. Substituting (4.24) in (4.23) and expanding the integrals then yields

$$N_{12} = (f_G - f_S)(\tau_2 - \tau_1) - f_S(t_2 - t_1) \\ + f_S\{(t_2 + \Delta t_2) - (t_1 + \Delta t_1)\} \qquad (4.25)$$

which rearranges to

$$N_{12} = (f_G - f_S)(\tau_2 - \tau_1) + f_S(\Delta t_2 - \Delta t_1) \qquad (4.26)$$

Assuming no refraction effects (i.e. assuming travel in a vacuum), the travel times Δt_2 and Δt_1 are equal to the respective ranges divided by the velocity of light, c. Hence (4.26) can be written

$$N_{12} = (f_G - f_S)(\tau_2 - \tau_1) + (f_S/c)(r_2 - r_1) \qquad (4.27)$$

which rearranges to

$$r_2 - r_1 = \{(N_{12} - (f_G - f_S)(\tau_2 - \tau_1)\} c/f_S \qquad (4.28)$$

Equation (4.28) is often referred to as the basic integrated Doppler count formula, as it relates the difference in range (sometimes called the *range-rate*) to the basic observables: the Doppler count, N_{12}, and the receiver times for starting and stopping the cycle counting, τ_1 and τ_2. It is the fundamental equation of satellite Doppler positioning.

In order to compute the position of the ground station P, we rewrite the range-rate in terms of the station and satellite coordinates as follows:

$$r_2 - r_1 = \{(X_P - X_2)^2 + (Y_P - Y_2)^2 + (Z_P - Z_2)^2\}^{1/2} \\ - \{(X_P - X_1)^2 + (Y_P - Y_1)^2 + (Z_P - Z_1)^2\}^{1/2} \qquad (4.29)$$

where the (usually known) coordinates of S_1 and S_2 are $(X_1, Y_1$ and $Z_1)$ and $(X_2, Y_2$ and $Z_2)$ respectively, and the unknown coordinates of the ground station are $(X_P, Y_P$ and $Z_P)$. Equations (4.28) and (4.29) are then combined to form the general observation equation for satellite Doppler positioning:

$$\{(X_P - X_2)^2 + (Y_P - Y_2)^2 + (Z_P - Z_2)^2\}^{1/2} \\ - \{(X_P - X_1)^2 + (Y_P - Y_1)^2 + (Z_P - Z_1)^2\}^{1/2} \\ - \{N_{12} - (f_G - f_S)(\tau_2 - \tau_1)\} c/f_S = 0 \qquad (4.30)$$

in which it is normal to consider N_{12} as the observable, and X_P, Y_P, Z_P and f_S as unknown parameters (everything else is assumed to be known). Note that f_S is considered unknown because the satellite oscillator will have drifted since its calibration, whereas the oscillator in the receiver (frequency f_G) can be calibrated by connecting the receiver to a frequency standard. Since (4.30) is an equation with four unknown parameters, we need to extend Fig. 4.11 to include five satellite positions, S_1, S_2,..S_5, and four integrated Doppler counts, N_{12}, N_{23}, N_{34} and N_{45}, so that we can form the four equations needed to obtain a solution. In practice, many more equations are formed (typically about forty during a single pass), and the method of least squares is used to estimate the four parameters. Note that (4.30) must be linearized before the application of least squares. Details of this process are outside the scope of this work, and readers are referred to, for instance, Cross (1983) for more details.

It is important to note (although it is not relevant to the multipass solutions in geodetic and engineering surveying) that, for a single satellite pass, the least squares normal equations resulting from the foregoing process are almost singular and their solution highly unstable. This is because a three-dimensional solution is barely possible from a single pass. In fact, if the orbit were a straight line, the point P could revolve around it in a circle without changing its distances from the satellite (rather like the 'danger circle' in classic resection theory). For this reason three-dimensional solutions are not attempted in the navigational mode (the usual solution, at least for navigation at sea, is to hold the height of the receiving station, i.e. the ship, fixed).

Finally it is mentioned here that for readers who find the algebraic explanation of the positioning process difficult to follow, there is a simple geometrical interpretation as follows. Essentially, according to (4.28), if we know the satellite oscillator frequency, each Doppler count leads to a difference in range from two known satellite positions. The locus of a difference of range in space is a hyperboloid (just as the locus of a range is a circle), so if we have three such Doppler counts we will have three hyperboloid surfaces and their intersection is the receiver position.

4.3.1.2 Relative positioning by TRANSIT. Generally in geodetic and engineering work a number of receivers are deployed simultaneously, and it is required to determine their relative positions. At the simplest level this is done by computing individual point positions and differencing the results. Errors due to imperfect orbit prediction (i.e. BE errors) and ionospheric refraction will have similar effects on the derived coordinates of close stations (up to several hundred kilometres apart), and will hence tend to cancel in the coordinate differences. Standard errors of coordinate differences approaching 0.5 m have been achieved by this technique. This can be improved to about 0.3 m by employing more sophisticated orbit and refraction modelling methods and by directly combining the observations at the various ground stations. The topic is treated in detail in Ashkenazi and Sykes (1980), and referred to again in 4.3.3.

4.3.2 The Global Positioning System (GPS)
Unlike TRANSIT, which was planned for intermittent offshore (low dynamic) use, GPS (sometimes referred to as NAVSTAR which is an acronym derived from NAVigation Satellite, Timing And Ranging) was conceived as a continuous multipurpose positioning and navigation system. When fully operational it will consist of eighteen satellites (plus three in-orbit spares) arranged in six orbit planes with inclinations of 55°, and with 60° spacing in longitude. The satellites will have a height of about 20 183 km and

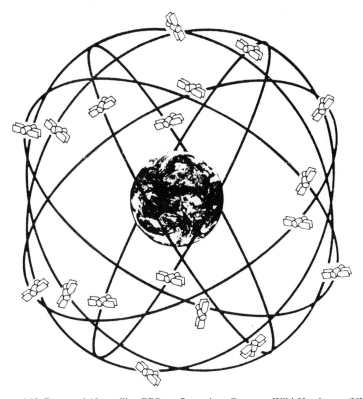

Figure 4.13 Proposed 18-satellite GPS configuration. Courtesy Wild Heerbrugg (UK).

periods of 12 h. The configuration is shown in Fig. 4.13 and is such that, for most points on the Earth, there will be at least four (and sometimes up to nine) satellites in view at any time.

The satellites themselves are far larger and more complex than those for TRANSIT. Each will transmit two signals (as with TRANSIT based on a single oscillator) which have been named L1 and L2 (because they are in the L-band of the electromagnetic spectrum) and have frequencies of 1574.42 MHz (154 × 10.23 MHz) and 1227.60 MHz (120 × 10.23 MHz) respectively. Both L1 and L2 carry a formatted data message (rather similar to the BE for TRANSIT) which contains the Keplerian elements and other information for satellite position computations. It also contains information to relate the GPS time system to Universal Time (previously GMT) and estimates of the state of the ionosphere to enable refraction corrections to be made. The satellites are tracked by five stations located at Kwajalein, Diego Garcia, Ascension, Hawaii and Falcon Air Force Station (AFS) in Colorado, which send their data to the master control station at AFS. There the data messages are computed for each satellite and sent to the most convenient 'ground antenna' (located at Kwajalein, Diego Garcia and Ascension) for upload to the satellite. Normally uploads occur every eight hours, but each satellite actually stores data for 14 days in case any problems arise with the tracking network. The orbital information currently has a standard error of about 20 m which would, in fact, degrade to about 200 m if not updated for 14 days.

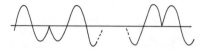

Figure 4.14 Binary biphase modulation.

Also modulated on L1 is a C/A (coarse acquisition, sometimes referred to as S, for standard) code and a P (precise) code, and modulated on L2 is just the P code. These codes are essentially pseudo-random binary sequences (i.e. sequences of zeros and ones) that are repeated every 1 ms and 7 days respectively. They are in fact realized by a technique known as binary biphase modulation which results in 180° changes of phase of the carrier as shown in Fig. 4.14. Note that the term pseudo-random refers to the fact that, although they have the stochastic properties of random sequences, the process that generates them is known so that they can, in principle, be recreated by a receiver. The P code has a frequency of 10.23 MHz (i.e. about 10^7 binary digits are transmitted every second with a physical spacing of about 30 m), whereas the C/A code has a frequency ten times smaller, with a spacing of about 300 m between the changes of phase.

4.3.2.1 *Principles of GPS positioning.* The GPS signals can be used, by geodetic receivers such as those in Fig. 4.15, for positioning in a number of ways, two of which will be considered here: *pseudo-ranging* and *carrier phase measurement*.

Pseudo-ranging is a technique whereby the range between a GPS satellite and a receiver is determined by use of the pseudo-random timing codes. Essentially the receiver generates an identical code (called a *replica code*), at an identical time, to that produced by a particular GPS satellite. It is then compared, by a process known as cross-correlation, with the satellite signal. Cross-correlation involves moving the receiver-generated binary sequence in a *delay-lock loop* until it is exactly in phase with the one received. The amount by which it must be moved is then the time of travel of the satellite signal to the receiver (since they were generated at the same time). Multiplication by the velocity of light then yields the distance.

GPS satellites carry extremely accurate caesium clocks, which are updated every time a new message is uploaded, so they can normally be considered to all intents to be exactly on the same time system (actually an atomic time scale known as *GPS time*). Receiver clocks are usually of a much lower quality, and the time of generation of the receiver random sequences is liable to contain a significant clock error. For this reason the functional models used to solve the pseudo-ranging positioning problem always contain this error. If Δt is the observed travel time and E the clock error, then the distance, r_{1P}, from a satellite S_1 to a receiver is given by

$$r_{1P} = (\Delta t - E)c \tag{4.31}$$

where, as before, c is the velocity of light. The observation equation is then

$$\{(X_P - X_1)^2 + (Y_P - Y_1)^2 + (Z_P - Z_1^2)\}^{1/2} = (\Delta t - E)c \tag{4.32}$$

in which Δt is the observable and there are four unknowns: X_P, Y_P, Z_P and E. Clearly observations to four satellites will yield the required receiver position (and the clock error).

Currently the accuracy of pseudo-ranging is such that it is generally reserved for

Figure 4.15 (*a*) Trimble 4000SX GPS receiver with antenna, power supply and data-logging computer. Courtesy GPS Survey Services Ltd.

Figure 4.15 (*b*) Texas Instruments TI4100 dual-frequency 6PS receiver with antenna and cassette tape recorder. (Texas Instruments).

Figure 4.15 (c) Wild Magnavox WM101 GPS receiver with antenna and internal tape recording system. Courtesy Wild Heerbrugg (UK).

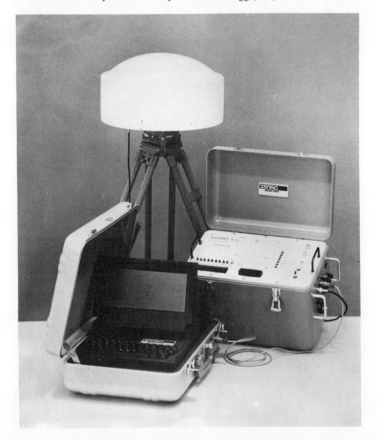

Figure 4.15 (d) ISTAC 2002 codeless GPS receiver with antenna and data logging computer. Courtesy Global Surveys Ltd.

Figure 4.16 GPS carrier signal after signal squaring.

navigation and approximate static positioning. Most geodetic and engineering work with GPS is based on the so-called *continuous phase observable*, and consequently it will be treated in detail here.

Essentially the phase observation process involves heterodyning the incoming carrier signal (either L1 or L2, or both) with a signal generated in the receiver, and making measurements with the resulting signal at specified times. The term 'continuous phase refers to the fact that, as well as measuring the instantaneous value of the phase (in the range zero to 360°), the receiver also counts the number of complete cycles from some arbitrary starting time (i.e. it increments an integer counter every time the phase changes from 360° to zero). Of course in order to access the received carrier the receiver must remove the code (which is modulated by 180° changes of phase). This can be carried out either by a process known as 'signal squaring' (which leads to a signal of the form shown in Fig. 4.16), or by using a prior knowledge of the pseudo random binary sequence to reconstruct the original carrier (i.e. the carrier before the code was added). The former, codeless approach, has the advantage of not requiring any knowledge of the code and of leading to a more precise measurement (because the wavelength is halved) but the process destroys the data message, and an external ephemeris (rather than a broadcast one) must be used. At present this creates considerable logistical difficulties, so most receivers adopt the latter technique, but should the codes become unavailable at any time in the future the codeless approach may be more widely adopted.

Actually the foregoing measurement process is identical to that used in satellite Doppler positioning (see 4.3.1.1), but since the 'counting' is continuous (rather than between specified epochs) the observation equations and processing techniques are rather different. Remondi (1985) has shown that the continuous phase observable $N[iaj]$, observed at station a and at epoch i from satellite j, is given by

$$N[iaj] = E_S[ij] + E_R[ia] + I[aj] + f_j \Delta t[iaj] \qquad (4.33)$$

where

$E_S[ij]$ is a term describing the satellite clock error at epoch i,
$E_R[ia]$ is a term describing the receiver clock error at epoch i,
$I[aj]$ is the *integer ambiguity*, which is the (arbitrary) integer value of the phase counter at the start of the observing period,
f_j is the frequency of the emitted signal, and
$\Delta t[iaj]$ is the time of travel of the signal from satellite j to receiver a, at epoch i.

The travel time $\Delta t[iaj]$ is given by

$$\Delta t[iaj] = (1/c)\{(X_a - X_j)^2 + (Y_a - Y_j)^2 + (Z_a - Z_j^2)\}^{1/2}$$
$$+ \Delta t^{\text{ion}}[iaj] + \Delta t^{\text{trop}}[iaj] \qquad (4.34)$$

where $\Delta t^{\text{ion}}[iaj]$ and $\Delta t^{\text{trop}}[iaj]$ are the signal delays in the ionosphere and troposphere respectively. Substitution of (4.34) into (4.33) leads to (4.35), which is the general continuous-phase observation equation for an observation at epoch i between satellite j

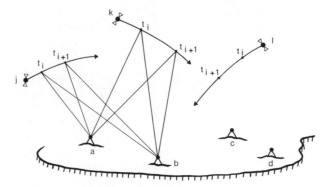

Figure 4.17 A number of GPS receivers simultaneously measuring phase to a number of satellites.

and receiver a.

$$N[iaj] - E_S[ij] - E_R[ia] - I[aj] - f_j(1/c)\{\{(X_a - X_j)^2 + (Y_a - Y_j)^2 + (Z_a - Z_j^2)\}^{1/2} - \Delta t^{\text{ion}}[iaj] - \Delta t^{\text{trop}}[iaj]\} = 0 \qquad (4.35)$$

4.3.2.2 *Relative positioning by GPS*. As with satellite Doppler, it is possible to compute individual point positions from pseudo-ranges and to difference them in order to derive relative positions, and hence azimuths, distances and height differences. Such a strategy is useful for precise kinematic surveying when the data can be transferred in real time by means of a telemeter link, but the quality is rarely such that it is suitable for geodetic work. Consequently a more sophisticated procedure based on the phase observable is usually adopted.

In a general relative positioning problem we may have m stations observing n satellites over p epochs as indicated in Fig. 4.17. This would lead to mnp observation equations, (4.35), with $3m$ station coordinate (assuming no datum defect), np satellite clock error, mp receiver clock error and mn integer ambiguity parameters to be estimated. In a typical case of 3 stations, 4 satellites and 1000 epochs this would mean 12 000 observation equations and 7021 parameters. The solution of such a system creates serious computational difficulties and, in practice, it is usual to reduce the number of parameters by differencing the observation equations (i.e. by subtracting them from each other). A number of strategies are possible based on combinations of differencing with respect to satellites, ground stations and time. Each removes a different group of parameters from the system, as follows:

(i) Differencing with respect to satellites removes the receiver clock error term $E_R[ia]$
(ii) Differencing with respect to ground stations removes the satellite clock error term $E_S[ia]$
(iii) Differencing with respect to time removes the integer ambiguity $I[aj]$.

It is important to note that, so long as the stochastic models are properly formed, identical answers are possible from all strategies, and the work saved in reducing the number of parameters is replaced by the work involved in dealing with non-diagonal weight matrices. Most commercially available software has, however, been written to run on portable microcomputers, so that computations can be carried out in the field, and consequently it is usual for it to be based on the assumption that the differenced observation equations are uncorrelated. This point is explored in more detail in Ashkenazi and Yau (1986) and Beutler et al. (1986).

The most commonly adopted strategy is to difference with respect to both satellites and ground stations to yield the so-called *double difference* observable. In the foregoing example this would lead to a system of 6000 observation equations with 15 parameters (probably less in practice when a datum was introduced). The workload in least-squares problems is, in general terms, related to the cube of the number of parameters, so the double differencing process clearly leads to a considerable saving and usually means that the computations can be carried out in the field on a portable microcomputer. Note that the least-squares procedure is usually adapted to force the integer ambiguity parameters to take integer values. Another commonly adopted strategy is to form *triple difference* observation equations, i.e. to difference with respect to time, satellites and ground stations. This process does not yield results of the highest quality (one reason being that the integer ambiguities are no longer solved for, and it is not possible to force them to integer values) but leads to an extremely simple system (with the only parameters being station coordinates) that is very useful for data validation. It is perhaps interesting to note that forming single differences with respect to time leads to an identical observation equation to that used in Doppler positioning.

4.3.2.3 *Status of GPS.* GPS phase 1 (concept validation program) began in 1973 and, in 1979, phase 2 (system testing) started with the launch of a number of Block I satellites. Currently (1988) seven of these are operational. These are arranged in two planes with an inclination of 63° (the optimum inclination for the original plan of 24 satellites) and enable GPS positioning on most parts of the Earth's surface for a few hours per day. The current plan is to begin the launch of the operational (Block II) satellites late in 1988 and to complete the 21-satellite constellation by 1992.

Currently both the C/A and P codes are generally available (the US Department of Defense has released details of the numerical processes employed to generate the pseudo-random sequences). There is some doubt as to whether or not this will be the case for the Block II satellites with the latest information, Baker (1986), suggesting that only the C/A code will be generally available. If this is the case then there will be some (not insurmountable) difficulties in accessing the L2 signal, as it carries only the P code. Also, there is a possibility that the orbit included in the data message may be deliberately downgraded (by a factor of ten). These measures are aimed at denying, for military reasons, the use of high-accuracy, real-time GPS point positioning to 'non-approved users'. It will not be possible, however, to restrict surveyors and engineers in their determination of post-processed relative positions. They can always produce their own orbits and use the signal squaring technique to access the carrier for phase measurement.

4.3.3 *Models for geodetic relative positioning*
The quality of relative coordinates derived from both TRANSIT and GPS observations can be considerably enhanced by the data processing methods employed. The three major factors are the treatment of refraction, orbit modelling and data filtering, and these will be treated briefly here.

Refraction occurs in two layers of the atmosphere: the ionosphere and the troposphere. Of the two, the ionospheric effect is the largest. In order to account for this, all geodetic TRANSIT receivers and some GPS receivers measure on both of the two broadcast frequencies (note that to improve results from single-frequency instruments, the systems broadcast estimates of the state of the ionosphere as part of their data message). Certain linear combinations of the measurements on these two frequencies are known to be free (to a first order–about 99%) of ionospheric refraction, and it is these combinations which are used as the observables in the respective observation equations. For more details of this, Clynch and Coco (1986) is recommended, and Clynch and Renfew (1982) discuss techniques for dealing with the residual (about 1%) effects. Tropospheric refraction is usually treated by measuring the local meteorological conditions (temperature, humidity and pressure) and applying appropriate atmospheric models such as those due to Black or Hopfield—see for instance Hopfield (1980).

There are a number of different ways to treat the orbits of TRANSIT and GPS satellites, and the topic is rather too complex to be treated here in detail. Essentially, the methods range from accepting the broadcast satellite positions as being perfect, to completely ignoring them and carrying out a separate orbit integration based on a selected set of force models. A midway treatment is to assume that an individual broadcast orbit has errors that can be described by some form of mathematical transformation (e.g. Helmert), so that in general we can write

$$X_S^E = F_X(X_S^B, Y_S^B, Z_S^B, \text{transformation parameters})$$
$$Y_S^E = F_Y(X_S^B, Y_S^B, Z_S^B, \text{transformation parameters}) \quad (4.36)$$
$$Z_S^E = F_Z(X_S^B, Y_S^B, Z_S^B, \text{transformation parameters})$$

where X_S^E, Y_S^E and Z_S^E are the required Earth-fixed coordinates of the satellite, and X_B^S, Y_S^B and Z_S^B are those that were computed from the information broadcast by the satellite. Substitution of (4.36) into the observation equations then introduces a new set of unknown parameters for each orbit, and the subsequent solution involves both the station coordinates and the orbit transformation parameters. This procedure is known as *orbit relaxation*, and is a simple example of the short-arc positioning technique introduced in 4.1.3. For a more detailed discussion of the treatment of TRANSIT and GPS orbits, readers are referred to Malyvac and Anderle (1982) and Beutler et al. (1986) respectively.

Finally, here the problem of data validation and data selection is mentioned. It is usual in geodetic and engineering work to filter the observed data extremely rigorously before accepting it for computation. A detailed discussion of this topic involves reference to a large number of statistical (and other) criteria and is clearly outside the scope of this work, but two points are worth mentioning. Firstly, much data selection is based on pure geometry. For instance, observations below a specified altitude (typically 15°) are commonly rejected (due to refraction errors near to the horizon), and in TRANSIT work an equal number of passes travelling northwards and southwards are insisted upon (as along-track orbit errors will then have compensating effects). Secondly, continuous-phase measuring techniques are subject to errors known as *cycle slips* which are caused by momentarily losing lock on a satellite. When the measurements resume, the fractional part of the phase will be correct (i.e. equal to that which it would have been without a loss of lock), but the integer counting will be in error. The problem of detecting and repairing cycle slips is currently receiving much attention (see for instance Remondi, 1985), but it is interesting to note here that if differences with respect to time are computed, then the problem will be isolated to a single measurement. Hence cycle slips do not present difficulties in TRANSIT work, and in GPS work they can be seen in individual residuals after a triple difference computation.

4.3.4 *Applications of TRANSIT and GPS positioning*
Relative positions with standard errors of, at best, 0.3 m have been achieved with the TRANSIT system. This figure appears to be independent of distance for station separations of up to several hundred kilometres, and it is generally thought that it cannot be substantially improved upon, due to limitations of the satellite oscillators, rather low frequency (and hence high ionospheric refraction) and difficulties in orbit prediction (due to the rather low orbits). TRANSIT has therefore only been able to compete, from an accuracy point of view, with terrestrial techniques over rather long distances, and hence for applications such as first- and second-order networks, large-scale photocontrol, and positioning offshore structures such as oil rigs. A special advantage, however, of TRANSIT is its relative freedom from systematic errors, and for this reason it has played an extremely important role in the strengthening of the overall scale and orientation of many existing national triangulation networks, especially in Europe and North America. For a discussion of its role in Britain see Ashkenazi *et al.* (undated). It should also be pointed out that to obtain the highest accuracy from TRANSIT, several days of observation are usually necessary.

The situation with regard to GPS is completely different. The wavelength of the L1 carrier is approximately 0.19 m, and observations with a resolution of 1° of phase (equivalent to 0.5 mm) are possible. The result is a system whereby the standard errors of derived coordinate differences are proportional to the distances between the stations (the major error sources being orbit prediction and atmospheric refraction). For very short distances (up to 1 km), standard errors of 1 mm have been achieved, and the 1 part per million (ppm) standard error can be maintained up to several hundred km so long as the correct equipment and processing models are used. Some authors, e.g. Mueller (1986), even claim that 0.1 ppm is achievable and that 0.01 ppm is a realistic goal for the future. Currently these accuracies need several hours of observing time, but it is expected that only minutes of observation time will be required with Block II satellites, future equipment and observing techniques. The general prospects for GPS, especially *vis à vis* inertial techniques, are reviewed in Cross (1986).

Of course the foregoing is something of an oversimplification, as the accuracy achievable depends on a large number of factors, including the type of equipment, whether or not dual-frequency measurements are made, the prevailing state of the ionosphere, the slope of the line joining the two stations (steep slopes present greater tropospheric modelling problems), the data validation process, the length of observation period, the number of satellites observed, the orbit modelling process, etc. Also, the ease with which the predicted future accuracy will be achieved (but probably not the accuracy itself) will depend on a number of political decisions that the US Department of Defence has yet to make. Nevertheless, the quoted figures give an overall impression of what can be obtained today and what might be expected in the future.

Accuracies of 1 ppm or better over a few minutes open up a vast range of engineering applications, and probably the major factor that will determine the rate at which GPS (or perhaps a rival system) replaces almost all existing engineering control systems will be the size, weight and cost of user equipment. This aspect is difficult to predict, but even with the current somewhat bulky and expensive equipment, a number of interesting engineering applications have already been found. They include the UK/France Channel Tunnel project (Calvert and Best, 1987); the monitoring of the subsidence of North Sea oil platforms, surveys to control large-scale urban mapping (Moreau, 1986); and the setting-out of the CERN large electron positron ring in Switzerland (Gervaise *et al.*, 1985). Austin (1986) is recommended for readers who would like to read a selection of papers on applications.

One application is worth examining in a little more detail because it exemplifies what may in the future be typical of the likely implementation of GPS for general engineering purposes. Merrell (1986) describes how the Texas State Department of Highways plan to establish eight regional reference points (RRPs) in the State of Texas. Each RRP is designed to service points up to 200 km away and will be unmanned. It will continuously collect GPS data which it will transfer by electronic mail to a central computer in the state capital at Austin. To position any point, an engineer will use a single receiver to collect GPS data and send it, in real time, via a mobile telephone communications link, to the central computer. The central computer will carry out relative positional computations using the most appropriate (nearest) RRP data, and send the computed position back to the engineer. In this way, the Highways Department expect to be able to deliver to their engineers almost immediate positions with standard errors of only a few centimetres anywhere in the state at any time.

One can imagine a time in the future when GPS, or similar satellite-based techniques, will replace terrestrial methods for the majority of engineering surveying positioning problems. The necessary accuracy can be achieved today but, as costs are generally rather high, GPS is being used only for special applications. This high cost is because occupation times are rather long (perhaps up to two hours), there is limited satellite coverage, and the receivers are rather bulky and expensive (in 1988 a GPS receiver typically costs four to five times as much as a 'total station' theodolite/EDM combination). Once the system is fully operational, extensive development in GPS receivers can be expected. It is predicted that they will eventually be of a size that can be 'handheld' and at a price that is easily affordable by most engineering surveying organizations. Also, it is highly likely that they will be integrated with other surveying systems (such as total stations) so that they can be used for positioning in places where GPS cannot normally operate (e.g. on streets bounded by very tall buildings).

4.4 Related space geodesy methods

There are a number of space geodesy techniques that, although they are not used by engineering surveyors, produce results that are of great indirect importance to satellite positioning for engineering purposes. In particular, they help in the determination of the Earth's gravity field and of the rotation of the Earth. Satellite laser ranging (SLR), satellite altimetry and Very Long Baseline Interferometry (VLBI) are three such techniques, and they will be briefly treated here.

4.4.1 *Satellite laser ranging*

SLR involves the transmission of a short laser pulse to a satellite carrying corner-cube reflectors. The pulse is retroreflected by the satellite and the return travel time measured. If Δt is the return travel time and t the epoch of emission, then the range r to the satellite at time $t + \Delta t/2$ is given by

$$r = c\Delta t/2 + C + R \qquad (4.37)$$

where c is the velocity of light in a vacuum, C a calibration correction (usually derived by ranging to terrestrial targets) and R a tropospheric refraction correction (computed from measured meteorological observations).

Laser ranging systems are generally rather large, permanently mounted instruments such as the one at the Royal Greenwich Observatory (Fig. 4.18), described in detail in Sharman (1982). There are, however, a number of mobile systems that have been deployed (in the USA, the Netherlands and West Germany) for special positioning work associated with geodynamics, especially in the Eastern Mediterranean and California—see for instance Tapley et al. (1985). At present (1988) it is estimated that about 20 permanent and 6 mobile systems are routinely operating.

Typically the lasers operate at about 10 Hz, i.e. ten pulses per second, with pulse lengths of about 150 ps (equivalent to about 45 mm). An accurate pointing device is necessary to ensure that the telescope points in the correct direction, and a computer-controlled tracking system is needed so that it follows the satellite as it passes overhead. Note that since laser light is in the visible part of the spectrum, laser rangers can operate only in cloudless skies. Ranging standard errors as low as 10 mm can now be achieved, although most systems are up to about five times worse than this.

There are currently about 12 satellites with corner cube reflectors, of which three are regularly used for geodetic purposes: LAGEOS (USA), Starlette (France) and Ajisai (Japan). These three are all rather similar passive spherical objects. Figure 4.19 is an example.

SLR has a number of important geodetic applications of which positioning is the most relevant here. By collecting and processing, using methods such as those of Sinclair and Appleby (1986), data from a variety of sites well distributed over the entire Earth, absolute and relative positions with standard errors of only a few cm can be determined. Such positions have recently confirmed theories of continental drift (see for example Christodoulidis et al., 1985), and now form the basis for the modern reference systems used in satellite positioning (see Mueller and Wilkins, 1986). Other applications include the monitoring of Earth rotation (it forms a major input into the new IERS referred to in 4.2.1), and the determination of the gravity field (by measurement of satellite orbit perturbations).

4.4.2 *Satellite altimetry*

Altimetry is the term used to describe the direct measurement of the height of a satellite. The technique adopted is for the satellite to emit a radar pulse and to measure its return travel time to the Earth's surface. Up to the present, only water (sea and lake) surfaces have consistently given good-quality returns, and most geodetic work has been carried out with this type of data. Returns have, however, been received from certain smooth surfaces such as the polar ice caps and deserts, and these have had important applications in earth sciences generally.

To date only three civilian satellites, Skylab, GEOS 3 and Seasat, have carried altimeters, although two others, ERS-1 (European Space Agency) and TOPEX (USA), are planned for the early 1990s. The Skylab and GEOS 3 results were of rather low accuracy and yielded relatively little geodetic information, but Seasat (launched in June

Figure 4.18 The satellite laser ranging system at the Royal Greenwich Observatory. Courtesy Science and Engineering Research Council.

1978), with its 0.1 m resolution satellite, has made a massive contribution to geodesy and geophysics, despite its failure after only 99 days in orbit.

Seasat had a height of about 800 km and an inclination of 108° and was tracked by laser ranging (it carried corner-cube reflectors) and Doppler methods (it emitted a continuous radio signal). Its altimeter, fully evaluated in Lame and Born (1982), had a rate of 1020 Hz (i.e. it emitted 1020 radar pulses per second) and its footprint (the size of the pulse when it hit the sea) had a diameter of 1.6 km.

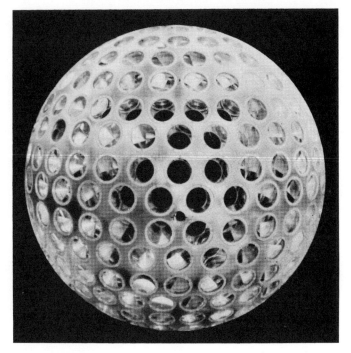

Figure 4.19 The LAGEOS laser ranging satellite.

The reason for Seasat's importance is that it furnishes a direct measurement of the sea surface (approximately the geoid), h_a, as shown in Fig. 4.20. Orbit computations give the distance OS, the radius r comes from ellipsoidal geometry, Δh (known as the sea surface topography) comes from oceanography, and h_t from tidal studies. Hence the geoid height N can be directly computed from (4.38):

$$OS = r + N + \Delta h + h_t + h_a \qquad (4.38)$$

Note that Δh and h_t are difficult to determine and are often ignored. Satellite altimetry is the most accurate method available for the determination of the geoid at sea (where gravity measurements present special problems) and hence forms an extremely important data source for almost all gravity models currently used for orbit prediction and geoid description—see, for example, Rapp (1983) for the determination of the geoid from Seasat, and Reigber et al. (1983) for an example of a combination method.

4.4.3 Very long baseline interferometry

VLBI is not a satellite system. It involves the measurement of the variation in the arrival times of radio signals from quasars at different observing sites (radio telescopes) around the world. Basically, data is recorded and accurately time-tagged at each site, and then cross-correlated, as explained for instance in Ashkenazi and McLintock (1982), with similar data at other sites. Extremely precise relative positions (typically 0.01 ppm) and Earth rotation parameters can be determined and, as with SLR, the data contributes to the IERS and coordinate system definition.

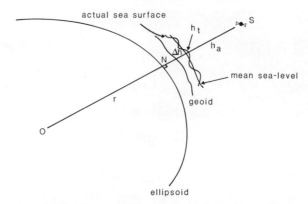

Figure 4.20 Basic geometry of satellite altimetry.

Relative positions can also be used for the study of deformations of the Earth due to continental drift and tidal phenomena. They are, however, of limited use on a global scale because there are very few radio telescopes around the world regularly devoting time to VLBI work. In the USA, however, a number of mobile systems have been constructed, and they are permanently deployed in California, monitoring the San Andreas fault system.

References

Ashkenazi, V., Crane, A.S. Preiss, W.J. and Williams, J.W. (undated) The 1980 readjustment of the triangulation of the United Kingdom and the Republic of Ireland OS(SN)80. *Ordnance Survey Professional Paper* **31**, 27 pp.

Ashkenazi, V. and McLintock, T. (1982) Very long baseline interferometry: an introduction and geodetic applications. *Survey Review* **26**(204), 279–288.

Ashkenazi, V. and Moore, T. (1986) The navigation of navigation satellites. *J. Navigation*, **39**(3), 377–393.

Ashkenazi, V. and Sykes, R.M. (1980) Doppler translocation and orbit relaxation techniques. *Phil. Trans. R. Soc. London* **A294**, 357–364.

Ashkenazi, V. and Yau, J. (1986) Significance of discrepancies in the processing of GPS data with different algorithms. *Bull. Géodésique* **60**(3), 229–239.

Austin, B. (1986) *Proc. Fourth Geodetic Symp. on Satellite Positioning*, University of Texas, 1509 pp.

Best, B. and Calvert, C.E. (1987) The Channel Tunnel grid 1986. *Proc. Second Industrial and Engineering Survey Conf.*, University College London, 182–192.

Baker, P.J. (1986) Global Positioning system policy. *Proc. Fourth Geodetic Symp. on Satellite Positioning*, University of Texas, 51–64.

Beutler, G., Gurtner, W., Rothacher, M., Schildknecht, T. and Bauersima, I. (1986) Determination of GPS orbits using double difference carrier phase observations from regional networks. *Bull. Géodésique*, **60**(3), 205–220.

Beutler, G., Gurtner, W. and Bauersima, I. (1986) Efficient computation of the inverse of the covariance matrix of simultaneous GPS carrier phase difference observations. *Manuscripta Geodaetica* **11**(4), 249–254.

Bomford, G. (1980) *Geodesy*. 4th edn, Clarendon Press, Oxford, 855 pp.

Christodoulidis, D.C., Smith, D.E., Kolenkiewicz, R., Klosko, S.M., Torrence, M.H. and Dunn, P. (1985) Observing tectonic plate motions and deformations from satellite laser ranging. *J. Geophys. Research* **90** (B11), 9249–9264.

Clynch, J.R. and Renfrew, B.A. (1982) Evaluation of ionospheric residual range error model. *Proc. Third Geodetic Symp. on Satellite Doppler Positioning*, University of Texas, 517–538.

Clynch, J.R. and Coco, D.S. (1986) Error characteristics of high quality geodetic GPS measurements: clocks, orbits and propagation effects. *Proc. Fourth Geodetic Symp. on Satellite Positioning*, University of Texas, 539–556.

Cross, P.A. (1983) Advanced least squares applied to position-fixing. *Working Paper* **6**, Department of Land Surveying, North East London Polytechnic, 205 pp.

Cross, P.A. (1986) Prospects for satellite and inertial positioning methods in land surveying. *Land and Mineral Surveying* **4**(4), 196–203.

Gervaise, J., Mayoud, M., Beutler, G., and Gurtner, W. (1985) Test of GPS on the CERN-LEP control network, *Proc. FIG Study Group Meeting on Inertial, Doppler and GPS Measurements for National and Engineering Surveys*, Universität der Bundeswehr, Munich, **202**, 337–358.

Hopfield, H.S. (1980) Improvements in the tropospheric refraction correction for range measurement. *Phil. Trans Roy. Soc. London* **A294**, 341–352.

Lame, D.B. and Born, G.H. (1982) Seasat measurement system evaluation: achievements and limitations. *J. Geophys. Res.* **87**(C5), 3175–3187.

Malyevac, C.W. and Anderle, R.J. (1982) Force model improvement for NOVA satellite. *Proc. Third Geodetic Symp. on Satellite Doppler Positioning*, University of Texas, 85–94.

Merrell, R.L. (1986) Application of GPS for transportation related engineering surveys. *Proc. Fourth Geodetic Symp. on Satellite Positioning*, University of Texas, 1165–1180.

Moreau, R. (1986) The Quebec 1985 TI 4100 test. *Proc. Fourth Geodetic Symp. on Satellite Positioning*, University of Texas, 1059–1072.

Mueller, I.I. (1986) From 100 m to 100 mm in (about) 25 years. *Proc. Fourth Geodetic Symp. on Satellite Positioning*, University of Texas, 6–20.

Mueller, I.I. and Wilkins, G.A. (1986) On the rotation of the earth and the terrestrial reference system: joint summary report of the IAU/IAG Working Groups MERIT and COTES. *Bull. Géodésique* **60**(1), 85–100.

Ordnance Survey (1980) Report of investigations into the use of satellite Doppler positioning to provide coordinates on the European Datum 1950 in the area of the North Sea. *Ordnance Survey Professional Paper* **30**, 32 pp.

Rapp, R.H. (1983) The determination of geoid undulations and gravity anomalies from Seasat altimeter data. *J. Geophys. Res.* **88**(C3), 1552–1562.

Reigber, Ch., Muller, H., Rizos, C., Bosch, W., Balmino, G. and Moynot, B. (1983) An improved GRIM3 earth gravity model. *Proc. Int. Assoc. of Geodesy Symp.* (*Hamburg*), Ohio State University, **1**, 388–415.

Remondi, B.W. (1985) Global Positioning System carrier phase: description and use. *Bull. Géodésique* **59**(4), 361–377.

Roy, A.E. (1982) *Orbital motion*. 2nd edn, Adam Hilger, Bristol, 495 pp.

Seppelin, T.O. (1974) The Department of Defense World Geodetic System 1972. *Int. Symp. on Problems Related to the Redefinition of the North American Datum*, Fredericton, New Brunswick, 43 pp.

Sharman, P. (1982) The UK satellite laser ranging facility. *SLR Technical Note* **1**, Royal Greenwich Observatory, 6 pp.

Sinclair, A.T. and Appleby, G.M. (1986) SATAN—programs for the determination and analysis of satellite orbits from SLR data. *SLR Technical Note* **9**, Royal Greenwich Observatory, 14 pp.

Tapley, B.D., Schultz, B.E. and Eanes, R.J. (1985) Station coordinates, baselines and earth rotation from LAGEOS laser ranging: 1976–1984. *J. Geophys. Res.* **90**(B11), 9235–9248.

Taylor, P.T., Keating, T., Kahn, W.D., Langel, B.A., Smith, D.E. and Schnetzler, C.C. (1984) Observing the terrestrial gravity and magnetic fields in the 1990's. *EOS* **64**(43), 609–611.

Vanicek, P. and Krakiwsky, E. (1986) *Geodesy: the concepts*. 2nd edn, Elsevier, Amsterdam, 697 pp.

5 Analysis of control networks and their application to deformation monitoring

A.H. DODSON

5.1 Introduction

The use of surveying control networks in civil engineering has become widely accepted over the last twenty years or so, particularly since the advent of modern computers which have the speed and storage capabilities required to make the rigorous calculation of such networks a realistic proposition in the construction environment. Nowadays, a comprehensive survey network design and analysis can be carried out with little effort on a desktop personal computer, and even networks of the size of the national primary triangulation of Great Britain can be computed rigorously on such a machine.

However, network analysis is appropriate not only to such large schemes, or prestigious projects. The statistical rigour and resulting power of the analysis techniques of least-squares network adjustment are applicable to small and large problems alike. In particular the ability of network analysis to detect errors and mistakes, and to estimate the accuracy of results, in addition to producing coordinates of the network stations, can be of equal importance to small and commonplace civil engineering tasks as well as to major projects.

A control scheme established for a particular project can be a very simple affair, such as a braced quadrilateral, or a complex network such as was required for the construction of the Gravelly Hill highway interchange near Birmingham, England (known as Spaghetti Junction—Ashkenazi, 1970). Indeed a simple traverse can be considered as a control network of sorts (and can be computed by least squares), but suffers from many defects, primarily a lack of redundant observations which undermine most of the advantages of a rigorous adjustment.

The first aim of this chapter is to introduce the reader to the theoretical basis of least-squares survey network computation and analysis, without dwelling on mathematical and statistical proofs and philosophical argument. However, the practical implications and problems are emphasized sufficiently so that the reader will become aware of the assumptions behind the method and will therefore be able to apply the analyses with confidence. Nevertheless, it must be stressed that the theory and practice of least-squares network analysis is a sufficiently broad topic to be the subject of whole books in itself, for example Mikhail and Gracie (1981) and Meissl (1982), and even, in Cooper (1987), the specific application of networks to civil engineering. In addition new ideas are constantly being proposed and discussed in scientific conferences and symposia. It should therefore be realized that this chapter can only give a basic introduction to the technique.

By its very definition, least-squares computation relies on statistical reasoning, and statistical tests form the basis of most of the resulting analysis. It is beyond the scope of this chapter to give a detailed description of the statistical theory involved in the analysis, but a reasonable knowledge of the statistical testing is assumed. Readers unfamiliar with the subject are referred to one of the many textbooks available, such as Chatfield (1983).

Although control networks are used for a variety of purposes in civil engineering, a growing use is in the monitoring of structural deformations and ground movement. With this in mind, the second aim of this chapter is to introduce the reader to the concepts involved in the analysis of networks which have been repeatedly observed in order to detect and quantify movements. Yet again, however, this topic is sufficiently involved to be the subject of a whole book (for example Caspary, 1987) and therefore only a brief coverage can be given here. More importantly perhaps, the various techniques for deformation analysis are the source of much discussion and have been the subject of an International Federation of Surveyors (FIG) study group and numerous international symposia (FIG, 1981; FIG, 1983). The reader should therefore be aware that the methods described herein are by no means the only ones used, although the problems which have to be addressed are of course universal.

The chapter begins with a brief introduction to the theory of errors before giving details of the adjustment (section 5.3) and analysis (section 5.4) of control networks by least-squares estimation. Section 5.5 briefly reviews the topic of the design of a control network before the subject of deformation monitoring is addressed in sections 5.6 and 5.7.

5.2 Theory of errors

5.2.1 *Types of error*

It is safe to say that all survey observations are subject to measurement uncertainty. The sources of error causing this uncertainty in engineering survey measurements are numerous and diverse. They include such factors as physical instability of a measurement station, atmospheric variation along a line of observation, instrument malfunction and plain human fallibility. For the purposes of this chapter it is reasonable to classify errors into three types which can be treated separately in the design, adjustment and analysis of engineering control networks. These types are random errors; systematic errors; and blunders.

5.2.1.1 *Random errors.* Random errors (sometimes, perhaps confusingly, called accidental errors), reflect the variability of repeated measurements due to sampling from statistical probability distributions. Variables (observations) which are subject to random errors are known as random or stochastic variables. In surveying it is usual to restrict the term random error to Normal probability distributions. For a single variable x the distribution is described by two parameters, the expectation (μ_x) and the variance (σ_x^2). The positive square root of the variance is known as the standard deviation (σ_x).

However, in surveying one is usually concerned with many different types of observation involving numerous random variables. In such cases a multidimensional normal distribution is appropriate, and then a third parameter, the covariance (σ_{xy}) between any two variables x and y, must be considered.

For a detailed explanation of the statistics of random variables and multivariant Normal distributions the reader is referred to any one of numerous statistical texbooks, or, more specifically for a surveyor's viewpoint, to Cooper (1974), Mikhail (1976) and Cross (1983). However, some important generalities need to be emphasized. Firstly, for *independent* variables x and y, the covariance $\sigma_{xy} = 0$. This implies that the covariance is a measure of the correlation between variables, and indeed the coefficient of correlation ρ_{xy} is given by

$$\rho_{xy} = \sigma_{xy}/(\sigma_x \cdot \sigma_y) \tag{5.1}$$

Secondly, the variance of a random variable is a measure of the repeatability of a measurement of that variable (under constant conditions) and thus gives an indication of the *precision* of the measurement. This is not the same as the *accuracy* of the measurement, the difference between them being discussed in 5.2.1.4.

Finally, since in surveying one is always dealing with small samples of observations from the overall population, it is only possible to obtain estimates of the population expectation and variance. The sample mean ($\bar{x}$) and standard error (s_x) are unbiased estimates (see Cooper, 1974, for definition) of the population expectation and standard deviation respectively.

5.2.1.2 *Systematic errors.* Systematic errors have a constant effect on repeated observations and therefore cannot be detected by repetitive measurements. A useful concept is to consider this type of error as being due to an error in the basic mathematical model used to represent the physical reality of the observation concerned. For example, in the measurement of a distance by steel tape, the model used for calculating the distance from the observation will include (explicitly or implicitly) values for the temperature and tension in the tape. Errors in these values will lead to a systematic error in the result. Other common examples of sources of systematic error are atmospheric refraction, theodolite collimation and circle graduation errors, and EDM frequency drift.

There are several ways of attempting to reduce the effects of systematic errors on surveying observations. It may be possible to calibrate the instrument concerned and hence to quantify the error, allowing corrections to be applied to subsequent measurements. Alternatively, appropriate observational procedures such as reversing a theodolite, or reciprocal (trigonometric) levelling may be adopted. Another approach would be to attempt to improve the mathematical model of the physical reality by the measurement of additional parameters, such as a temperature profile to model atmospheric refraction. Nevertheless, however good the model, instrument calibration and observational procedure, in reality there will always be some systematic error present in any measurement, and its effect will produce a shift in the mean of a sample of observations from the true population expectation. Due to its nature it will not affect the standard error of the observations and is not therefore included in any *precision* estimate obtained from the standard error. However, it must be considered in an overall *accuracy* assessment (see 5.2.1.4).

5.2.1.3 *Blunders.* Blunders (mistakes or gross errors) are an inevitable consequence of human fallibility, but the use of careful observational procedures and in particular the

ANALYSIS AND APPLICATION OF CONTROL NETWORKS 149

use of independent checks on observations should largely reduce their effect. However, despite the use of checks and precautions, blunders do still sometimes escape notice and find their way into network analysis computations. It is therefore essential that analysis techniques provide a suitable means for blunder detection. The ability of a network to do this is quantified by *reliability* criteria (5.2.1.4 and 5.4.1).

5.2.1.4 *Precision, accuracy and reliability*. The concepts of precision, accuracy and reliability have been a matter of debate for many years. They may, however, be considered as directly relating to the types of error present in observations. Precision is a quality representing the repeatability of a measurement and therefore only includes an assessment of random error, whereas accuracy (is considered) to be an overall estimate of the error in a quantity, including systematic effects. It therefore follows that precision may be regarded (and is often referred to) as the 'internal accuracy'. It should be noted here that repeatability is not even a true measure of precision unless the complete measurement process is repeated. A case in point is the observation of several rounds of theodolite directions which will not lead to a truly representative value for precision unless, for example, the theodolite is taken down and re-centred and levelled between each round, since random errors, such as centring, are not re-sampled.

However, in general terms the comparison of results obtained using completely different measurement techniques (and therefore subject to different systematic errors) will lead to a good estimate of accuracy, provided that blunders have been eliminated.

The ability of a measurement scheme to detect and hence eliminate blunders leads to the concept of *reliability*. An observation which is reliable is unlikely to contain an undetected blunder, and conversely a blunder is unlikely to be detected in an unreliable observation. The concept of reliability will be discussed later in the context referred to section 5.4 and to Cross (1987c).

For a fuller description of the quality assessment of a survey network the reader is referred to section 5.4 and to Cross (1987c).

5.2.2 *The principle of least squares*

In surveying it is common, and indeed good, practice to measure more quantities than the minimum required to achieve the desired result. In other words, redundant observations are made in order to provide a check for blunders as well as to allow some assessment of the precision of the measurements undertaken and the results obtained. However, once an overdetermined system of observations exists, there comes the problem of how to compute the results. This is resolved by recourse to an adjustment of one sort or another, and most commonly this takes the form of a least-squares adjustment. Two questions therefore arise. What is the purpose of an adjustment and why is the least-squares adjustment most often used?

The objectives of an adjustment are given in Bomford (1980) as: (i) to produce unique values for the unknowns, which (ii) will be of maximum probability, and (iii) to indicate the precision with which the unknowns have been determined.

Most surveyors would not disagree with these objectives, but not many know why the least-squares method provides the most widely accepted solution. What is 'least squares'?

Given an overdetermined set of m observations in n unknowns, it is unlikely, due to the existence of observational errors, that the observations will be consistent, and hence any subset of n linearly independent observations will produce different values for the n

unknowns. It is therefore not possible to produce a unique solution from the observations unless an additional criterion is introduced. The least-squares method (which was first used by Gauss) defines this criterion by constraining the sum of the squares of the residual errors to be a minimum, i.e.

$$V^\mathrm{T} V \to \text{minimum} \tag{5.2}$$

where $V = \hat{l} - l$ = vector of residual errors
 l = vector of observations
and $\hat{l}$ = vector of adjusted observations, which are totally internally compatible.

The adjusted observations are not necessarily (and very unlikely to be) the true values, and therefore the residual errors are not true observational errors. However, a property of least-squares estimation, which is one of the main reasons for its widespread use, is that 'on average' the adjusted observations are the true observations. In other words, least-squares estimation leads to an unbiased estimate of the true observations. In addition, it can be shown that least-squares estimation provides the minimum variance for the solution unknowns (e.g. station coordinates) and for any quantities derived from these unknowns (e.g. adjusted distances, directions) and it is hence referred to as the Best Linear Unbiased Estimate (BLUE). These properties are all independent of any assumption about the probability distribution of the observational errors. However, if these errors are Normally distributed (as is generally assumed to be the case in surveying), then the least-squares method also leads to the maximum likelihood ('most probable') solution.

The preceding description implicitly assumes observations of equal precision. However, the same results hold true for independent observations of differing precision, if a diagonal weight matrix W is included:

$$V^\mathrm{T} W V \to \text{minimum} \tag{5.3}$$

where the elements of W consist of the inverse of the variances of the observations. Both Gauss and later Laplace assumed independent observations which lead to a diagonal W. However, later it was shown that the principles hold true for correlated observations, i.e. for a full matrix W. In this case W is the inverse of the covariance matrix of the observations.

Cross (1983) gives a full description and justification, together with a short historical background, of the method of least squares.

5.3 Adjustment of survey observations by least-squares method

The previous section described the basic principles of least-squares estimation without reference to the parameters which it is specifically required to estimate through the adjustment process. When dealing with survey observations, these parameters are most likely to be the coordinates of the survey stations (hence enabling the computation of derived quantities such as distance and bearings), but may also include unknowns related to the observations (e.g. systematic observational biases—see 5.4.4) and to the reference coordinate system. We shall initially consider the simple case involving only station coordinates. For further simplicity, the following sections will cover only the two-dimensional plane situation, which is adequate for many engineering purposes. However, the least-squares adjustment computation can be carried out in any reference frame, and the more involved three-dimensional adjustment is equally valid, provided observations are related to a common reference system. It must be appreciated, however, that due to the variation of the direction of the vertical (which is of the order of 30″ per kilometre), the combination of observations related to local verticals (i.e. most survey measurements) must be carried out with careful consideration. For a detailed explanation of three-dimensional adjustment the reader is referred for example to Grist (1984) and Hein and Landau (1983).

Before any estimate of the desired parameters can be attempted, a mathematical model linking the parameters with the observations must be formulated. This model is known as a functional model. Two special cases of the functional model lead to sets of equations known as 'condition equations' and 'observation equations' (Cooper, 1987). In modern survey practice it is very unusual to use an adjustment by 'condition equations' primarily because of the difficulty in formulating the equations for large networks. Thus an 'observation equation' adjustment is normally carried out, this method also being commonly known as the 'variation of coordinates' or 'adjustment by parameters' method. It should be noted that identical results will be obtained irrespective of whether the condition or observation equation method is used. However, for the reasons given above only the observation equation method will be considered further here.

5.3.1 Observation equations

For each observation in the network, an observation equation is formulated in terms of the parameters required. For example, the equation for the simple case of a (two-dimensional) measured distance expressed in terms of the unknown coordinates is given by

$$l_{ij} = [(X_j - X_i)^2 + (Y_j - Y_i)^2]^{1/2} \tag{5.4}$$

However, in general (as can be seen above), these equations are non-linear and, in order for a practical least-squares estimation to be carried out, must be linearized by using a first-order Taylor's expansion. This leads, for the above example, to

$$\delta l_{ij} = -\frac{(X'_j - X'_i)}{l'_{ij}} \delta X_i - \frac{(Y'_j - Y'_i)}{l'_{ij}} \delta Y_i$$

$$+ \frac{(X'_j - X'_i)}{l'_{ij}} \delta X_j + \frac{(Y'_j - Y'_i)}{l'_{ij}} \delta Y_j \tag{5.5}$$

where $\delta X_i, \delta Y_i, \delta X_j, \delta Y_j$ represent corrections to initial, approximate values of X'_i, Y'_i, X'_j, Y'_j respectively, and δl_{ij} is given by

$$\delta l_{ij} = l_{ij} - l'_{ij} \tag{5.6}$$

with l'_{ij} being the distance calculated from the approximate coordinates.

For an observed value of the distance, l^o_{ij}, within an overdetermined network the linearized observation equation becomes (abbreviating the coefficients above to K, L, M and N):

$$K\delta X_i + L\delta Y_i + M\delta X_j + N\delta Y_j = l^o_{ij} - l'_{ij} + v \tag{5.7}$$

where v is the least-squares residual error, which is equivalent to the difference between the adjusted and observed values of the distance, neither of which is, in general, the true value.

Similar equations can be formulated for all types of observations, that for a clockwise horizontal angle φ_{ijk} measured at station i between stations j and k being given by

$$\left[\frac{(Y'_j - Y'_i)}{(l'_{ij})^2} - \frac{(Y'_k - Y'_i)}{(l'_{ik})^2}\right]\delta X_i + \left[\frac{(X'_k - X'_i)}{(l'_{ik})^2} - \frac{(X'_j - X'_i)}{(l'_{ij})^2}\right]\delta Y_i$$

$$-\frac{(Y'_j - Y'_i)}{(l'_{ij})^2}\delta X_j + \frac{(X'_j - X'_i)}{(l'_{ij})^2}\delta Y_j + \frac{(Y'_k - Y'_i)}{(l'_{ik})^2}\delta X_k$$

$$-\frac{(X'_k - X'_i)}{(l'_{ik})^2}\delta Y_k = \varphi^o_{ijk} - \varphi'_{ijk} + v \qquad (5.8)$$

Equations for other types of observation are given, for example, in Cooper (1987).

In general the set of observation equations (or linearized functional model) can be expressed by the matrix equation

$$Ax = b + V \qquad (5.9)$$

where A is the coefficient matrix, (which is also commonly known as the configuration matrix or design matrix)
x is the vector of unknown corrections to the approximate coordinates
b is the vector of observed minus computed (from approximate coordinate) values
V is the vector of residual errors.

It must be noted here that the linearization process used to formulate the observation equations is valid only if the approximate values adopted are close enough to the true values for the second- and higher-order terms in the Taylor expansion to be neglected. This is not generally *known* to be the case (although it may be so), and therefore an iterative solution should be carried out. Provided reasonable approximate coordinates are used, it is unusual for more than one or two iterations to be needed. In addition, it is of interest to note that some geodesists are questioning the validity of linearized models, and the use of non-linear adjustment is being investigated (Teunissen, 1985). Such techniques are, however, beyond the scope of this book.

5.3.2 *Least-squares solution*
The least-squares solution of an overdetermined set of observation equations is obtained by setting

$$V^T V \to \text{minimum} \qquad (5.2)$$

i.e. $V_1^2 + V_2^2 + V_3^2 \quad .. \quad .. \quad .. \quad V_n^2 \to \text{minimum}$

The application of this criterion leads to the so-called Normal equations (see for example Cooper, 1987):

$$A^T A \hat{x} = A^T b \qquad (5.10)$$

and hence to the solution for the least-squares estimate of the parameters, $\hat{x}$:

$$\hat{x} = (A^T A)^{-1} A^T b. \qquad (5.11)$$

However, the above oversimplifies the solution. In general, the observations will be of different types and precision, and may well be correlated. Hence the least-squares estimates are properly defined as those which minimize the quadratic from $V^T W V$, where W is the 'weight matrix', the inverse of the covariance matrix of the observations (also known as the stochastic model), i.e.

$$W = C^{-1} \tag{5.12}$$

The covariance matrix C is defined as containing the variances of the observations on the diagonal and the appropriate covariances as the off-diagonal elements. In many instances, it may be assumed that observations are uncorrelated (covariance = 0) and in this case W becomes the diagonal matrix

$$W = \begin{bmatrix} \frac{1}{\sigma_1^2} & 0 & & & 0 \\ & \frac{1}{\sigma_2^2} & & & \\ & & \frac{1}{\sigma_i^2} & & \\ & \text{Symm.} & & & 0 \\ & & & & \frac{1}{\sigma_n^2} \end{bmatrix} \tag{5.13}$$

where σ_i^2 is the variance of the ith observation.

The quadratic form criterion is thus simplified to the equivalent of

$$W_1 V_1^2 + W_2 V_2^2 + \ldots \quad \ldots \quad W V_n^2 \to \text{minimum} \tag{5.3}$$

However, even in this simple case the allocation of appropriate variances (or weights) can be a difficult problem (see 5.4.2). In the more complex situation of correlated observations, the allocation of covariances substantially adds to the difficulty. The use of modern space systems usually leads to correlated observations, and for a discussion of this problem the reader is referred to Cross (1987b). For the purposes of this chapter it will be assumed that observations are uncorrelated and therefore a diagonal weight matrix exists.

The introduction of the weight matrix W leads to a new set of Normal equations and corresponding solution, for the least-squares estimates, namely

$$A^T W A \hat{x} = A^T W b \tag{5.14}$$

and

$$\hat{x} = (A^T W A)^{-1} A^T W b \tag{5.15}$$

Apart from the solution vector $\hat{x}$, the matrix $(A^T W A)^{-1}$ is of great importance, since it can be shown that the covariance matrix of the unknowns $(C_{\hat{x}})$ is given by

$$C_{\hat{x}} = \sigma_0^2 (A^T W A)^{-1} \tag{5.16}$$

$C_{\hat{x}}$ (which is symmetrical) is more explicitly expressed as

$$C_{\hat{x}} = \begin{bmatrix} \sigma^2_{x_1} & \sigma_{x_1 x_2} & & & \sigma_{x_1 x_n} \\ & \sigma^2_{x_2} & & & \\ & & & \sigma^2_{x_i} & \\ & \text{Symm.} & & & \\ & & & & \sigma^2_{x_n} \end{bmatrix} \qquad (5.17)$$

where $\sigma^2_{x_i}$ is the variance of the ith unknown and $\sigma_{x_i x_j}$ is the covariance of the ith and jth unknowns. Note that the subscripts of the variances and covariance should properly be $\hat{x}_i$, $\hat{x}_i \hat{x}_j$, etc., but the $\hat{}$ symbols have been omitted for clarity. The quantity σ^2_0 is known as the 'unit variance' or 'variance factor'. Correspondingly, σ_0 is known as the 'standard error of an observation of unit weight'. It can be shown that if the stochastic model for the observations has been chosen correctly, then σ^2_0 should equal unity. An unbiased estimate for σ^2_0 can be calculated from

$$\hat{\sigma}^2_0 = \frac{V^T W V}{n - k} \qquad (5.18)$$

where n is the number of observations and k is the number of unknowns, i.e. $(n - k)$ is the number of degrees of freedom.

In order to calculate $\hat{\sigma}^2_0$ it is therefore necessary to compute the residual vector V by back-substitution of the solution vector $\hat{x}$ into the observation equations. However, the residuals are of interest in themselves, particularly in the detection of outliers (see 5.4.1.1). At present our concern is with $\hat{\sigma}^2_0$, which can be used as an aid to the assessment of the validity of the assumed stochastic model since it is an indication of by how much, *on average*, the weight matrix has been over- or underestimated. This is discussed further in 5.4.2.

If $\hat{\sigma}^2_0$ is unity, it follows that

$$C_{\hat{x}} = (A^T W A)^{-1} \qquad (5.19)$$

However, in the case of it not being unity, the use of the estimated $\hat{\sigma}^2_0$ to premultiply $(A^T W A)^{-1}$, to obtain $C_{\hat{x}}$ may be dangerous, since very large or small $\hat{\sigma}^2_0$ values may be due to the estimation process rather than a poor assumed stochastic model. For example, purely fortuitously, observations could lead to zero residuals (by compensating errors) and hence to a zero value for $\hat{\sigma}^2_0$. This would mean that the $C_{\hat{x}}$ (which would become the null matrix if $\hat{\sigma}^2_0 (A^T W A)^{-1}$ is calculated) would imply perfect estimates of the unknowns (i.e. zero variances). This would obviously not be true. The estimated value should be tested statistically against its assumed value (i.e. 1) before it is used in the above manner. Further details and examples are given in Cooper (1987).

Once the covariance matrix $C_{\hat{x}}$ has been determined, it plays a significant role in the analysis of the network, and in particular in the assessment of deformations. It is the ability of the least-squares method to provide comprehensive error analysis in addition to the estimates of the parameters that makes it such a powerful tool in the adjustment of survey observations.

5.3.3 *Rank deficiency and constraints*

In general, the least-squares estimate of parameters cannot be obtained from the solution equation given above since the normal equation coefficient matrix will be singular.

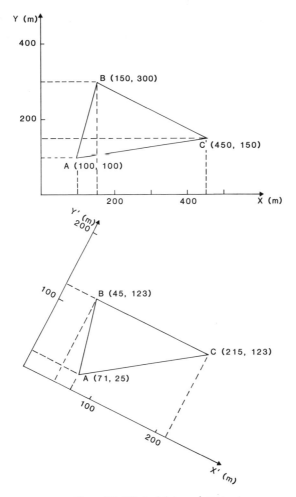

Figure 5.1 Effect of datum change.

In other words, the design matrix A has a column rank deficiency. Once this deficiency has been removed, the normal equations can be solved and the estimate $\hat{x}$ obtained. The reason for the rank deficiency is that the coordinate datum is not completely defined by the observations. For a two-dimensional network the datum definition requires four defined elements; two position, one scale and one rotation. The need for these elements can be seen in Fig. 5.1, where two identical sets of measurements (i.e. the angles of a triangle) will give completely different estimates of the coordinates of the stations, depending on the definition of the origin of the coordinate system (two position elements), the orientation of the coordinate axes (one rotation element) and the scale of the axes (one scale element). It will be evident that for a three-dimensional network the datum requires seven defined elements (three position, three rotation and one scale).

However, certain types of measurement implicitly define datum elements. For example, a distance measurement included in a network will provide scale to that network. Similarly a (gyro-theodolite) azimuth will provide orientation and an absolute satellite-Doppler measurement will provide position. Hence the datum deficiency of a network will depend on the types of measurements present in that network. Nevertheless, in most engineering networks a datum deficiency (= rank deficiency of matrix A) will exist and the least-squares estimate $\hat{x}$ can then only be obtained by the introduction of constraint equations.

Constraint equations are introduced by adding to the observation equations additional equations which remove the rank deficiency. For example, arbitrary position observations may be added to remove a positional defect. The addition of two station positions (i.e. four coordinate values) in a two-dimensional network will remove all possible datum deficiencies, since two station positions in themselves define scale and orientation. If more equations are added than are necessary to remove the rank deficiency, then the network is 'over-constrained'. It is usual to add only the minimum number of constraint equations necessary to remove the deficiency, and the solution is then termed a 'minimum constraint' solution. It is important to realize that the introduction of arbitrary constraint equations will lead to arbitrary estimates from the solution. This is of particular importance in the analysis of networks, and in the application of networks to deformation monitoring, and will be returned to later in the chapter (5.4.3 and 5.6.2).

When adding constraint equations it is possible to do so either in a stochastic or a non-stochastic sense. That is to say, the arbitrary observations may be 'fixed' or given appropriate variances (and covariances) in the stochastic model. In the latter case, it may be possible to reduce the arbitrary nature of the constraint if the measurement added is approximately realistic and the variance attached to it represents the nature of the approximation. It should be noted, however, that if the solution has only minimum constraints, then the value of the constraint observation (and its variance) will remain unaltered by the adjustment process.

An alternative approach to that described above is to remove rank deficiency by the so called 'free-network' adjustment, also known as the method of 'inner constraint'. In this method, constraints are imposed on the average positional, rotational and scale changes throughout the network (assuming full rank defect). This results (Cooper, 1987, for example) in a solution which minimizes $\hat{x}^T \hat{x}$, in other words the sum of the squared corrections to the approximate coordinates is a minimum. This method of adjustment has been widely used and is often reputed to remove the problem of datum definition. However, it is important to appreciate that the datum is effectively defined by the approximate coordinates used, and as such is affected by any change in those coordinates. In addition it should be noted that the centroid of the approximate coordinates (and their average orientation and scale) are maintained after the adjustment process.

5.4 Analysis of control networks

The analysis of control networks centres around the assessment of the precision, reliability and accuracy of the estimated parameters. These quantities, which were described in general terms in 5.2.1.4, will now be discussed in more detail.

5.4.1 *Reliability and tests for outliers*

The reliability of a network can, in simple terms, be understood as the ability of that network to detect errors in the observations. This concept must therefore be linked to the detection and rejection of 'outliers'. The ideas of reliability are complex and this section will deal only with general principles and a few specific tests which are most commonly used. For a fuller description the reader is referred to, for example, Caspary (1987).

5.4.1.1 *Detection of outliers.* In order to assess whether an observation is an outlier, it is usual to compute the ratio of the residual of the observation to its standard error, i.e.

$$\hat{Z}_i = V_i/\sigma_{V_i} \tag{5.20}$$

the standard error being given by the square root of the ith diagonal element of the matrix C_V:

$$C_V = \sigma_0^2 W^{-1} - AC_x A^T \tag{5.21}$$

The statistics $\hat{Z}_i$ is then compared in a one-dimensional test with a value computed from the appropriate (usually normal) distribution for a specified percentage rejection level α, where α might be for example 0.05 (equivalent to 5% rejection level). An observation which does not pass the test is then subject to further investigation and possibly rejection.

This simple approach has two major drawbacks. Firstly, a one-dimensional (Normal) distribution test is not strictly statistically valid, indeed even a Student t-test is not, since V_i and σ_{V_i} are usually computed from the same sample and are therefore dependent. Several authors have proposed alternative methods, notably Baarda's Data Snooping Technique (Baarda, 1968) and Pope's Tau Method (Pope, 1976), the latter in particular being commonly used. Details and comparisons of several methods, including those mentioned, are given in Caspary (1987).

Secondly, the simple method described is unlikely to be efficient in outlier detection if more than one error is present in the observations. (It should be noted that this is also the case for Pope's method.) A discussion of a method for multiple error detection may be found in Cross and Price (1985). Nevertheless, simple outlier rejection based on the one-dimensional Normal distribution has commonly been, and will no doubt continue to be, used.

Whatever test is used for outlier detection, it is important to realize that two possible errors can be made. These are referred to as Type I and Type II errors and are defined as follows.

(i) *Type I error*: The rejection of an observation which is not an error (in general terms the rejection of an hypothesis which is true).
(ii) *Type II error*: Acceptance of an observation which is an error (in general, the acceptance of an hypothesis which is not true).

5.4.1.2 *Internal reliability.* The concept of reliability can be envisaged in terms of the probability of gross errors remaining undetected by the outlier rejection procedure. If the chances of this happening for a particular observation are small, then the observation is said to be reliable, or to have high reliability. If all observations are reliable, then the network as a whole is also reliable (i.e. is globally reliable). This idea of reliability is usually referred to as 'internal reliability'. It may be quantified in terms of the largest gross error which will remain undetected in an observation, also known as the marginally detectable error, Δ_i, given by

$$\Delta_i = \delta_i \tau_i \sigma_i \tag{5.22}$$

where σ_i is the standard error of the ith observation,

$$\tau_i = \sigma_i / \sigma_{V_i} \tag{5.23}$$

and δ_i is a dimensionless factor computed from the probability density function of the Normal distribution, dependent on the significance level α, the probability of making a Type I error, and β, which is the probability of making a Type II error. $(1 - \beta)$ is known as the 'power' of the test.

Hence if values for α and β are chosen, the size of the marginally detectable error for each observation can be computed. (Note: no observations are necessary for this computation, which can be carried out at the network design stage—see 5.5). Fuller details of this computation can be found in Cooper (1987). As an example, for $\alpha = 5\%$ and $\beta = 15\%$

$$\delta_i = 1.960 + 1.037 = 2.997 \approx 3.0 \tag{5.24}$$

where 1.960 is obtained from Normal distribution tables (one-tail) test for $(1 - \alpha/2) = 0.975$, and similarly 1.037 corresponds to $(1 - \beta) = 0.850$.

Hence, if an observation has been made with a standard error of 5 mm, and τ_i has a value of 2.0, then, when using 5% as a rejection criteria for outliers, there will be a 15% chance that an error of 30 mm $(\Delta_i = 3 \times 2 \times 5 \text{ mm})$ remains undetected.

It should be noted that the Tau factor τ_i (not to be confused with the Pope Tau test) is a measure of internal reliability in itself. It can be shown that τ must lie between unity and infinity, and the closer it is to infinity the less reliable the observation. However, quantification of 'less' is difficult. Some authors, such as Ashkenazi (1981) have used different but related ratios as measures of reliability.

Where observations are found to be unreliable it will usually be necessary to strengthen the network in the region concerned, as the inclusion of undetected gross errors in an adjustment is normally unacceptable.

The type of reliability described above is known as a local internal reliability. For quantitative assessment of global internal reliability, the reader is referred for example to Cooper (1987) or Caspary (1987).

5.4.1.3 *External reliability*. External reliability refers to the effect of undetected gross errors in the observations on the quantities to be estimated from them. The size of this effect is often more important than the size of the undetected error itself. The effect on the estimated parameters of a marginally detectable error in the ith observation is simply given by

$$\Delta \hat{X}_i = (C_{\hat{x}} A^T W e_i) \Delta_i \tag{5.25}$$

where e is a null column vector except for a unit element in the ith row.

The corresponding effect on any quantity φ (where $\varphi = f(x)$) computed from the estimated parameters can also be found from

$$\Delta \varphi = f(\Delta x_i) \tag{5.26}$$

This measure of external reliability can, however, be tedious to compute, since the effect of all marginally detectable errors on all possible quantities φ could be considerable. Overall (global) external reliability measures are also sometimes used, for example (from Caspary, 1987):

$$\text{trace}(WAC_{\hat{x}}A^TW) \qquad (5.27)$$

but further consideration of these is outside the scope of this chapter.

5.4.2 Estimation of the weight matrix

One of the most difficult problems to be addressed in the analysis of a network is the choice of the stochastic model of the observations, or equivalently the weight matrix. Even assuming the simple case of a diagonal matrix, the relative weights of the various observations need to be determined. Perhaps the most commonly used approach to this problem is to adjust with minimum constraint a subset of the observations, all assumed to be of equal weight, and consisting of only one type of measurement (say theodolite angles). The resulting value obtained for σ_0^2 may be considered as representing the proportion by which the initially assumed weights were on average incorrect, since it is known that for correctly chosen weights, σ_0^2 should equal unity (see 5.3.2). The assumed weights can therefore be adjusted proportionately.

Subsequently, additional subsets of observations are included in the adjustment and the procedure repeated, amending only the weights of the newly added observations each time. It should be noted, however, that since the value of σ_0^2 reflects the overall error in the assumed weights (including those already determined), the alteration of the weights of additional subsets of observations is not in direct proportion to the value of σ_0^2. The process must therefore be carried out iteratively until a value for unity is achieved.

Alternatively the weights of each subset can be determined from independent adjustments of the subsets individually; however, this assumes sufficient observations of each type are available for an adjustment to be possible. Indeed, even the former approach requires sufficient observations of at least one type of measurement to be present. If this is not the case, the method of variance component estimation can be employed. This approach will not be discussed here, but more details are given for example in Cross (1987c).

5.4.3 Precision estimates

Once one can be confident that blunders have been detected and that the covariance matrix of the observations has been adequately estimated, it is reasonable to make use of the covariance matrix of the estimated parameters $(C_{\hat{x}})$ to assess the precision of the network. There are various methods for doing this and, as with reliability, it is possible to use global or local quantities. However, before going further it is necessary to understand the concept of 'estimable' (or invariant) quantities.

5.4.3.1 Estimable quantities.

In principle an estimable quantity is one which can be estimated (calculated) from the observations themselves, *without* the introduction of any (arbitrary) constraints (see 5.3.3). For example, in a network consisting only of angle and distance measurements, the coordinates of a station are not estimable since such a network has a positional datum defect. However, the distance between any two points (not necessarily one which has been measured) is an estimable quantity, since it can be calculated from the angles and distances observed irrespective of the choice of datum position and orientation. However, in a network which has a scale defect, distance is not an estimable quantity, although distance ratios are.

The importance of estimable quantities is not so much in the quantity itself but in the precision estimate attached to such a quantity, since this again will be independent of the choice of arbitrary datum constraints whereas that for an inestimable quantity will not. Hence it is of little value to consider the precision of inestimable quantities since the precision will be entirely dependent on the datum chosen, and will therefore change if a new datum is defined. Nevertheless, such quantities are frequently used and it is therefore important that their value is understood. This point is reinforced later in 5.4.3.3.

5.4.3.2 *Global precision estimates.* There are several commonly used global precision estimates giving single-figure measures which in some way express the precision of a network as a whole. Although these estimates are not commonly used in engineering networks, examples are given here for completeness.

(i) The trace of the covariance matrix $C_{\hat{x}}$. This estimate is datum dependent, but the minimum of all possible tr $(C_{\hat{x}})$ is obtained from the covariance matrix resulting from the 'inner constraints' adjustment (see 5.3.3), and this value is perhaps most often used as a global precision estimate.
(ii) The mean variance of all network point coordinates; this again is datum-dependent.
(iii) The largest eigenvalue (λ_{max}) of $C_{\hat{x}}$. This represents the square root of the semi-major axis of the largest point error ellipse (see 5.4.3.3) in the network. The ratio of $\lambda_{max}/\lambda_{min}$ is also used as an indication of the homogeneity of the network.

5.4.3.3 *Local precision estimates.* Perhaps the most obvious local precision estimates are the variances of the estimated parameters, usually the coordinates of the network stations. These are given directly by the diagonal elements of the covariance matrix $C_{\hat{x}}$. However, more graphical representations are obtained by computing the error ellipses (or ellipsoids in three dimensions) of which two types may be distinguished: (i) absolute error ellipses, and (ii) relative error ellipses.

An *absolute error ellipse* for a point P can be computed from the submatrix (of $C_{\hat{x}}$) which contains the variances of the coordinates of P and the associated covariance, namely σ_x^2, σ_y^2 and σ_{xy}. It can easily be shown (e.g. Bomford, 1980) that the maximum and minimum values of the variance of the point position are given by

$$\sigma_{max, min}^2 = \sigma_y^2 + \sigma_{xy} \cot(\psi) \tag{5.28}$$

where ψ is given by

$$\tan(2\psi) = 2\sigma_{xy}/(\sigma_x^2 - \sigma_y^2) \tag{5.29}$$

The two solutions for ψ (which are 90° apart) provide the corresponding maximum and minimum values of the variance.

The absolute error ellipse is plotted (see Fig. 5.2) with the semi-major axis equal to σ_{max}, the orientation of the ellipse with respect to the x, y axes being determined by ψ. It should be appreciated that the ellipse is not the locus of the standard error (σ_p) in any direction, but lies within this locus which is the pedal curve of the ellipse, shown dotted in Fig. 5.2. It is, however, the error ellipse which is most generally used in practice, and the pedal curve is not usually drawn.

ANALYSIS AND APPLICATION OF CONTROL NETWORKS

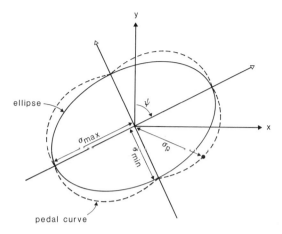

Figure 5.2 Absolute point error ellipse.

There is a 39.4% probability (see for example Cross, 1983, for full explanation) that the least-squares estimate of the position of a point will lie within the error ellipse centred on the point's true position. This is not the same as saying there is a 39.4% probability of the true position being within the error ellipse centred on the estimated position. However, as the estimated position is all that is usually known, the latter ellipse is what is generally drawn, and is referred to as the 39.4% (or one-sigma) confidence ellipse. Since such a low percentage confidence is not often useful, it is more appropriate to draw the 2-, $2\frac{1}{2}$- or 3-sigma ellipse.

Although absolute error ellipses are useful in giving an overall picture of network precision as shown by Fig. 5.3, they are datum-dependent since they are calculated with respect to inestimable quantities, i.e. coordinates. This datum dependence is demonstrated by the propagation in their size as the positions under consideration get further from the chosen datum, whether defined by minimum constraint (as in Fig. 5.3) or inner constraint (Fig. 5.4).

Given the appropriate submatrices of $C_{\hat{x}}$ (i.e. σ_x^2, σ_y^2 and σ_{xy}) for two points in a network, differences for the pair can be calculated from

$$\sigma_{x_j - x_i}^2 = \sigma_{x_i}^2 + \sigma_{x_j}^2 - 2\sigma_{x_i x_j} \tag{5.30}$$

$$\sigma_{y_j - y_i}^2 = \sigma_{y_i}^2 + \sigma_{y_j}^2 - 2\sigma_{y_i y_j} \tag{5.31}$$

$$\sigma_{(x_j - x_i),(y_j - y_i)} = \sigma_{x_i y_i} - \sigma_{x_i y_j} - \sigma_{x_j y_i} + \sigma_{x_j y_j} \tag{5.32}$$

These two variances and a covariance can then be used (in a similar manner to σ_x, σ_y, σ_{xy} for absolute error ellipses) to compute the semi-major and semi-minor axes of the *relative error ellipse* for the pair of points. This ellipse is conventionally drawn on the line joining the two points, as shown in Fig. 5.5. From the error ellipse, the standard error of the estimated length of the line ($\hat{l}_{ij}$) between the points, as well as the standard error of the estimated bearing ($\hat{\alpha}_{ij}$) between them, can be computed. In Fig. 5.5 the tangents to the ellipse, which are parallel and perpendicular to the line joining i and j, have been drawn. The resulting distances PP_1 and PP_2 are related to the required quantities simply by

$$PP_2 = \sigma_{\hat{l}_{ij}}, \tag{5.33}$$

162 ENGINEERING SURVEYING TECHNOLOGY

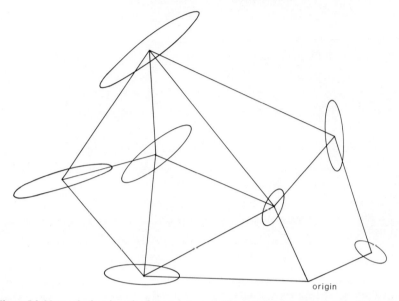

Figure 5.3 Network showing absolute point error ellipses (minimum constraint adjustment).

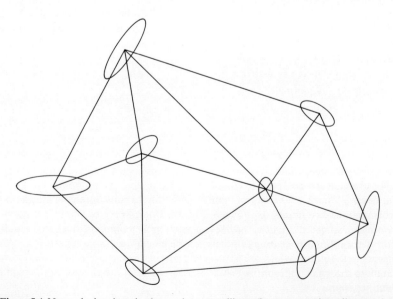

Figure 5.4 Network showing absolute point error ellipses (inner constraint adjustment).

ANALYSIS AND APPLICATION OF CONTROL NETWORKS 163

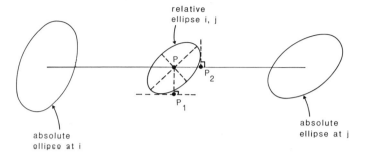

Figure 5.5 Relative error ellipse.

and
$$PP_1 = \hat{l}_{ij}\sigma_{\hat{a}_{ij}} \tag{5.34}$$

Relative error ellipses are datum-independent since they refer to coordinate differences. They are therefore particularly useful in assessing network precision.

The covariance matrix $C_{\hat{x}}$ contains all the required information to compute the *variance* of any quantity which can be calculated from the estimated parameters $\hat{x}$. The straightforward relationship for any derived quantity is given by

$$\sigma_\varphi^2 = f^T C'_{\hat{x}} f \tag{5.35}$$

where f represents the vector of linearization coefficients relating the estimated parameters to the quantity required, φ, and $C'_{\hat{x}}$ is the appropriate submatrix of $C_{\hat{x}}$. For example, for the variance of a distance $\hat{l}_{ij}$, f^T is given by eqn (5.5):

$$f^T = -\frac{(\hat{x}_j - \hat{x}_i)}{\hat{l}_{ij}} - \frac{(\hat{y}_j - \hat{y}_i)}{\hat{l}_{ij}} + \frac{(\hat{x}_j - \hat{x}_i)}{\hat{l}_{ij}} + \frac{(\hat{y}_j - \hat{y}_i)}{\hat{l}_{ij}}, \tag{5.36}$$

and $C'_{\hat{x}}$ by:

$$C'_{\hat{x}} = \begin{bmatrix} \sigma^2_{x_i} & \sigma_{x_i y_i} & \sigma_{x_i x_j} & \sigma_{x_i y_j} \\ & \sigma^2_{y_i} & \sigma_{y_i x_j} & \sigma_{y_i y_j} \\ & & \sigma^2_{x_j} & \sigma_{x_j y_j} \\ \text{Symm.} & & & \sigma^2_{y_j} \end{bmatrix} \tag{5.37}$$

Any derived quantity may be used, whether observed or not, although estimable quantities are probably the most useful.

5.4.4 Accuracy and systematic biases

The assessment of the accuracy of a network can be achieved only if all the systematic errors in the measurements can be quantified in addition to the precision considerations previously mentioned. If system biases are suspected and a realistic model for their effect can be proposed, then they can be included in the functional model of the network. It is important to realize, however, that such biases can be determined in a minimum constraint adjustment only if there is at least one observation in the network which is unaffected by the bias. For an inner constraint adjustment, biases will be determined with respect to values defined by the adopted approximate coordinates.

An example of bias estimation is given by the 1980 readjustment of the Ordnance Survey national triangulation (Ashkenazi et al., 1985). Here scale bias parameters for EDM measurements were included in the adjustment, and were determined with respect to satellite Doppler-derived positions which were assumed to be free from scale error. The values obtained for the bias associated with microwave Tellurometer distance measurements was between 4 and 10 ppm. This should be compared with the estimated precision of the previous adjustment, OSSN 70, which gave an average scale standard error of 2.5 ppm (Ashkenazi and Cross, 1972).

A full discussion of systematic error modelling is beyond the scope of this chapter, but interested readers are referred to Ashkenazi (1981). It is, however, of great importance to realize the limitations of precision analysis, and to appreciate the difference between precision and accuracy.

5.5 Network design

Much can be done to design a network to ensure that it will achieve its desired aim, before any measurements are made. The basis on which the design can be carried out is seen from the form of the matrix equation which gives the covariance matrix of the estimated parameters, namely

$$C_{\hat{x}} = \sigma_0^2 (A^T W A)^{-1} \tag{5.38}$$

The only term in this equation which requires any observational data is σ_0^2, which can be assumed to be unity for design purposes, since this will be its value once the weight matrix has been appropriately estimated. It therefore follows that all precision estimates based on $C_{\hat{x}}$ may be computed without any observations being made, provided the intended positions of the observation stations are approximately known and the measurements which are to be made have been identified. This situation represents the most usual design problem, that is, to decide where to position observation stations and which measurements to make in order to satisfy defined (precision) criteria. However, this is only one of four orders of design, which are commonly defined as follows.

(i) *Zero-order design* (A, W fixed; x, C_x free). The datum problem; the choice of an optimal reference system for the parameters and their covariance matrix. A common solution is to obtain the minimum trace of C_x (see 5.4.3.2). If estimable quantities are used in the definition of the required precision criteria, then the solution to this order of design problem is immaterial.

(ii) *First-order design* (W, C_x fixed; x, A free). The configuration problem; this is the situation introduced above where the station positions and observational scheme are to be designed given observational accuracies and a required precision.

(iii) *Second-order design* (A, C_x and x fixed; W free). The weight problem; this is the determination of the observational accuracies required for a given scheme to meet defined precision criteria.

(iv) *Third-order design* (C_x fixed; A, W and x partly free). The improvement problem; really a combination of first- and second-order design, required for example when strengthening an existing network to meet improved precision criteria.

The design required usually not only needs to solve the problem of meeting precision criteria, but must also be the minimum-cost solution, often referred to as the optimum design. This introduced cost element can be very difficult to quantify, but possible designs are usually assessed subjectively taking regard of previous experience.

Once the design problem has been formulated, there are two basic approaches to the solution. Firstly, and most commonly, there is the computer simulation, or pre-analysis method, whereby proposed networks are analysed in turn to see whether they meet the required criteria, being subjectively modified by operator intervention, and using his experience, if the proposed scheme is either too strong or not strong enough. This method has been successfully used on many engineering projects (for example see Ashkenazi et al., 1982), but it has the drawback of possibly (or even probably) missing the optimum solution. In contrast, the analytical approach attempts to mathematically formulate the design problem in terms of equations or inequalities and then to explicitly solve for the optimum solution. This latter method has so far been of only limited success, but is still being developed. For further details of analytical design, the reader is referred to Grafarend and Sanso (1985); however, Cross (1987a) gives a concise description of both the simulation and analytical techniques.

Finally, when considering design for deformation analysis (see sections 5.6, 5.7) it is important to take account of the sensitivity of the resulting network to the particular deformation expected. This is because the purpose of such a network is usually not only to detect possible movements, but also to try and establish the general mechanism of the motion taking place. In other words, it is required to test theoretical models of deformation against the result of the network analyses.

5.6 Networks for deformation monitoring

Deformation monitoring, whatever the structure involved, consists of four stages: specification, design, implementation and analysis. At first sight the specification of the requirements of the monitoring scheme might seem straightforward; for example, a given magnitude of movement is required to be detected. However, the problem is more involved than it might appear. To what confidence level is the movement to be detected? Are all points to be monitored to the same accuracy? Is the direction of movement important? Is 'absolute' or relative movement required, and, if 'absolute' what stable origin can be assumed? It is only after questions such as these have been answered that a design can commence. It will be apparent that the specification will usually need to be decided upon by the engineer, rather than the surveyor, but taking due consideration of the points such as those mentioned. However, in order for the survey scheme adopted to provide sufficient information to meet the specifications, it will very often be necessary to adopt a rigorous network analysis approach, and here of course the surveyor's role will be paramount. It should be noted that, although conventional survey network analysis techniques can be effectively adapted for deformation monitoring, the observations in the network need not be limited to conventional measurements. For example offset (Gervaise, 1976) and optical (laser) alignment (Williams, 1987) measurements might be included.

The two remaining aspects of a monitoring scheme which will be considered in detail are the design and analysis stages. Before doing so, it is important to briefly consider the implementation stage which, although it is of utmost importance, and is probably the most costly element, usually involves standard measurement techniques. On many occasions particularly high-precision observations are required, and special attention must be paid to such matters as centring, targeting and levelling of instruments. Nevertheless, such problems will not be discussed further here.

5.6.1 *Design of a monitoring network*

As has already been mentioned in section 5.5, a network may be designed to meet specific criteria, before any observations are actually made. In the specific case of a deformation monitoring network, the design may not only be required to meet precision (e.g. variances of point positions or derived quantities) and reliability criteria, but also to be sensitive to the deformation pattern which is expected to take place. If such a pattern of a deformation can be formulated in a mathematical model, then network designs can be quantitatively assessed as to their capability to identify the true deformation. Such an ability is sometimes referred to as the 'sensitivity' or 'testability' of the network. This concept of sensitivity is relatively new and not yet widely used; however, two methods of assessment are given in Niemeier *et al.* (1982) and Fraser (1982), and brief details are outlined in Cooper (1987). In essence, the techniques involve an F-test being carried out, at the required significance level, to test an assumed value of deformation against the covariance matrix of that deformation C_d derived from the designed network covariance C_x. The size of deformation which can just be detected at this significance level can then be calculated.

Since a postulated deformation model between two epochs of observations represents in effect a systematic difference between the two sets of measurements, the sensitivity assessment of a network can be regarded as being related to the detection of systematic errors (Niemeier *et al.*, 1982).

5.6.2 *Analysis of deformation monitoring networks*

The analysis of a network observed for the monitoring of deformation will usually initially consist of a conventional analysis with the appropriate tests for the detection of outliers, and the computation of the estimated parameters and associated covariance matrix and unit variance. However, a monitoring network will be repeatedly measured at various epochs, and a comparison of successive network adjustments must be carried out in some way in order to detect any deformations which have taken place. It is essential to realize that the straightforward difference in the two sets of coordinates does not in itself provide sufficient information to assess whether points have moved or not, since some consideration must be given to the accuracy with which the coordinates have been determined. The same magnitude of difference at two stations may represent a significant movement in one case and not in the other. Any analysis must therefore be carried out in conjunction with all the available accuracy estimates. There are many approaches to this problem, several of which are described and compared in Chrzanowski and Chen (1986). However, before proceeding to give details of techniques, it is necessary to differentiate between two-epoch and multi-epoch analysis. In general, it is not only the comparison of two sets of measurements which is required, although this may be sufficient for many engineering purposes, but the complete analysis of all the observational data to hand. This may be done rigorously by using a multi-epoch analysis, or subjectively by inspecting the cumulative results of successive, rigorous, two-epoch analyses. Since the latter is less complex, it will be addressed first.

5.6.2.1 *Two-epoch analysis.* The starting point for a two-epoch analysis will be the results of two single-epoch adjustments, namely

$$\hat{x}_1, \hat{x}_2; W_1, W_2; C_{\hat{x}_1}, C_{\hat{x}_2}; \hat{\sigma}^2_{0_1}, \hat{\sigma}^2_{0_2}. \tag{5.39}$$

The aim of the analysis is to identify stable reference points in the network (if any), and detect single-point displacements which will later be used to aid in the development of an appropriate deformation model (see 5.7). It is necessary to stress how crucial is the detection of outliers in each of the single epoch adjustments, since errors which escape detection are likely to be assessed as deformations later in the analysis. The reliability of deformation monitoring networks is therefore of utmost importance.

The differentiation of monitoring networks into two classes is useful.

(i) *Absolute network*: a network in which one or more points can be considered stable, i.e. outside the deforming zone or structure, thus providing a reference datum against which changes in coordinate values can be assessed.
(ii) *Relative network*: a network in which all points are assumed to be subject to deformation, thus having no stable reference datum.

The first stage of a two-epoch analysis of an absolute network is to assess the stability of the reference points by assuming them to form a relative network and testing whether any points have moved. This may be achieved by carrying out a global congruency test see (5.6.2.1). Similarly in analysing a relative network, the first step is usually to establish whether any group of points in the network has retained its shape between the two epochs, again by use of the global congruency test. If such a group can be identified, then these points may be used as a datum, thus providing an absolute network for the analysis of the other stations. If in either case no group of stable points can be identified, then the resulting relative network must be assessed only in terms of datum invariant criteria (5.4.3.1).

5.6.2.1 *Global congruency test, combined adjustment and detection of significant movement*. From the results of the two single-epoch adjustments it is possible to calculate the displacement $\hat{d}$ and the associated covariance matrix $C_{\hat{d}}$ from

$$\hat{d} = \hat{x}_2 - \hat{x}_1 \tag{5.40}$$

$$C_{\hat{d}} = C_{\hat{x}_2} + C_{\hat{x}_1} \tag{5.41}$$

(assuming $\hat{x}_1$ and $\hat{x}_2$ are uncorrelated), and in addition the quadratic form given by

$$\Omega = \hat{d}^t C_{\hat{d}}^{-1} \hat{d} \tag{5.42}$$

It is readily shown (Caspary, 1987) that a suitable test of the hypothesis that the points under consideration have remained stable, i.e. $E(\hat{d}) = 0$, is

$$T = \frac{\Omega}{h\hat{\sigma}_0^2} \tag{5.43}$$

where h is the rank of $C_{\hat{d}}$ $(2n-4)$ for a 2D network
$\hat{\sigma}_0^2 = (r_1 \sigma_{0_1}^2 + r_2 \sigma_{0_2}^2)/r$
$r = r_1 + r_2$
$r_i =$ degrees of freedom in the adjustment of the ith epoch.

T is tested against the Fisher distribution (F-test) with $F_{h,r}$ at an appropriately chosen level of significance. If the test is successful (the hypothesis is not rejected), then the two epochs are assumed congruent, i.e. the points involved have remained stable. If the test is unsuccessful, at least one point has moved, and must be removed from the group of reference points. Several methods exist for identifying which point (or points) should be removed. The simplest of these methods is to identify the point which has the greatest contribution to Ω. This point is then eliminated from the reference group and the global congruency test repeated. The process is repeated until a stable group of points is identified. The contribution of any point i to Ω can be calculated (Cooper, 1987) by partitioning $\hat{d}$ and $C_{\hat{d}}^{-1}$:

$$\hat{d} = \begin{bmatrix} \hat{d}_s \\ \hat{d}_i \end{bmatrix}, \qquad (5.44)$$

$$C_{\hat{d}}^{-1} = \begin{bmatrix} W_s & W_{si} \\ W_{is} & W_i \end{bmatrix} \qquad (5.45)$$

then

$$\Omega_i = D_i^T W_i D_i \qquad (5.46)$$

where

$$D_i = W_i^{-1} W_{si} \hat{d}_s + \hat{d}_i$$

Having determined, by means of the global congruency test, a group of points which have remained stable, it is now necessary to calculate coordinates for these stations, as well as for the other, unstable points. There are different solutions to this problem. Firstly, it would be possible to adopt the first epoch estimates for the stable group and use these in a computation of the second epoch observations. However, this is not sensible since the measurements between the stable points in the second set are being ignored. It is also not entirely reasonable to adopt the separate estimates $\hat{x}_1$ and $\hat{x}_2$, since this would result in stable points having changing coordinates. The most preferable solution is to carry out a *combined adjustment* of the observations from both epochs, with only one set of unknown coordinates being estimated for the stable group of points, and two (one for each epoch) being estimated for the moving points. In fact, the required solution may be obtained without actually carrying out the combined solution since the displacements and covariance matrix of the unstable points can be obtained directly from the information available from the single epoch solutions (Caspary, 1987). The difference in the resulting coordinates for the moving points, (namely the displacements), together with the associated covariance matrix, can then be used in an assessment of the significance of the detected movements.

The movement $\hat{d}_i$ calculated for an unstable point can be tested for significance by comparing it with the appropriate elements ($C_{\hat{d}_i}$) of the associated covariance matrix. The test statistic which is most commonly used is

$$T = \frac{\hat{d}_i^T C_{\hat{d}_i}^{-1} \hat{d}_i}{\hat{\sigma}_0^2} \qquad (5.47)$$

and it is tested against the Fisher distribution $2F_{2,r}$ at the chosen level of significance. (Note that the definitions of $\hat{\sigma}_0^2$ and r are as on p. 167.)

In a similar fashion to the computation of absolute point error ellipses (5.4.3.3) it is possible to compute a point displacement ellipse by using the appropriate sub-matrix of $C_{\hat{d}}$ in place of the sub-matrix of $C_{\hat{x}}$. This ellipse may then be plotted along with the displacement vector for a graphical representation of the significance of the movement. In Fig. 5.6 this has been done with a 90% confidence ellipse. In this example, the movement is significant at the 10% level, since the displacement vector extends outside the ellipse.

The size of a given $(1 - \alpha)$ percentage confidence ellipse is obtained by multiplying the axes of the standard ellipse (obtained by the procedure described above) by $\sqrt{2(F_{2,r})_\alpha}$

Examples of the use of this type of approach are given for example in Caspary (1987), FIG (1981) and FIG (1983).

ANALYSIS AND APPLICATION OF CONTROL NETWORKS 169

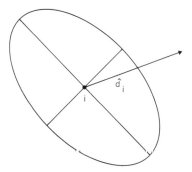

Figure 5.6 Point displacement ellipse.

5.6.2.2 An alternative approach. The global congruency test approach aims to define a group of stable points which will provide a datum for the assessment of the movement of the other points under consideration. An alternative to this is to choose an (arbitrary) datum station and orientation and compute each epoch using a minimum constraint adjustment. In order for the analysis of other station movements (by a similar methods to 5.6.2.1) to be meaningful, it is necessary that the 'fixed' origin and orientation are stable between measurement epochs. The stability of these points may be monitored externally, or may be tested statistically by the calculation of the significance of any systematic origin or orientation bias between the two epochs under consideration. This approach is described in more detail in Dodson and Ashkenazi (1984), and an example of its use is given in Ashkenazi *et al.* (1982).

5.6.2.3 Multi-epoch analysis. Multi-epoch analysis has not been widely studied, and anything other than a brief statement about it is outside the scope of this book. However, it is important to know of the principles of such an analysis, whilst bearing in mind that much of the same information can be obtained from successive two-epoch analyses combined with the estimation of deformation models (see 5.7).

Multi-epoch methods aim to combine the surveying observations taken in three-dimensional space with the fourth dimension of time, by assigning velocities and accelerations to the deformation model. This process can be carried out either by a combined adjustment of all epochs (which can lead to excessive computational effort), or by use of a sequential method. Further details of such approaches are given in Caspary (1987).

5.7 Determination of deformation parameters

As well as determining the deformation of individual points in the survey network, it is often desirable to be able to estimate the overall mode of deformation of the body under consideration, i.e. to derive a mathematical model of the deformation taking place. In order to attempt this, it is necessary to have some original estimate of the model which can be compared with the observed deformations, perhaps subsequently leading to a refinement of the original model. It is worth noting also that appropriately chosen models can be tested through derived datum-invariant quantities, thus avoiding the problems involved with datum definition described in 5.6.2.

The model to be tested is usually based on three components of deformation: (i) translation and (ii) rotation (representing rigid body motion), and (iii) pure deformation (strain). It will be apparent that only (iii) is datum-invariant.

The non-translational deformation tensor (in three dimensions) may be defined as

$$E = \begin{bmatrix} E_{xx} & E_{xy} & E_{xz} \\ E_{yx} & E_{yy} & E_{yz} \\ E_{zx} & E_{zy} & E_{zz} \end{bmatrix} + \begin{bmatrix} 0 & -W_{zz} & W_{yy} \\ W_{zz} & 0 & -W_{xx} \\ -W_{yy} & W_{xx} & 0 \end{bmatrix} \quad (5.48)$$

$$\text{Strain} \text{Rotation}$$

and hence measured displacements may be modelled in terms of polynominals. For the simpler 2-dimensional case such a model would be

$$\hat{d}_i = \begin{bmatrix} \Delta x_i \\ \Delta y_i \end{bmatrix} = \begin{bmatrix} a_0 + a_1 x_i + a_2 y_i \\ b_0 + b_1 x_i + b_2 y_i \end{bmatrix} \quad (5.49)$$

where a_0 and b_0 are constants (translation)

$a_1 = E_{xx}$

$a_2 = E_{xy} - W_{zz}$

$b_1 = E_{xy} + W_{zz}$

$b_2 = E_{yy}$

and this type of model can therefore be written in general terms as

$$B\boldsymbol{\delta} = \hat{d} + V_d \quad (5.50)$$

where $\boldsymbol{\delta}$ is the vector of unknown model parameters,
B is the coefficient matrix, and
V_d is the residual vector of the model fit.

The least-squares solution of the above equation for $\boldsymbol{\delta}$ can be tested for significance against a covariance matrix C_δ which can also be derived (see for example Dodson and Ashkenazi, 1984). In addition, the corresponding strain ellipses can be computed for each point. A generalized approach for the solution is given in Chrzanowski et al. (1982). It is not usual in fitting a deformation model to assume all the points are on one continuous body. Often the object being monitored is considered as comprising of separate discontinuous blocks, each of which is continuous within itself. Separate displacement functions are thus derived for each block. An example for a simple two-block case is shown in Fig. 5.7. The relevant displacement functions for points in Blocks 1 and 2 would be:

Block 1: (Homogeneous strain + rigid body translation)

$$\Delta x_1 = a_0 + E^1_{xx} x + E^1_{xy} y \quad (5.51)$$
$$\Delta y_1 = b_0 + E^1_{xy} x + E^1_{yy} y \quad (5.52)$$

Block 2: (Homogeneous strain only)

$$\Delta x_2 = E^2_{xx} x + E^2_{xy} y \quad (5.53)$$
$$\Delta y_2 = E^2_{xy} x + E^2_{yy} y \quad (5.54)$$

where subscripts and superscripts 1 and 2 refer to block numbers.

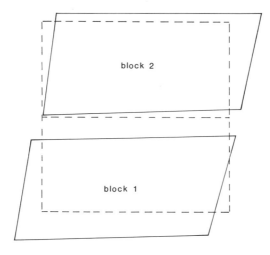

Figure 5.7 Two-block displacement model.

An alternative approach to the derivation of strain parameters based on a finite-element type analysis has also been used (Welsch, 1983). In this method the object is divided into triangular elements, and equations are set up to express boundary conditions between the elements. Again, a knowledge of the expected deformation is helpful for the choice of elements. If adjacent elements show similar strain patterns after initial strain parameter estimation, then these elements may be combined and a recomputation carried out.

5.8 Applications to deformation monitoring

Interest in the use of the techniques and procedures previously described has grown and developed considerably since the advent of high-accuracy electromagnetic distance measuring instruments and electronic data-capture systems. The latter in particular have facilitated the processing of the measurements and hence allow a rapid analysis of the results to be carried out. To date, many of the civil engineering applications have concerned dam monitoring, whether concrete (e.g. Egger and Keller, 1976) or earth embankment structures (e.g. Ashkenazi and Dodson, 1982). This concentration on dams and reservoirs has mainly been due to the potentially catastrophic consequences of structural failure, in conjunction with the presence of recent legislation and regulations requiring the frequent monitoring of such structures. In the United Kingdom the legislation takes the form of the Reservoirs Act 1975 (details of the Act and its implications are given in Denyer-Green, 1987), and in Switzerland for example (Egger, 1987) the Swiss National Committee on Large Dams has published an 'Executive Decree' regarding dam regulations.

However, surveying monitoring schemes have by no means been limited to dams and reservoirs. The procedures have been used for a variety of structures, including, for example, nuclear accelerators (e.g. Ashkenazi et al., 1983; Gervaise, 1976). In addition the monitoring of ground movements has been undertaken in several different environments, including tectonically active areas (e.g. Baldi et al., 1983), open-cast mines (e.g. Niemeier *et al.*, 1982) and subsidence areas (e.g. Chrzanowski and Szostak-Chrzanowski, 1984).

With the growing accessibility of network-analysis software packages and increased computational power, surveying network schemes can provide a precise, flexible and statistically rigorous means of monitoring and analysing the movement of virtually any man-made or natural feature, and their use for such purposes is certain to increase.

References

Ashkenazi, V. (1970) Adjustment of control networks for precise engineering survey. *Chartered Surveyor*, January, 1–7.
Ashkenazi, V. (1981) Least squares adjustment: signal or just noise. *Chartered Land and Minerals Surveyor* **3**(1) 42–49.
Ashkenazi, V., Crane, S.A., Price, W.J. and Williams, J.W. (1985) The 1980 readjustment of the triangulation of the United Kingdom and the Republic of Ireland. *Ordnance Survey Prof. Paper, New Series*, **31**, 27 pp.
Ashkenazi, V. and Cross, P.A. (1972) Strength analysis of Block VI of the European triangulation. *Bull. Géodésique* **103**, 5–24.
Ashkenazi, V., Dodson, A.H., Crane, S.A. and Lidbury, J.A. (1983) Setting-out and deformation measurement for a nuclear accelerator. *Proc. 3rd Int. Symp. on Deformation Measurements by Geodetic Methods*, Akademiai Kiade, Budapest, 743–762.
Ashkenazi, V., Dodson, A.H., Jones, D.E.B. and Samson, N. (1982) Measuring reservoir ground movements. *Civil Engineering*, February, 35–43.
Baarda, W.W. (1968) A testing procedure for use in geodetic networks. *Netherlands Geodetic Commission Publications on Geodesy, New Series* **2**, **5**.
Baldi, P., Achilli, V., Mulargia, F. and Broccio, F. (1983) Geodetic surveys in the Messina Straits area. *Bull. Géodésique* **57**, 283–293.
Bomford, G. (1980) *Geodesy*. 4th edn., Clarendon Press, Oxford, 855 pp.
Caspary, W.F. (1987) Concepts of network and deformation analysis. *School of Surveying Monograph* **11**, University of New South Wales, 183 pp.
Chatfield, C. (1983) *Statistics for Technology*. 3rd edn., Chapman and Hall, New York, 381 pp.
Chrzanowski, A. and Chen, Y.Q. (1986) Report of the ad-hoc committee on the analysis of deformation surveys. Paper 608.1, *Proc. 18th FIG Int. Congr.*, Toronto, 19 pp.
Chrzanowski, A., Chen, Y.Q. and Secord, J.M. (1982) A generalised approach to the geometrical analysis of deformation surveys. *Proc. 3rd Int. Symp. on Deformation Measurements by Geodetic Methods*, Akademiai Kiade, Budapest, 349–370.
Chrzanowski, A. and Szostak-Chrzanowski, A. (1984) A comparison of empirical and deterministic prediction of mining subsidence. *Proc. 3rd Int. Symp. on Land Subsidence*, Venice, 10 pp.
Cooper, M.A.R. (1974) *Fundamentals of Survey Measurement and Analysis*. Granada, London, 107 pp.
Cooper, M.A.R. (1987) *Control Surveys in Civil Engineering*. Blackwell Scientific, Oxford, 381 pp.
Cross, P.A. (1983) Advanced least squares applied to position fixing. *Working paper No 6*, Department of Surveying, North East London Polytechnic, 185 pp.
Cross, P.A. (1987a) CAD, geodetic networks and the surveyor. *Land and Minerals Surveying* **5** (9) 466–476.
Cross, P.A. (1987b) A method for the combination of GPS networks and fixed terrestrial control. *XIX General Assembly, IUGG*, Vancouver.
Cross, P.A. (1987c) The assessment and control of the quality of industrial and engineering survey information. *Proc. 2nd Industrial and Engineering Survey Conf.*, University College, London, 66–83.
Cross, P.A. and Price, D.R. (1985) A strategy for the distinction between single and multiple gross errors in a geodetic network. *Manuscripta Geodaetica* **10**, 172–178.
Denyer-Green, B.P.D. (1987) The Act and its implications. *Proc. Seminar on The Reservoir Act 1975—The Surveyor's Role*, Nottingham, 21 pp.
Dodson, A.H. and Ashkenazi, V. (1984) Ground movement or measurement error? *Proc. 3rd Int. Conf. on Ground Movements and Structures*, Vol. **3**, UWIST, 10 pp.
Egger, K. (1987) Behavioural control of existing reservoirs and dams. *Proc. Seminar on The Reservoir Act 1975—The Surveyor's Role*, Nottingham, 8 pp.

Egger, K. and Keller, W. (1976) New instrumental methods and their applications for geodetic deformation measurements on dams. *Trans. 12th Int. Congr. on Large Dams*, Mexico City, 27 pp.

Fédération International des Géomètres (FIG) (1981) *Proceedings, 2nd International Symposium on Deformation Measurements by Geodetic Methods (1978)*, ed. L. Hallermann, Konrad Wittwer, 604 pp.

Fédération International des Géomètres (FIG) (1983) *Proceedings, 3rd International Symposium on Deformation Measurements by Geodetic Methods (1982)*, eds. I. Joo and A. Detrekoi, Akademiai Kiade, Budapest, 900 pp.

Fraser, C.S. (1982) The potential of analytical close-range photogrammetry for deformation monitoring. *4th Canadian Symp. on Mining Surveying and Deformation Measurements*, Banff.

Gervaise, J. (1976) Geodesy and metrology at CERN: a source of economy for the SPS programme. *CERN Report* **76–19**, CERN, Geneva.

Grafarend, E.W. and Sanso, F. (1985) *Optimisation and Design of Geodetic Networks*. Springer, Berlin, 606 pp.

Grist, S.N. (1984) Computing models in spatial geodesy. PhD thesis, University of Nottingham, 250 pp.

Hein, G. and Landau, H. (1983) A contribution to 3-d operational geodesy, Part 3: OPERA—A multipurpose program for the adjustment of geodetic observations of terrestrial type. Deutsche Geodätische Kommission, Munich.

Meissel, P. (1982) *Least squares adjustment: a modern approach*. Technical University of Graz.

Mikhail, E.M. (1976) *Observations and Least Squares*. Dun-Donnelley, New York.

Mikhail, E.M. and Gracie, G. (1981) *Analysis and Adjustment of Survey Measurements*. Van Nostrand Reinhold, New York, 340 pp.

Niemeier, W., Teskey, W.F. and Lyall, R.G. (1982) Monitoring movement in open pit mines. *Austral. J. Geodesy, Photogrammetry and Surveying*, **37**, 1–27.

Pope, A.J. (1976) The statistics of residuals and the detection of outliers. *NOAA Technical Report* **NOS65 NGSI**, National Oceanic and Aeronautical Administration, Washington DC.

Teunissen, P.J.G. (1985) The geometry of geodetic inverse linear mapping and non-linear adjustment. *Netherland Geodetic Commission Publications on Geodesy* **8**(1), 177 pp.

Welsch, W.M. (1983) On the capability of finite element strain analysis as applied to deformation investigations. Paper 608.5, *Proc. 17th FIG, Congr.* Sofia, 12 pp.

Williams, D.C. (1987) Zone plate and related systems. *Seminar on the Reservoir Act 1975—The Surveyor's Rule*, Nottingham, 12 pp.

PART B

DEVELOPMENTS IN REMOTE SENSING AND PHOTOGRAMMETRIC TECHNOLOGY

The first part of the book considered the field survey instrumentation currently in use for point positioning in relation to large-scale mapping, control surveys for mapping, high-precision measurement and setting out, together with methods of analysing such data. By contrast, this second part of the book deals with the technological developments in the fields of airborne and spaceborne remote sensing (Chapter 6), aerial photogrammetry (Chapter 7) and close-range photogrammetry (Chapter 8).

Remote sensing (Chapter 6) is an activity which uses a varied range of technologies, both for data acquisition and for digital processing of such data. The primary use of remotely sensed imagery is qualitative, generally involving the interpretation of the imagery to identify some spatial characteristic of the terrain. Historically, the photographic camera was the first device used to capture an image of the terrain, or other object, from a distance. In recent years it has been supplemented by a number of other imaging devices, of which thermal and multispectral scanners and side-looking radar are the most important. The most significant development has, however, been in the digital image processing techniques used to enhance and extract information from digital imagery. Thus remote sensing, particularly from space, is largely concerned with the interpretation and thematic mapping of phenomena of a wide areal extent, often to a low degree of positional accuracy.

On the other hand, aerial photogrammetry (Chapter 7) is primarily concerned with topographic mapping, where the location of detailed point and line feature information is required to a high degree of accuracy. Consequently, very high demands are set, both for the planimetric accuracy of the objects to be mapped, and for the accuracy of the height information. Despite the advances in electronic surveying instrumentation which have been discussed previously, as far as topographic mapping is concerned, photogrammetry using aerial photography continues to be the primary method for medium- to small-scale mapping. Only at very large scales (1: 1000 and larger), and over small areas, can ground surveying methods compete economically with aerial photogrammetry.

Aerial photogrammetry is well established as the primary method to use for national mapping and for the mapping needs of large engineering projects. However, in many developed countries (such as the United Kingdom), the systematic national mapping is now complete, and the use of photogrammetry is restricted to revision, an activity to which the technique is not well suited, given the random nature of development and the need for non-metric mapping information. Thus photogrammetry is not in as widespread use as previously, and is somewhat in decline in many such countries. However, elsewhere, especially in developing countries where basic topographic mapping is often of limited areal extent, aerial photogrammetry is still a technique of prime interest and importance.

In the specific case of large engineering projects aerial photogrammetry also retains its importance. Even if large-scale maps exist, as in the case of the United Kingdom at 1:1250 and 1:2500 scales, they contain no height or contour information. These data can only be supplied in a timely manner by aerial photogrammetric methods. This is especially the case with digital terrain modelling information where methods such as progressive sampling allow the data to be acquired in a manner, and at a density, in keeping with terrain roughness.

Close-range photogrammetry (Chapter 8) refers to the subdiscipline which deals with the specific measurement problems occurring when the distance from the imaging source to the object is of the order of a few tens of metres at most (and generally much

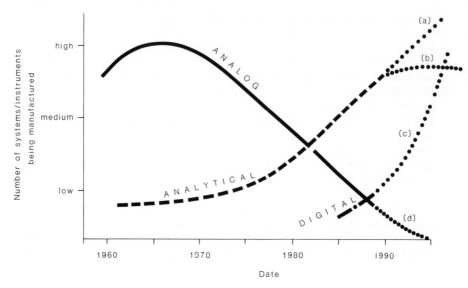

Figure B1 Development of analog, analytical and digital photogrammetric instrumentation and future scenarios.

closer). It is not a new topic, and in fact was developed before aerial photogrammetry. However, the development of analytical techniques and specialist instruments designed specifically for close-range measurement has opened up a whole new range of applications, particularly in engineering. Its use is nevertheless not restricted to engineering, and the method is in widespread use in a number of other fields such as architecture, archaeology and medicine. While the method overlaps with more direct methods of high-precision measurement (such as those based on electronic theodolites), there are quite unique circumstances when close-range photogrammetry is of particular benefit. The first occurs when remeasurement may be required at a later date. The ability to store the 'object' on film offers a major advantage over alternative methods. Similarly, if the object is moving, for instance in monitoring laboratory samples, the ability of photography to 'freeze' the motion of the object is also of significance.

In the case of both aerial and close-range photogrammetry, the nature of the instrumentation used for precise measurement has changed considerably in the last 30 years (Fig. B1). The analogue stereoplotter, based on the use of optical or mechanical projection and manufactured during the 20 year period between 1955 and 1975 is now in decline. The extent of this decline is such that today only one or two manufacturers are currently producing such instruments, and only in small numbers. Although the analytical stereoplotter was originally invented in the late 1950s, it did not have a major impact in the civilian sector until the mid-1970s. However, since then the technology has developed considerably (with significant real-term reductions in price and increases in performance) and such instruments are now manufactured by all major photogrammetric equipment manufacturers. The impact of 'all-digital' acquisition and measurement systems is still in its infancy, although many expect that the current technological limitations to the widespread development of such instruments will be overcome in the next decade.

For the future, a number of different possible scenarios can be identified. The first and easiest to identify is the further decline in the production of analogue instrumentation (Fig. B1d), the second (Fig. B1c) is the increasing interest in and development of all digital measurement systems, although the rate of development is much less easy to predict. Furthermore, it remains to be seen whether the development of all-digital systems will be at the expense of analytical instruments (Fig. B1b) in the short term (5–10 years), or whether the two approaches will continue to develop in parallel (Fig. B1a) and eventually merge in the longer term (10 years or more).

6 Remote sensing for topographic and thematic mapping

T.J.M. KENNIE

6.1 Introduction

The creation of maps by aerial photogrammetry and the interpretation of information from aerial photography are two well-established techniques in surveying and civil engineering. In recent years, however, developments in satellite and airborne sensors, together with improvements in computing speed and reductions in the costs of computer memory, have led to the creation of the relatively new discipline of remote sensing. Although still in an embryonic state, particularly in the field of image processing, remote sensing is nevertheless proving to be a cost-effective method of investigating regional engineering phenomena, such as surface drainage, and more localized phenomena such as landslides, building heat loss and so on. It is also being used, although to a much more limited extent, for topographic and hydrographic mapping.

This chapter is concerned with the fundamental principles of this new technology. Section 6.2 outlines the basic physical principles of the subject, while section 6.3 describes the characteristics of the airborne and satellite sensors which are used to acquire the remotely sensed data. The full potential of this type of data can only be fully exploited if appropriate hardware and software is available to enhance and classify the raw data. Section 6.4 therefore discusses the requirements of the digital image processing systems used in remote sensing. Finally, section 6.5 reviews a selection of the current applications of remote sensing relevant to the land surveyor and civil engineer.

6.2 Basic principles of remote sensing

Remote sensing basically refers to measurement at a distance. More precisely, it is concerned with the detection and measurement of variations in electromagnetic (EM) energy. Not all regions of the EM spectrum are of equal importance, since some are more affected by the attenuating effects of the Earth's atmosphere than others; Fig. 6.1 illustrates these attenuating regions and identifies those areas of the EM spectrum most significant to remote sensing.

6.2.1 *Characteristics of EM radiation*

The visible and reflected infrared regions of the spectrum represent the most important 'atmospheric windows', since they enable solar radiation reflected from the Earth to be

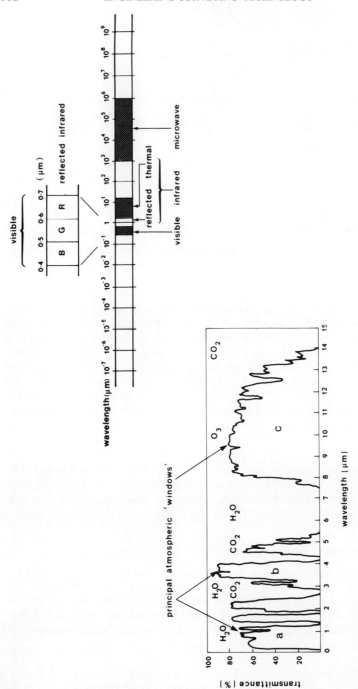

Figure 6.1 (Top) The electromagnetic spectrum (EM) and the principal regions relevant to remote sensing. (Bottom) Regions of the EM spectrum attenuated by the Earth's atmosphere. (a) Visible/near infrared; (b) far infrared (emitted); (c) far infrared (thermal).

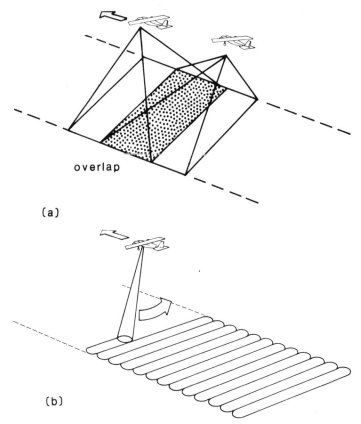

Figure 6.2 Modes of data acquisition. (*a*) Simultaneously using an aerial camera; (*b*) sequentially using a linescanner.

measured. At longer wavelengths, for example in the thermal infrared (8–14 μm) region, the measurement of radiation is of greater importance. The amount of energy emitted by a body can be quantified by means of the Stefan-Boltzmann law:

$$W = C\Sigma T^4 \qquad (6.1)$$

where W is the radiant emittance in $W\,m^{-1}$
  Σ is the emissivity of the object (a variable quantity which describes how efficiently an object is radiating in comparison with a theoretically perfect radiator)
  C is the Stefan-Boltzmann constant, equal to $5.7 \times 10^{-8}\,W\,m^{-2}\,K^{-4}$ and
  T is the absolute temperature of the object.

The application of this simple concept enables the temperature of an object to be inferred by remotely detecting the radiance from the feature being examined. However, before precise determinations of temperature can be achieved, a knowledge of the spatial variation in the emissivity of the surfaces being sensed is vitally important. More details of the basic physical principles of the subject can be found in Curran (1987).

6.2.2 *Characteristics of remote sensing systems*

Depending on the nature of the radiation being measured, it is possible to record the reflected or emitted EM energy, using either a lens and photographic emulsion or a scanner and crystal detector element. The geometrical distinction between the two approaches is illustrated in Fig. 6.2. The scanning technique enables EM energy outside the sensitivity range of photographic emulsions to be measured.

All measurements within both the visible and infrared regions (reflected and emitted) are normally referred to as *passive* in nature, since the radiation being recorded occurs naturally. Although it is possible to measure passive radiation at longer wavelengths, for example in the microwave region, it is generally easier to actively generate EM radiation of this wavelength and record the reflected radiation from the terrain. Systems of this type are referred to as *active*, and side-looking radar is a typical example of a system based on this technique.

Before discussing the specific characteristics of the various data acquisition systems in detail, mention should be made of the distinction between the terms *photograph* and *image*. A 'photograph' generally refers to an image which has been detected by photographic techniques and recorded on photographic film. An 'image', in contrast, is a more general term which is used to describe any pictorial representation of detected radiation data. Therefore, although scanner data may be used to create a photographic product, it would normally be referred to as an 'image', since the original detection mechanism involved the use of crystal detectors, rather than a lens and photographic emulsion.

6.3 Data acquisition

A wide and varied range of sensing devices has been developed for the acquisition of remote sensing imagery. In general terms, instruments can be classified according to two criteria, the mode of operation and their spectral range. Thus it is possible to differentiate between:

(i) Instruments which expose the entire format at a single instant and record visible and reflected (near) infrared radiation by photographic (or non-photographic) processes (6.3.1)
(ii) Instruments which operate by scanning the image and record radiation in the visible, reflected infrared and emitted infrared regions of the spectrum by means of crystal detector elements, (6.3.2 and 6.3.3)
(iii) Instruments which operate by measuring the time taken for actively generated pulses of microwave radiation to be reflected from the Earth's surface (6.3.4).

In all three instances, the sensors may be mounted either in airborne or in spaceborne platforms, most commonly in fixed-wing aircraft and unmanned satellites. Each has its own specific and largely complementary features. For example, airborne platforms enable small localized phenomena to be recorded at high spatial resolution. By contrast, spaceborne platforms enable wide synoptic views to be obtained, but at much lower spatial resolutions. In the latter case this may range from kilometres in the case of meteorological satellites, to a few tens of metres in the case of the most recent Earth resources satellites.

6.3.1 *Frame imaging systems*

A frame imaging system refers to a sensing device which exposes the entire format of an image at a single instant. The most common example is the aerial camera where

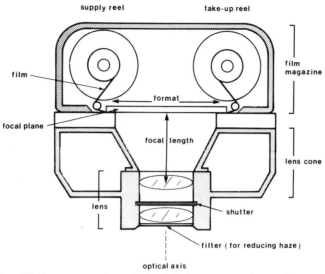

Figure 6.3 Components of a typical aerial photogrammetric mapping camera.

each frame is exposed by a camera shutter and the latent image is recorded on a photographic emulsion. Alternatively, the recording media may be non-photographic, and these are discussed further in section 6.3.1.2. The main distinguishing feature, however, is that in all cases the image produced has a perspective geometry and can therefore be analysed using conventional photogrammetric measuring techniques such as those described in Chapter 7 and discussed briefly in section 8.5.

6.3.1.1 *Photographic systems.* Vertical aerial photography obtained using a high-precision mapping camera is by far the most commonly available source of remotely sensed data. The design and construction of a modern photogrammetric mapping camera is given in Fig. 6.3.

The main components are:

(i) The main camera body which houses the motor drives and other mechanical components which are necessary to operate the film transport mechanism and the shutter
(ii) The lens cone, which will normally be interchangeable, the lens, a diaphragm to regulate its aperture and a between-the-lens shutter
(iii) The magazine containing the unexposed and exposed spools of film, and the film flattening device which is usually of the vacuum suction type
(iv) The mount on which the camera will be supported and which enables the operator to level the camera, or to rotate it around its vertical axis to eliminate the effects of aircraft crabbing.

The current types of aerial camera produced by manufacturers such as Wild (Switzerland), Zeiss Oberkochen (West Germany) and Zeiss Jena (East Germany) and used for photogrammetric purposes are highly developed, and are characterized by:

(i) A relatively large format—230 × 230 mm is standard, although larger formats up to 230 × 460 mm have been used for civilian mapping from space platforms (Table 6.9)

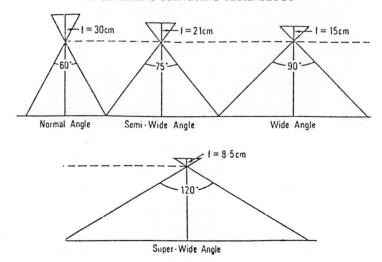

Figure 6.4 Angular coverage of aerial photogrammetric cameras of varying focal length.

Table 6.1 Geometric characteristics of typical designs of photogrammetric mapping cameras

	Normal-angle	Semi-wide-angle	Wide-angle	Super-wide-angle
Focal length	12" (30 cm)	8¼" (21 cm)	6" (15 cm)	3½" (8.5 cm)
Angular coverage	56°	75°	93°	120°
Base:height ratio	0.3	0.45	0.6	1.0
Photo scale (assuming $H = 3000$ m)	1:10 000	1:14 300	1:20 000	1:34 300
Ground resolution (40 lp/mm)	0.8 ft (0.25 m)	1.05 ft (0.35 m)	1.6 ft (0.5 m)	2.8 ft (0.85 m)
Ground area covered	2.3 × 2.3 km	3.3 × 3.3 km	4.6 × 4.6 km	8.4 × 8.4 km

(ii) Very low geometric distortions (5–7 μm in the negative plane)
(iii) Large-aperture lenses ($f/4$ to $f/5.6$ at fullest opening) with high light-gathering powers at considerable angular values from the optical axis
(iv) Between-the-lens shutters of high efficiency (over 90% at large apertures) which also permit the use of short exposure times (0.5–1 ms) to cut down image motion
(v) Angular coverage of single exposures ranging from 60° (normal angle) to 120° (superwide angle) (Fig. 6.5) and Table 6.1
(vi) Use of a range of different emulsion types which are sensitive to specific wavebands.

For the purpose of mapping, it is necessary to ensure that the individual photographs overlap in their coverage of the terrain, primarily to enable stereo-viewing or stereo-measurements to be made. Although a 50% overlap in the forward direction of flight would be sufficient to provide a stereoscopic view for a number of reasons, illustrated by Fig. 6.4, it is common practice to specify that the overlap should be at least 60%. In

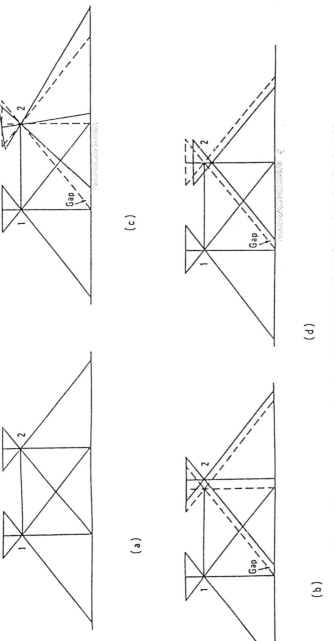

Figure 6.5 Factors causing gaps in stereoscopic coverage with 50% forward overlap (Petrie, 1987).

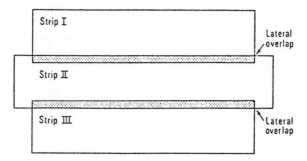

Figure 6.6 Lateral overlap between strips of aerial photography.

order to avoid gaps between adjacent strips of photography, it is again common procedure to specify a lateral overlap of between 15 and 20% (Fig. 6.6).

The ratio of the air base B to the flying height H is a parameter of particular importance for photogrammetric measurement, since it has a direct influence on the accuracy of heighting. Large $B:H$ ratios, while preferable from a geometric standpoint, are also associated with super wide-angle coverage. The less favourable aspect of using such a camera is the loss of fine detail in the corners due to relief displacement and the 'dead' areas which occur in urban areas with tall buildings or regions of high relief.

Many photogrammetric cameras also include Image Movement Compensation (IMC) systems to further enhance the resolution of the photography. In essence, such systems operate by moving the film in its image plane at a rate corresponding to the aircraft speed. The importance of such a system at large scale can be highlighted by the following typical case. A standard high-resolution photogrammetric mapping camera producing 1:3000 scale photography will have a ground resolution of about 6 cm. If one assumes an exposure period of 1/250 second from a survey aircraft flying at 200 kph, then the aircraft will have moved a distance of 22 cm during the exposure interval. Thus the inherent high resolution of the large-scale photography will be degraded by the slight blurring of the image due to aircraft movement while the shutter is open. Meier (1985) provides a more exhaustive treatment of the theory of IMC, and Fig. 6.7 summarizes the effect of IMC for a number of typical photographic scales and aircraft speeds.

Small-format (typically 70 or 35 mm) low-precision cameras have also been used on occasions for interpretive work over small sites (Heath, 1980; Tomlins, 1983; Graham and Read, 1984). Coverage may involve the production of either vertical or oblique photography. In the latter case, the camera axis is intentionally tilted. Using the camera in this manner helps overcome the problems associated with the very narrow angular coverage of such camera systems when acquiring vertical photography. The oblique view may, however, mask features of interest, and the variation in scale across the format may make the accurate measurement of dimensions very difficult.

In instances when small-format cameras are being used, it is common to combine several (typically four) cameras to acquire photography of the same ground area, each camera having a different film/filter combination. The most common photographic films are black-and-white panchromatic (monochrome), black-and-white infrared, true-colour and false-colour infrared. Table 6.2 summarizes the characteristics of each type. Alternatively, filters may be placed over the each camera lens to restrict the

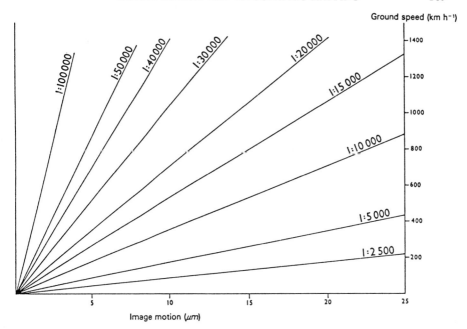

Figure 6.7 Image motion for a range of photographic scales and aircraft speeds (Meier, 1985).

responses to a narrow waveband, the results being recorded on a conventional panchromatic emulsion. These produce the so-called 'multispectral photography'. Such photography may be interpreted either by examining hardcopy or by using an 'additive viewer' to interactively examine the various film/filter combinations (Beaumont, 1977).

6.3.1.2 *Non-photographic systems.* It is also possible to record an image produced from a frame imaging system using non-photographic techniques. The most commonly used devices are video cameras, return beam vidicon (RBV) cameras, and cameras which use charge-coupled devices (CCD). With each of these devices, the image data are recorded either in an analogue manner on video tape, or digitally on high-density digitial tape. Hardcopy imagery can subsequently be generated, either by replaying the video tape and directly photographing the television screen on which the image is being displayed by use of a hardcopy video printer, or digitally by means of a raster scan laser-based film output device.

Video cameras mounted in an aerial platform have been used for a number of engineering applications, including the monitoring of traffic flow (Mountain and Garner, 1981), the recording of stages in the development of engineering projects, and the assessment of the impact of construction activities such as pipelines on the terrain.

The imaging system used in a video camera is based on a conventional lens and shutter arrangement. However, unlike photographic methods, the video image is formed by recording the charge falling on to a photosensitive screen (Fig. 6.8) rather than on to a silver halide emulsion. Incoming light falling on to the charge face of the vidicon tube induces a positive charge proportional to the brightness of the incoming

Table 6.2 Characteristics of various photographic emulsions

Film type	Structure	Sensitivity	Advantages	Limitations	Cost
Black-and-white panchromatic	Single-layer emulsion	0.4–0.7 μm	Good definition and contrast, wide exposure latitude. Boundaries are often clearly portrayed. Subtle textural variations can be identified and geometric patterns are more obvious	Sometimes difficult to interpret ground features due to the inability of the eye to distinguish between subtle differences in grey tones	Low
Black-and-white infrared	Single-layer emulsion	0.4–0.9 μm	It provides specific enhancement of different forms of vegetation and clearly defines bodies of water and ground moisture distribution. Helps to penetrate haze, giving a sharper more contrasting image	Contrast may be excessive and detail is often lost in areas of shadow	Low
True-colour	Multi-layered emulsions. Top layer is sensitive to blue light, the 2nd layer to green and blue light and the 3rd to blue and red light. A blue absorbing filter is introduced between the 1st and 2nd layers.	0.4–0.7 μm	Good contrast and tonal range. Wide exposure latitude for negative film types. Materials are more identifiable and hence moisture conditions are also more identifiable	Less good definition than panchromatic film	Intermediate to high
False-colour infrared	Multi-layered emulsion incorporating a reflected infrared sensitive layer. Top layer is sensitive to blue and reflected infrared light, 2nd to green and blue light and 3rd to red and blue light. The film is normally used with a filter to block out blue light.	0.4–0.9 μm	Helps to penetrate haze, provides accurate identifiable data on vegetation, rocks, soils, water bodies and moisture distribution	Lower resolution than colour film. Exposure difficult to determine and processing expensive	High

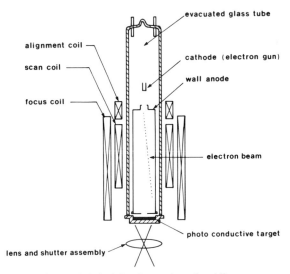

Figure 6.8 Principle of operation of a vidicon.

radiation, and hence a latent image is formed on the charge face. In order to 'develop' this latent image, an electron beam scans the charge face in a raster fashion, and in so doing discharges the initial positive charge.

The lenses used in video cameras normally have very short focal lengths (typically 8–35 mm), and the image format sizes are also very small in comparison with conventional photographic methods. Successive frames of imagery are recorded at rates which are typically between 25 and 30 Hz on to either U-Matic or VHS video tape.

Video cameras can be designed for a range of spectral sensitivities, although those which record the visible part of the EM spectrum, either in black-and-white or in colour, are the most common. Meisner and Lindstrom (1985) and Everitt and Nixon (1985), however, discuss the modifications which they designed for a conventional colour video camera in order to increase its spectral sensitivity and so enable the production of false-colour infrared imagery.

The *return beam vidicon* (*RBV*) camera operates in a manner similar to the more common vidicon. However, unlike the vidicon, the signal is continuously reflected back to the aperture of the electron gun. The main advantages of this design are the ability to produce higher-resolution images, and the ability of the photoconductive surface to remain in focus for longer periods.

The best-known example of an RBV camera system is that which operated on board Landsat 3 (section 6.3.2.2). Two RBV cameras with a spectral range of 0.505–0.750 µm (green to near infrared) were operated, and had a ground pixel equivalent to 30 m. Unfortunately, technical difficulties on board the satellite reduced the operational life of the sensor; nevertheless, considerable image coverage of the Earth's surface was acquired during this period.

A *charge coupled device* (CCD) camera differs from the previous two examples by having a rectangular, areal array of sensor elements situated in the imaging plane of the camera. Like the RBV, a latent image is formed on the surface of each array element, and the charge pattern formed is subsequently measured. Although the geometric stability of the image is superior to that produced by a RBV camera (since a secondary scan is not required), at present the spatial resolution is significantly lower. Current arrays reported in the literature range from 100 × 100 elements (Hodgson et al., 1981) to 288 × 385 elements (Traynor and Jeans, 1982), although an 800 × 800 element array has been reported for astronomical applications. Increasing interest has also been shown in their use as 'digital framing' or 'solid-state' cameras (Petrie, 1983; Burnside, 1985) to replace the conventional 230 × 230 mm format photographic camera. At present however, the size and density of arrays which are needed and the vast storage requirements (particularly at high resolution) preclude their widespread application.

CCDs are also appropriate to many other fields of surveying instrument design, and are discussed elsewhere in this book in relation to theodolites (1.5.1.3) and photogrammetric instrumentation (7.6.1, 8.7).

6.3.2 *Optical/mechanical line scanners*

The geometrical distinction between the instantaneous, photographic (or non-photographic) frame imaging technique and the sequential line scanning approach has been illustrated previously in Fig. 6.2. The other specific difference between such systems is the ability of the latter to measure radiation in regions of the EM spectrum which cannot be recorded on photographic emulsions, below 0.4 μm and above 0.9 μm wavelengths, in the ultraviolet and the longer infrared regions of the spectrum.

The basic principle of a line scanner involves the detection of the reflected or emitted radiation from objects present on the ground during an angular scan across the terrain, along a line perpendicular to the direction of flight. Scanning is performed by either a mechanically rotating or oscillating mirror. The area on the ground over which the radiation is measured is defined by the instantaneous field of view (IFOV) of the scanner. This also represents the size of the picture elements or pixels along each scan line. The IFOV is often used as a crude measure of the spatial resolution of the scanner. However, IFOV does not take into account other factors, particularly scene contrast, which can have a major impact on the detectability of features.

Two distinct designs of line scanner can be identified—those which record only a single waveband, generally in the thermal infrared part of the EM spectrum (in which case they are referred to as thermal scanners), and those which record images in several wavebands, the multispectral scanners (MSS).

6.3.2.1 *Thermal infrared (IR) line scanners.*

Thermal IR line scanners operate by recording emitted IR radiation corresponding to the atmospheric window in the 8–14 μm regions of the EM spectrum. This region also coincides with the peak value of emitted energy from the Earth's surface.

The design of a typical Thermal IR linescanner is illustrated in Fig. 6.9. The incoming scanned radiation is focused on to a crystal detector, often of mercury cadmium telluride (HgCdTe), which transforms the radiation into an electronic signal. The analogue signal is recorded on video tape, and may be used to modulate a light beam moving across a cathode ray tube to produce a visible image. This 'quick look' image is also generally recorded on a continuous strip of moving film. It is also common practice to record the data digitally for later analysis in an image processing system.

In order to quantify the recorded signals, a blackbody reference is used to calibrate the raw data, and temperature variations as small as 0.2°C can be determined using such equipment. For absolute temperature measurements, however, factors such as wind speed and the spatial variations in the emissivity of different surface materials are

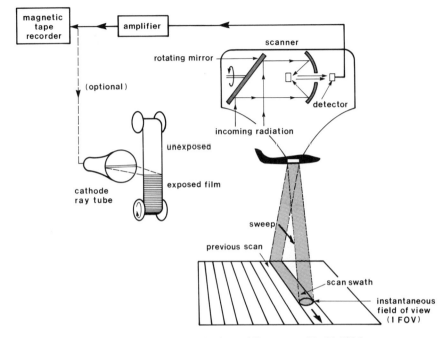

Figure 6.9 Components of a thermal linescanner (Rudd, 1974).

vitally important. Fig. 6.10 is a typical example of the type of image which may be created using this type of scanner, and which could, for example, be used for monitoring heat loss from buildings.

6.3.2.2 *Multispectral line scanners.* In general terms, multispectral scanners (MSS) can be classified into those of a mechanical design using, for example, rotating mirrors, or alternatively, those of the linear array design to be discussed in section 6.3.4. The following discussion is restricted to the principal features of the former design, although many of the characteristics are common to both.

The main components of an MSS are illustrated in Fig. 6.11. The line scanning mechanism is essentially the same as that used by thermal IR linescanners. The main difference lies in the separation of the incoming radiation into several spectral channels by means of a diffraction grating or prism. This enables the reflected or emitted radiation to be measured in several discrete wavebands for each individual pixel in the image. Both airborne and spaceborne platforms can be used to house MSS sensors.

The *Landsat* satellite system (Table 6.3), formerly known as the Earth Resources Technology Satellite (ERTS) was designed in the late 1960s by the US National Aeronautical and Space Administration (NASA), primarily to assess and monitor land resources. The first satellite was launched in 1972 and until 1985 the system was operated as a public service. The operational activities were then transferred to EOSAT, a joint company formed by the Hughes Aircraft Co. and RCA.

Three distinct sensors have been carried on board Landsat (up to 1987, 5 satellites have been in operation):

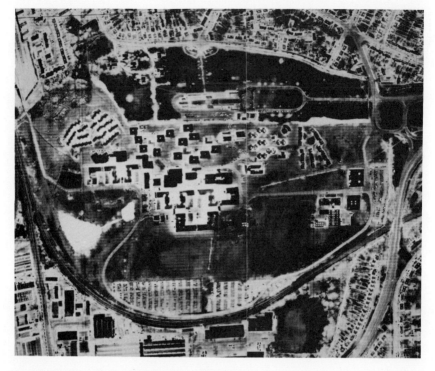

Figure 6.10 Thermal linescan image of the University of Surrey campus. Courtesy Clyde Surveys Ltd.

(i) A 4- or 5-waveband multispectral scanner (MSS)
(ii) A more advanced 7-waveband MSS termed the Thematic Mapper (TM)
(iii) A 3-waveband RBV camera (as discussed in section 6.3.1.2).

The former type of MSS has been carried on board all Landsat satellites. Scanning is performed using an oscillating mirror which scans six lines perpendicular to the orbital direction (Fig. 6.12). The swath width of a single image is 185 km, and the IFOV of the sensor produces a ground pixel equivalent to 56×79 m in the four visible and near infrared channels. The spectral range of these channels is from 0.5–1.1 μm. A fifth thermal channel covering the 8–14 μm spectral range has also been provided, and has operated on occasions, although with an enlarged IFOV giving a ground resolution equivalent to 235 m.

The *Thematic Mapper* (TM) sensor has only been operated on board Landsat 4 (launched 1982) and Landsat 5 (launched 1984). It is essentially an advanced MSS sensor which offers an improved spatial performance (with an IFOV equivalent to 30 m) and spectral range (7 wavebands) compared to the MSS. Nevertheless, in spite of these improvements, the high cost of TM data has proved to be a major limitation which has restricted the more widespread adoption of this type of data for engineering projects. The details of the specific wavebands covered by the TM are provided in Table 6.4, and the comparative spatial resolution of the two sensors is illustrated by Fig. 6.13.

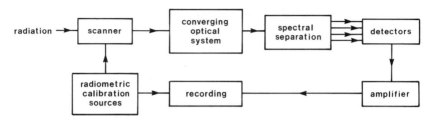

Figure 6.11 Principle of operation of a multispectral scanner.

Table 6.3 Characteristics of a selection of satellite scanners

	Line scanners		Pushbroom scanners	
	Landsats 1–5 Multispectral Scanner (MSS)	Landsats 4, 5 Thematic Mapper (TM)	Modular Optoelectronic Multispectral Scanner (MOMS)	Le Système Probatoire d'Observation de la Terre (SPOT)
Operated by (country)	EOSAT (USA)	EOSAT (USA)	DFVLR (W. Germany)	SPOT Image (France)
Date of launch	1972	1982, 84	1983	1986
Orbital altitude (km)	900	705	300	830
Ground resolution (m)	80	30	20	10, 20
Spectral range	0.5–12.6	0.45–12.5	0.575–0.975	0.5–0.89
No. of wavebands	5	7	2	3
Further reading	NOAA (1978–1986)	NOAA (1978–1986)	—	Chevrel et al. (1981)

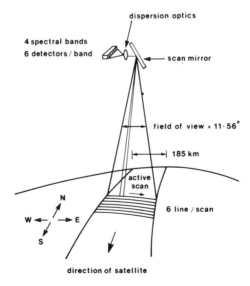

Figure 6.12 Landsat Multispectral Scanner (MSS) (after NASA).

Table 6.4 Landsat Thematic Mapper: spectral channels

Band number	Spectral range (μm)	Application
1	0.45–0.52	Water penetration
2	0.52–0.60	Measurement of visible green reflectance
3	0.63–0.69	Vegetation discrimination
4	0.76–0.90	Delineation of water bodies
5	1.55–1.75	Differentiation of snow and cloud
6	10.40–12.50	Thermal mapping
7	2.08–2.35	Geological mapping

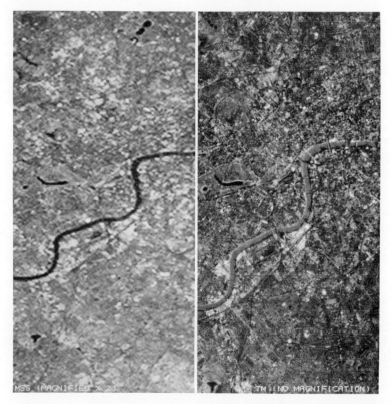

Figure 6.13 Comparative spatial resolution of (left) Landsat Multispectral Scanner (MSS) and (right) Landsat Thematic Mapper (TM) imagery. Courtesy National Remote Sensing Centre.

Figure 6.13 shows two typical single-waveband, black-and-white photographic images produced from Landsat digital data. Only a limited amount of information can be gleaned from the visual interpretation of single-band imagery. However, it is also possible to photographically combine three different wavebands to form a hardcopy colour 'composite' image. In the case of MSS imagery it is only possible to produce a 'false'-colour composite, while with TM it is also possible to produce a 'true'-colour composite image.

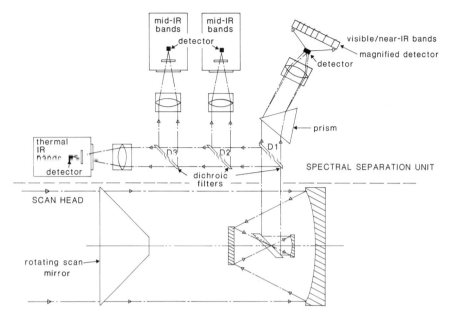

Figure 6.14 Optical components of Daedalus AADS 1268 Airborne Thematic Mapper.

Further benefits can be obtained if the imagery is examined interactively using an image processing system. However, to date, the high cost of such systems has meant that few surveying and engineering companies have invested in this technology. With the continued reductions in the cost of computer processing power, this is less likely to be a major limiting factor in the future.

The *Daedalus AADS 1268 Airborne Thematic Mapper* (ATM) is a typical airborne multispectral line scanner. Scanning is performed by a rotating mirror which operates at speeds of between 12.5 and 50 scanlines per second. The separation into 11 spectral channels is performed by three dichroic filters and a diffracting prism (Fig. 6.14). Table 6.5 summarizes the radiometric specifications of the scanner. The instrument has an IFOV of 2.5 m rad, which approximates to a ground resolution of a metre at a flying height of between 400 and 500 m. Data in the form of digitized analogue video signals for each channel are recorded on to high density digital tape (HDDT). These data are subsequently transferred on to computer compatible tape in order to read the data into an image processing system. Figure 6.15 is a typical example of the imagery which has been obtained with this type of scanner.

6.3.3 *Optical/mechanical frame scanners*

A more complex optical/mechanical design of scanner which has been used for acquiring remote sensing imagery is the frame scanner. In this instance, entire frames of data are captured, rather than single lines as described in section 6.3.2. Sensors of this design normally operate in the thermal part of the spectrum and may therefore be referred to as thermal video frame scanners (TVFS). Sensors of this type are particularly useful for engineering projects because they are portable and can be used in light aircraft

Table 6.5 Spectral channels for the Daedalus 1268 AADS ATM

Spectral band number	Wavelength (μm)	Spectral region
1	0.42–0.45	Blue
2	0.45–0.52	
3	0.52–0.60	Green
4	0.605–0.625	Red
5	0.63–0.69	
6	0.695–0.75	Near-IR
7	0.76–0.90	
8	0.91–1.05	
9	1.55–1.75	
10	2.08–2.35	Mid-IR
11	8.50–13.00	Far (thermal) IR

(Kennie *et al.*, 1986). This also reduces the operational costs of using the sensor. One particular TVFS system which has been used for remote sensing applications is the Barr and Stroud IR18.

6.3.3.1 *Barr and Stroud IR18 TVFS*. Unlike the line scanners previously described, which are designed specifically for airborne use, TVFS systems such as the Barr and Stroud IR18 (Fig. 6.16) are designed for both ground and airborne operations. Since such a system may therefore be used from both static and dynamic platforms, some mechanism for scanning in two orthogonal directions is necessary, and is described in detail in Amin (1986). Thus, incoming radiation is initially focused on to a frame scanning mirror oscillating at 50 Hz. The radiation reflected by this mirror is then directed on to a spherical focal plane. The line scanning is performed by a hexagonal prism which rotates at a rate of 39 000 rpm, scanning four lines of the image simultaneously. The image reflected by the line scan mirror is then focused by the relay lens onto a bank of four SPRITE (signal processing in the element) HgCdTe detector elements. The outputs from the detectors are stored in analogue form in a CCD storage unit and then read out in a TV-compatible PAL video format.

The main disadvantage of this design of scanner is the lack of a calibrated blackbody reference for quantitative temperature measurements. For some engineering applications, this is a severe limitation, but for others, for example where anomalous thermal patterns may indicate regions of subsidence (Edmonds *et al.*, 1986), a knowledge of the relative temperatures within a scene can be just as important. Quantitative temperatures can, however, be derived from the imagery by correlating known ground temperatures with corresponding image points. This technique has been used to assess the temperature variations throughout a road network with the aim of identifying those areas most prone to ice formation (Stove *et al.*, 1987). Figure 6.17 is an example of the type of imagery which was generated for this project.

6.3.4 *Linear array (pushbroom) scanners*
The linear, multilinear or 'pushbroom' scanner differs considerably from the optical-mechanical design. In fact, although such devices are referred to as scanners, the main

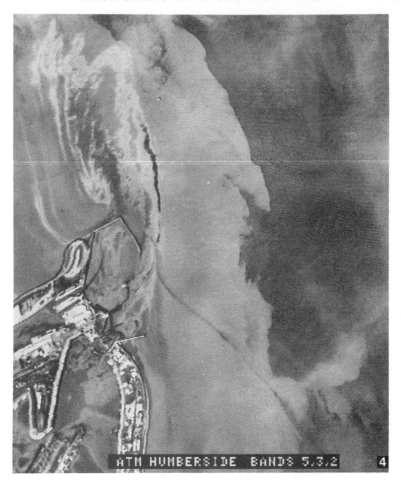

Figure 6.15 Airborne multispectral scanner image of the River Humber produced using the Daedalus AADS 1268 Airborne Thematic Mapper. Courtesy Hunting Technical Services Ltd.

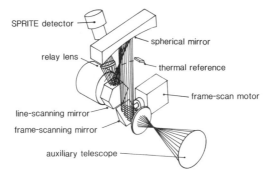

Figure 6.16 Components of the Barr and Stroud IR18 Thermal Video Frame Scanner (TVFS). Courtesy Barr and Stroud Ltd.

Figure 6.17 Density sliced Barr and Stroud IR18 TVFS image of part of the M9 motorway in Scotland, illustrating temperature variation over the road surface (original in colour). Courtesy ERSAC Ltd.

distinguishing feature of this design is the lack of any means of actually scanning the scene of interest. By contrast, each line of the image is formed by measuring the radiances as imaged directly on to a one-dimensional linear array of CCD detectors located in the instrument's focal plane (Fig. 6.18). Each line is then scanned electronically and the radiance values recorded on magnetic tape. As before, successive lines of the image are produced by the forward motion of the aircraft or satellite. Care has to be taken in the calibration of the individual detector elements to ensure that no degradation in the radiometric quality of the imagery occurs. Several satellite and airborne pushbroom scanners are currently in use, and the following sections describe a selected number of them.

6.3.4.1 *Le Système Probatoire d'Observation de la Terre (SPOT)*. In common with Landsat, SPOT (Table 6.3) is also a commercial satellite remote sensing system, the original development being funded by the French Government and aerospace industries. SPOT was launched on 22 February 1986, and is the first of what is hoped

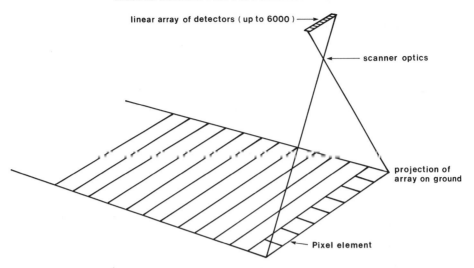

Figure 6.18 Pushbroom design of scanner.

will be a series of four satellites which will provide coverage until the turn of the century.

The sensors on board the satellite consist of two high resolution visible (HRV) pushbroom scanners. The HRV instrument is designed to operate in either a panchromatic or a multispectral mode. In the former case, the IFOV of the sensor produces a 10-m pixel, whereas in the latter case the IFOV produces a 20-m pixel. Figure 6.19 illustrates a typical panchromatic image.

The two HRV instruments are designed to be operated in either a vertical (nadir) or an oblique (off-nadir) position. Nadir pointing enables a swath of up to 117 km to be imaged (Fig. 6.20), whilst off-nadir viewing enables areas up to 425 km away from the satellite ground track to be imaged (Fig. 6.21). The ability to obtain coverage in this manner provides much greater flexibility in data acquisition, and enables the time interval between successive views of the same area on the ground to be reduced from 26 days to 2–3 days, assuming cloud-free conditions. This ability to point the sensor up to 27° off nadir also enables a range of scenes to be acquired by programming the sequence of sensor pointings from the ground (Fig. 6.22).

One of the most interesting features of SPOT imagery from a surveying and engineering point of view is the ability to provide stereoscopic coverage by using the lateral overlap between successive runs, as illustrated by Fig. 6.23. The use of stereoscopic imagery for the production of digital terrain models is discussed in section 6.5.1.

SPOT data are available to users in several different forms, ranging from 1:400 000 photographic negatives to geometrically corrected digital data. Figure 6.24 summarizes the different levels of correction which can be applied to the raw data. The operation and distribution of the satellite data is controlled by SPOT Image, supported by local distributors throughout the world. In the United Kingdom, for example, distribution rights are shared between the National Remote Sensing Centre (NRSC) and Nigel Press Associates (NPA).

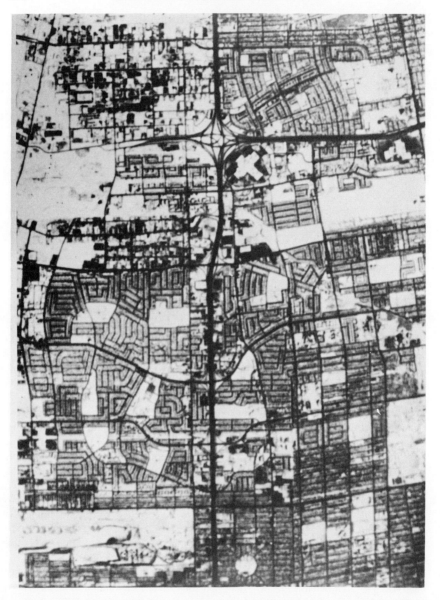

Figure 6.19 SPOT—panchromatic image of part of Montreal. Courtesy SPOT Image.

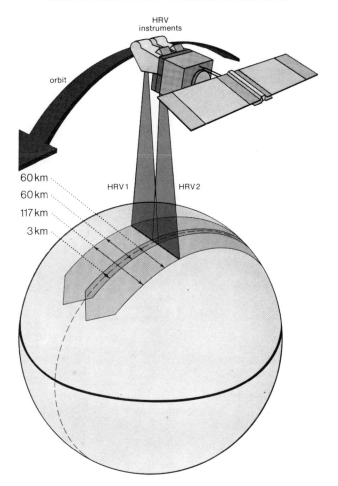

Figure 6.20 SPOT—Swath width of scanner in nadir pointing mode of operation. Courtesy SPOT Image.

6.3.4.2 *Modular optoelectronic multispectral scanner (MOMS).* MOMS is a further example of a satellite-based pushbroom scanner (Table 6.3). The sensors were designed by the West German MBB Company for the German Aerospace Research Establishment (DFVLR). Unlike SPOT, it is not an operational system, and was flown on board the Space Shuttle in June 1983 as a development system. The scanner design is more complex than for SPOT (Fig. 6.25), and consists of four linear arrays of 1728 pixels which enable a continuous line of 6912 pixels to be scanned using a dual-lens optical arrangement. Operating at the altitude of the Space Shuttle, this was equivalent to a ground pixel of some 20 m. A second-generation MOMS sensor currently under development will not only have an extended range of spectral bands but will also enable stereoscopic imagery to be obtained.

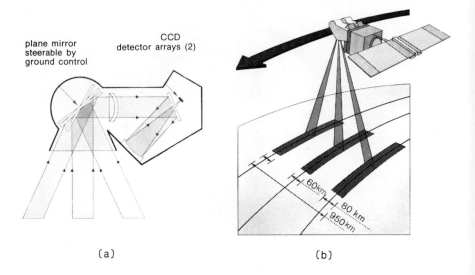

Figure 6.21 SPOT—(a) steerable mirror for off-nadir viewing; (b) ground coverage in off-nadir mode of operation. Courtesy SPOT Image.

6.3.4.3 *Pushbroom imaging spectrometers.* The requirement to be able to measure the many small variations in the spectral response of surface materials (particularly minerals) has been one of the primary factors leading to the development of very narrow waveband imaging systems. Such devices are similar in many respects to MSS. However, whereas in an MSS the spectrum of a surface material is recorded by a relatively small number of widely spaced samples, an imaging spectrometer records a much more detailed spectral pattern by simultaneously recording up to several hundred contiguous spectral measurements (Fig. 6.27). The primary application of such instruments is in the field of mineral exploration (Goetz *et al.*, 1985) but other applications include the assessment of damage to vegetation and forestry, as caused for example by acid rain.

The *Moniteq Programmable Multispectral Imager* (*PMI*) operates in a pushbroom mode, wide coverage being obtained by five registered units which enable a region up to 1925 pixels wide to be imaged. The instrument can be configured to specific applications by varying its spectral and spatial resolution. Figure 6.26 illustrates the variations which can be accommodated. Thus up to 288 narrow waveband spectral channels can be utilized by use of the transmission grating spectrometer.

The concept of the imaging spectrometer can also be utilized on board a spacecraft, and such a device, the *High Resolution Imaging Spectrometer* (HIRIS), is currently under development by the NASA Jet Propulsion Laboratory (JPL). It is planned that such an instrument could be operated on board the Space Shuttle in the early 1990s. A prototype airborne system, known as the Airborne Imaging Spectrometer (AIS), is

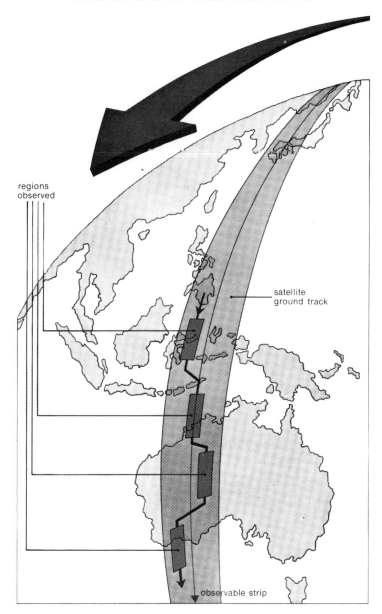

Figure 6.22 SPOT—programmed sensor pointing. Courtesy SPOT Image.

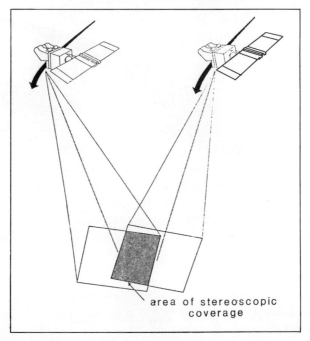

Figure 6.23 SPOT–Geometry of a typical pair of lateral stereoscopic images.

currently being tested. At present the instrument has a spectral range between 1.2 and 2.4 μm in 128 contiguous wavebands.

6.3.5 *Microwave sensors*
Both passive (sensing naturally-occurring radiation) and active (sensing reflected, system-generated radiation) designs of microwave sensor have been developed. Passive scanning microwave radiometers are not, however commonly used for engineering applications at present because of the difficulty of measuring naturally-occurring microwave radiation. Of much greater importance are the active microwave sensors such as side-looking radar (SLR).

6.3.5.1 *Side-looking radar (SLR)*. Side-looking radar was initially developed for military purposes during the early 1950s, although it was some 20 years later before it became available for civilian use. The main attractions of SLR were, firstly, the sensor's ability to produce imagery both during the day and at night and, secondly, the ability to acquire imagery of the terrian through haze, smoke, cloud and rain. Two distinct designs of SLR are used; these are referred to as *real aperture radar* and *synthetic-aperture radar*.

The principle of operation of a typical real-aperture radar system is illustrated by Fig. 6.28. It can be seen that the sensor measures the time interval between the transmission and reflection of a microwave pulse from objects in the terrain. Since the reflections from an individual microwave pulse are recorded sequentially, this creates, along a single line, a microwave image of the terrain. Successive strips of imagery are

REMOTE SENSING FOR TOPOGRAPHIC MAPPING 207

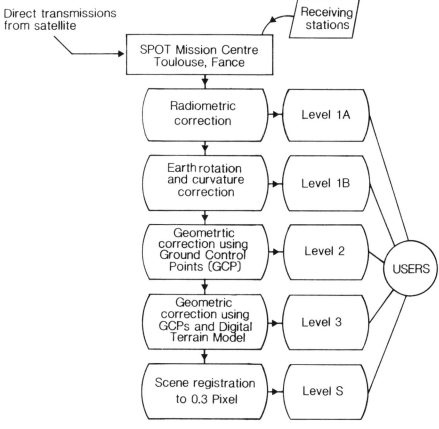

Figure 6.24 Forms of SPOT data illustrating different levels of correction applied to various products.

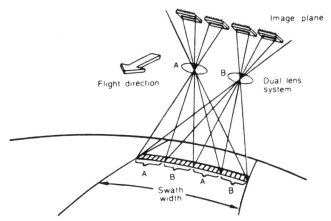

Figure 6.25 Optical and geometrical arrangement of MOMS scanner. Courtesy DFVLR.

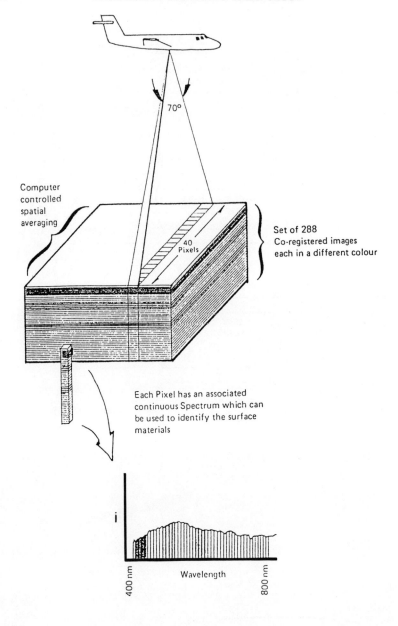

Figure 6.26 Moniteq PMI: (*a*) full spectral mode; (*b*) full spatial mode. *i*, intensity.

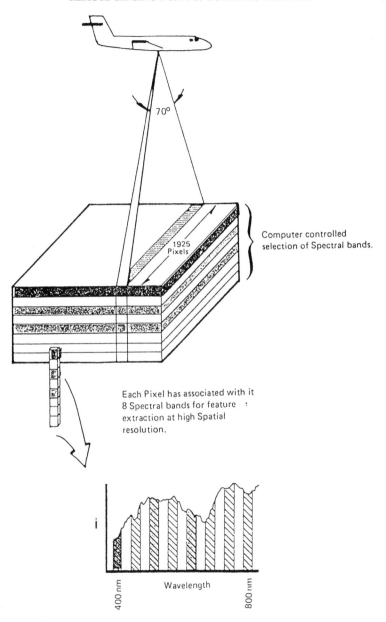

(b)

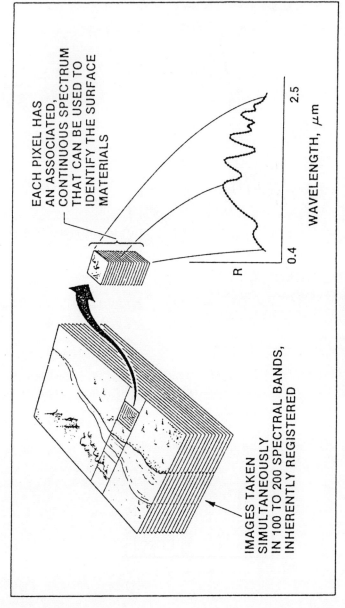

Figure 6.27 Principle of airborne imaging spectrometry.

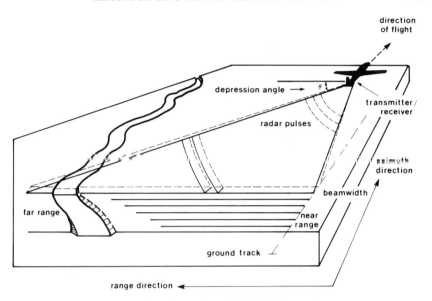

Figure 6.28 Principle of operation of a side-looking radar system.

then produced by the forward motion of the aircraft, thus creating, line by line, a continuous microwave image. Figure 6.29 is an example image produced from a side looking airborne radar (SLAR) system.

The image which is produced by a SLR system is very different from that produced by a conventional optical system or passive microwave radiometer. The view obtained is a record of the range to objects on the terrain and is based on the Earth's reflective properties at microwave wavelengths (8–300 mm). Consequently, the nature and intensity of the reflections are influenced by factors such as ground conductivity and surface roughness, which are much less significant at shorter wavelengths.

The spatial resolution of a real-aperture system is determined by two factors—the pulse length and the antenna beamwidth. The first of these factors determines the resolution in the range direction, whereas the second influences the discrimination between features in the direction of flight, in the azimuth direction (Fig. 6.28). A critical consideration is the antenna length, since this directly influences the beamwidth of the sensor and thus controls the resolving power of the sensor in the azimuth direction. The resolution of a real-aperture radar is consequently limited by the size of antenna which can be fixed to the side of the aerial platform. Although this restriction can be a serious deficiency when high-resolution imagery is required, these systems nevertheless have the ability to produce imagery in real time, without the need for sophisticated optical or computer processing. This has been a particularly important factor in the development of such systems for military use. The limitations in the resolution of real-aperture radars can, however, be overcome by using suitable signal processing techniques. Sensors of this design are referred to as synthetic-aperture radar (SAR) systems.

A detailed description of the principles of SAR is beyond the scope of this chapter, and readers should refer to Jensen *et al.* (1977) and Trevett (1986) for further

Figure 6.29 Typical SLAR image created by the SAR-580 system.

information on the technique. In very simple terms, the antenna used in SAR systems is synthesized electronically in order to increase its effective length. Thus a relatively short antenna, mounted on an airborne or spaceborne platform, is able to act as if it were a much longer antenna, with a consequent increase in resolving power in the azimuth direction. This improvement is achieved by processing not only the time interval between the outgoing and reflected signals but also the change in frequency (the Doppler shift) between the two signals. Figure 6.30 illustrates how this technique was used on board the Shuttle Imaging Radar-A (SIR-A) sensor.

Several airborne SAR systems have also been flown, and such systems have been used to generate imagery over extensive parts of South America (Von Roessel and De Godoy,, 1974), West Africa and Indonesia. The European Space Agency (ESA) sponsored the SAR-580 campaign providing the most recent airborne SAR imagery in the United Kingdom (Trevett, 1985). The campaign also provided valuable information for the design of the forthcoming European Radar Satellite (ERS-1). Up to 1988, three SAR sensors have been operated, for civilian use, from satellite platforms, with at least a further two planned for the near future. Table 6.6 summarizes the principal characteristics of these systems.

REMOTE SENSING FOR TOPOGRAPHIC MAPPING 213

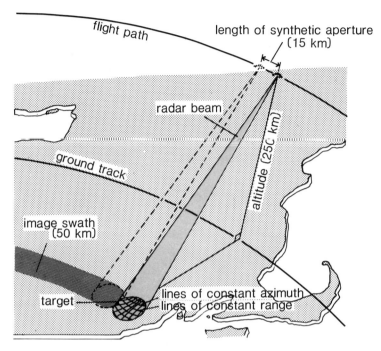

Figure 6.30 Shuttle imaging radar (SIR)–a synthetic aperture radar (Elachi, 1982).

Table 6.6 Characteristics of a selection of satellite-based SAR systems

	Past			Future	
	Seasat	Shuttle Imaging Radar (SIR)-A	SIR-B	European Radar Satellite (ERS-1)	Radarsat
Date of launch	1978	1981	1984	1989	1991
Operated by	NASA	NASA	NASA	European Space Agency	Canada
Altitude (km)	800	245	255, 274, 352	675	1000
Wavelength (mm)	230	230	540	540	540
Look angle (°)	20	50	57	—	30–45°
Resolution (m)	25	40	30	25	30
Swath width (km)	100	50	20–25	80	150
Further reading	Elachi (1980)	Elachi (1982)	—	Braun and Velten (1986)	Annett (1985)

It is important, however, to appreciate the limitations of SLR. These difficulties arise from two principal sources: background clutter on the image which can make interpretation quite difficult, and the geometry of the 'illuminating' signal relative to the terrain. The latter may lead to difficulties in the detection of linear features such as roads. Further details of the effects of these difficulties for topographic mapping can be found in section 6.5.1. In spite of these limitations, SLR sensors are important, particularly for monitoring in cloud-covered tropical regions and also in temperate and northern latitudes. For the future, as the technology for digitally processing SAR imagery improves, it is likely that such data will have a much more significant role to play in operational remote sensing activities than at present.

6.4 Digital image processing (DIP)

Digital image processing is central to the effective use of modern remote sensing data. Although a limited amount of information can be gleaned from the interpretation of a photographic image produced from Landsat imagery, for example, the full potential of the data can only be realized if the original data, edited and enhanced to remove systematic errors, are used with a suitably programmed image processing system.

6.4.1 *Hardware*

Until relatively recently, most DIP of remote-sensing data was carried out on a large expensive dedicated system such as that illustrated in Fig. 6.31. This enables data to be read into the system from a magnetic tape reader, to be displayed on a colour TV monitor and, after suitable processing, to be output to a colour filmwriter for the production of high-quality imagery. Typical of the modern dedicated image processing systems, which also incorporate local microprocessor support for image display and arithmetic processing, is the GEMSYS 35 system manufactured by GEMS of Cambridge (Fig. 6.32).

A trend in recent years has been the move towards smaller and cheaper image processing based on the new generation of 32-bit superminicomputers and 16-bit IBM-compatible personal computers (Fearns, 1984; Tait, 1987). Table 6.7 lists a selection of the remote sensing systems currently available.

6.4.2 *Software*

The range of algorithms which are used to restore and classify remote-sensing data is extremely wide and varied. Furthermore an appreciation of the choice of the most appropriate technique, and the advantages and limitations of each technique, is a subject of some complexity and beyond the scope of this chapter. Consequently, only a very brief description of the most commonly used techniques will be provided. For further details, Bernstein (1978) and Bagot (1985) should be consulted.

6.4.2.1 *Image restoration.* Several preliminary transformations are normally carried out on raw satellite data to reduce the effects of Earth rotation and curvature and variations in the attitude of the satellite. If sufficient ground control points are available, the imagery can be fitted to the control using least-squares techniques. This procedure is also necessary if the remote-sensing data are to be combined with existing map data. Geometric preprocessing may also be required for airborne imagery, for example, if multitemporal sources of data are being compared.

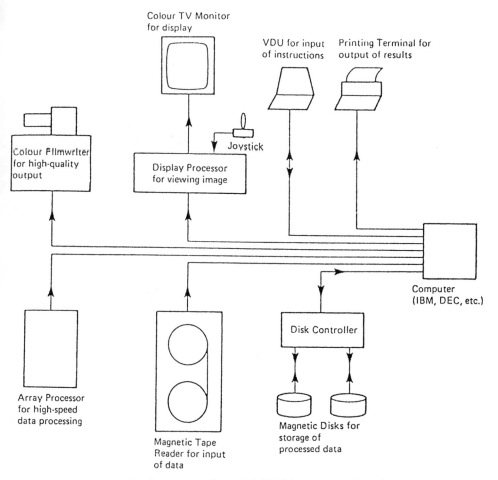

Figure 6.31 Components of a typical digital image processing system.

Variations in the output from the detectors used in an MSS may cause a striping effect on the final image. Such an effect can be eliminated by a process termed 'destriping'. In some cases, complete scan lines may be missing from the data, and again preprocessing can be carried out to overcome this type of problem. Radiometric preprocessing to eliminate the effects of atmospheric variations may also be necessary, particularly with airborne imagery.

6.4.2.2 *Image enhancement. Contrast stretching* is a technique which is used to brighten an image and produce one which allows the full potential range of the output device to be realized. It is often carried out automatically, and is generally one of the first operations performed by a user when viewing a new image. A range of different techniques can be used, and the choice generally depends on the distribution of the grey values or digital numbers within the scene. Figure 6.33, for example, illustrates one particular intensity histogram distribution.

Figure 6.32 GEMSYS 35 image processing system. Courtesy GEMS of Cambridge.

Table 6.7 A selection of remote sensing systems currently available

	Name of system	Company
Dedicated image processing system	Gemsys 33 I^{2}S Model 75 Vicom VDP Series Dipix Aries II	GEMS, Cambridge, UK International Imaging Systems, California, USA Vicom Systems, San Jose, USA Dipix Systems, Ottawa, Canada
Personal computer-based systems	Diad Systems LS10 Microimage ERDAS ImaVision	Diad Systems, Edenbridge, Kent, UK CW Controls, Southport, Lancashire, UK Terra Mar, California, USA Earth Resources Data Analysis System, Atlanta, USA RBA Associates, Ottawa, Canada

Two distinct forms of contrast stretching are normally used; the linear (single or two-part) and the more complex non-linear (e.g. normalization or gaussian). For example, Fig. 6.34 illustrates the effect of a simple linear stretch on the data shown in Fig. 6.33. This graph represents the 'transfer function' between the input and output digital numbers. If this transfer function is used on the original input data distribution, the effect would be to stretch the input range of 30 to 180 so that the output range varied linearly between 0 and 255.

Spatial filtering is carried out in order to either enhance boundaries between features (edge enhancement or high-pass filtering) or to smooth and eliminate boundaries

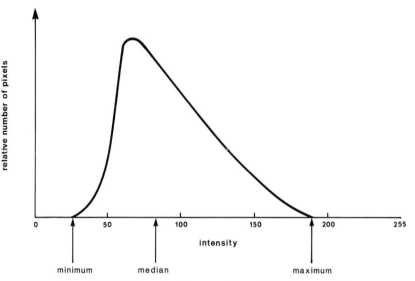

Figure 6.33 Pixel intensity histogram (adapted from Bagot, 1985).

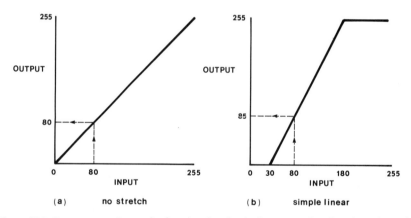

Figure 6.34 Contrast stretch-transfer function for simple linear stretch using data shown in Fig. 6.33.

(smoothing or low-pass filtering). The former approach is most suitable for enhancing geological features, particularly lineaments. The latter may be used to reduce random noise in an image, and Fig. 6.35 illustrates the concept.

Image ratioing is the process of dividing the radiance values of each pixel in one waveband by the equivalent values in another waveband. Ratioed images reduce radiance variations caused by local topographic effects, and consequently may be used to enhance subtle spectral variations between surface materials that are difficult to detect on single-band images. Figure 6.36 illustrates the concept in more detail. In this case the DN values for the same terrain class can be seen to vary, depending on the

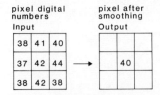

Figure 6.35 Concept of 3 × 3 low pass (smoothing) spatial filter. The central pixel value (40) is the mean of the nine surrounding values.

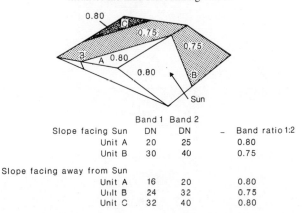

Figure 6.36 Use of image ratioing to eliminate systematic effects caused by varying illumination conditions (adapted from Taranik, 1978).

illumination conditions (sunlit or shaded slopes). The ratioed values remove this systematic bias in the original data.

Principal components analysis is a statistical technique which is used to improve the discrimination between similar types of ground cover. It is based on the formation of a series of new orthogonal axes which reduce the correlation existing between successive wavebands in a multispectral data set.

For example, Fig. 6.37 is a scatter plot showing the radiance values for two features A and B in two different wavebands X and Y. In the original wavebands (Fig. 6.37a) the variation between features A and B is very small in comparison to the original range of the data. Principal components analysis operates by generating a new set of axes along the lines of maximum variance; this helps exaggerate the difference between features such as A and B. Thus, in Fig. 6.37b, although the data range has been compressed, the variation between features A and B is now almost half the total range of the data and consequently eases the identification of these feature classes.

6.4.2.3 *Image classification*. *Density slicing* is the simplest form of automated classification. It enables a single band (generally thermal IR data) to be colour-coded according to the grey level of the individual pixels. Thus, for example, pixel values 0–20 may be coloured blue, 20–25 red and so on. This not only aids interpretation, but if the density slices are related to temperature levels, the resulting image can be used to quantitatively assess the variations in temperature across the scene (Fig. 6.17).

Supervised methods can be used in cases where ground data exists about the terrain

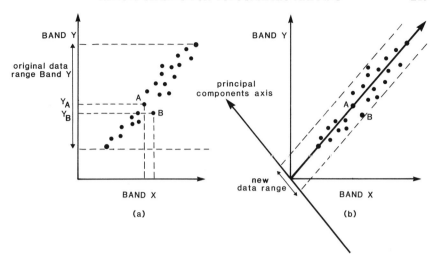

Figure 6.37 Graphical illustration of principal components analysis: (*a*) original wavebands; (*b*) new principal components wavebands.

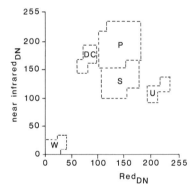

Figure 6.38 Supervised classification–parallelopiped or box classification of five land cover types (Curran, 1986). *W*, water; *DC*, deciduous woodland; *P*, pasture; *S*, bare soil; *U*, urban areas.

under investigation. It is possible to use this data to 'train' the computer to perform a simplified form of automated interpretation. For example, assume a particular group of pixels are known to represent an area of deciduous woodland. By examining the maximum and minimum numerical values of the radiances in each of the spectral bands being used (7 for Landsat TM, although only the least correlated bands would be used in practice) a 'spectral pattern' for deciduous woodland could be defined. If the computer then compared this multidimensional spectral pattern with the spectral pattern of every pixel in the scene, then it would be possible to display only those areas which satisfied the criteria set by the training sample. Consequently, all deciduous woodland over a large area could be automatically determined.

Generally, several different land cover types would be analysed and classified using this technique. Figure 6.38 illustrates a hypothetical set of data and the classifier which could be used to discriminate each of the five spectral classes. The process outlined above refers to the simplest form of classifier, the 'box' or parallelopiped classifier. Other more sophisticated techniques, such as maximum likelihood, can also be used.

In *unsupervised classification* of multispectral data, no attempt is made to 'train' the computer to interpret common features. Instead, the computer analyses the pixels in the scene and divides them into a series of spectrally distinct classes based on the natural clusters which occur when the radiances in n bands are compared. The user then analyses other evidence to determine the nature of each of the distinctive classes which have been identified by the computer. Density slicing, for example, is a simple form of unsupervised classification. Unsupervised classification techniques are generally more expensive in terms of processing requirements than supervised methods.

6.5 Applications of remote sensing

The applications of remote sensing cover a wide range of scientific and engineering disciplines and many of these are beyond the scope of this chapter. Consequently two general topics relevant to the surveyor and engineer will be discussed. The first relates to the use of satellite imagery for topographic mapping and the second to the use of airborne and satellite imagery to generate thematic mapping information appropriate to engineering projects.

6.5.1 *Small-scale topographic mapping from satellite sensors*

The creation of topographic maps is based almost totally on the use of remotely sensed data. Only at very large scales can ground surveying compete economically. The majority of maps created in this manner are based on aerial photography obtained using a high precision mapping camera and measured using a photogrammetric stereoplotter. The principles of these techniques are well known and considered in detail by Burnside (1982) and Wolf (1983), and to a more limited extent in Chapters 7 and 8.

In recent years as the spatial resolution of satellite sensors has improved, the potential of spaceborne cameras, scanners and radar systems for small-scale topographic mapping has been considered. This section examines the progress which has been achieved in this field.

6.5.1.1 *Satellite sensors suitable for topographic mapping.* A wide range of satellite based sensors has been considered as potentially suitable for cartographic applications. Three distinct groups of sensors can be identified; photographic sensors, scanning sensors and synthetic aperture radar.

Several limited-term missions have been flown to examine the capability of satellite-based photographic cameras for mapping; examples include the Skylab Itek camera, and more recently the Spacelab Zeiss RMK 'metric camera' and the Itek Large Format Camera (LFC).

Unlike photographic sensors, the scanning systems are generally housed on board Earth resources satellites providing repeat coverage of the terrain. The most significant sensors in this category are Landsat MSS, Landsat TM and SPOT. Landsat RBV has also been used for mapping, and indeed was primarily designed for this application; however, because of the limited amount of data obtained from the sensor, it will not be considered.

Table 6.8 Requirements for standard map series (Doyle, 1982)

Map scale	Positional accuracy (m)	Elevation accuracy (m)	Typical contour interval
1:1 000 000	300	30	100
1:500 000	150	15	50
1:250 000	75	8	25
1:100 000	30	6	20
1:50 000	15	3	10
1:25 000	8	2	5

The cloud penetration benefits of *synthetic aperture radar* (SAR) have already been discussed, and this is a particularly important feature for space systems imaging in temperate latitudes, such as the UK, or in tropical regions, such as South America and West Africa. Three particularly important systems in this context have been Seasat and the Shuttle Imaging Radars A and B (SIR-A and SIR-B) (Table 6.6). The experience from both of these systems will be invaluable in maximizing the success of future systems such as ERS-1 and Radarsat.

The level of success of all these systems for topographic mapping is dependent on two factors; first, the geometrical accuracy of the data which are obtained, and second, the level of completeness of detail which can be interpreted from the data.

6.5.1.2 *Geometrical accuracy testing.* Before examining in detail the results which have been obtained from the various sensors described above, it is useful to consider the accepted accuracy requirements for various standard map scales. Comparing these with the geometrical accuracy results obtainable from satellite systems will then enable the potential of the different sources of data to be assessed. Table 6.8 (Doyle, 1982) summarizes these specifications.

The procedure for testing the geometric accuracy of satellite imagery normally involves a comparison of the coordinates of a series of well-defined control points on the imagery with their known ground coordinates (from an existing map).

Various mathematical models can be used to perform the transformation from image coordinates to ground coordinates. These range from simple linear conformal or similarity transformation (4 parameters, scale change, rotation and translation in X and Y), to affine transformations (6 parameters, as linear conformal but with different scale changes in X and Y) and polynomial transformations with different numbers of terms (see section 7.3). The choice of transformation depends largely on the type of imagery and the level of pre-processing which has been performed on the imagery.

A summary of the results which have been reported for the three photographic sensors discussed previously is provided in Table 6.9. Comparison of these results with those for the mapping specifications given in Table 6.8 indicates that, geometrically, this type of data could be suitable for planimetric mapping at scales between 1:50 000 and 1:100 000.

Table 6.10 summarizes a selection of the results which have been obtained by applying different transformations to MSS imagery obtained from Landsats 2 and 3. Comparative results using TM imagery of two test sites in the United States are provided by Welch *et al.* (1985), and are illustrated in Fig. 6.39. Similar results were reported by Fusco *et al.* (1985) and Isong (1987) (Table 6.11).

Table 6.9 Characteristics of spaceborne photographic missions

	NASA Skylab S-190B	ESA Metric Camera	NASA Large Format Camera
Focal length (cm)	46	30	30
Format (cm)	12.5 × 12.5	23 × 23	23 × 46
Coverage (km)	110 × 110	190 × 190	225 × 450
Scale	1:950 000	1:820 000	1:1 000 000
Flying height (km)	385	250	300
Planimetric accuracy (m)	20–25	20–30	20–30
Reference	Petrie (1985)	Menuquette (1985)	

Table 6.10 Selected accuracy tests for Landsat MSS imagery

Image source	No. of control points	Type of transformation	Root mean square errors X (m)	Y (m)	Reference
Landsat MSS Bulk Image	181	Uncorrected	215	195	Wong (1975)
"	"	Linear conformal (similarity)	94	67	"
"	"	20-term polynomial	37	44	"
Landsat MSS Precision Images	35	Linear conformal (similarity)	64	81	Mohammed (1977)
"	"	Affine	46	52	"
"	"	12-term polynomial	40	49	"

The results obtained in virtually all cases are remarkably good, and indicate that sub-pixel level accuracies can be achieved with relatively low-order processing. The results from Welch *et al.* (1985) also indicate the higher geometrical characteristics of Landsat 5 TM imagery.

These results indicate that, geometrically, Landsat MSS could be suitable for planimetric mapping at scales of up to 1:100 000 and for Landsat TM up to 1:50 000. Preliminary results using simulated SPOT imagery (Welch, 1985), indicate that, geometrically, it could be used for planimetric mapping at scales of up to 1:50 000.

Comprehensive reviews of the geometric and cartographic potential of satellite SAR systems for topographic mapping can be found in Petrie (1978), Dowman and Morris (1982), and Leberl *et al.* (1985).

A summary of the geometrical accuracy results obtained from four differently processed Seasat images of the area is provided in Ali (1986). The results indicate that optically processed Seasat images contain substantial scale errors compared to their digitally processed counterparts. These results also indicate that the effect of using higher-order polynomials is more noticeable on the optically processed images than on those produced digitally. The imagery produced by the German Aerospace Agency (DFVLR) produced the best results during the test process. In both cases, the effect of terrain relief on the accuracy of the results was noticeable.

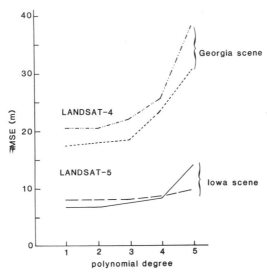

Figure 6.39 Landsat TM: comparison of RMSE values determined at check points for Landsat TM datasets in Georgia and Iowa as a function of the degree of polynomical used for rectification (adapted from Welch *et al.*, 1985).

Table 6.11 Summary of results of various geometric transformations applied to Landsat Thematic Mapper imagery of an area around London, England (Isong, 1987)

Transformation used	Terrain coordinates	
	Vector standard errors (m)	
	Control points $n = 20$	Check points $n = 20$
Linear conformal (similarity)	19.81	20.90
Affine	19.60	21.10
4-term polynomial	18.84	20.76
6 ″ ″	18.71	20.49
8 ″ ″	18.78	21.97
10 ″ ″	19.49	22.13

Leberl and Domik (1985) report the results which have been achieved with Seasat and SIR-A imagery using an analytical stereoplotter and the SMART (stereomapping with radar techniques) bundle adjustment software system. Table 6.12 summarizes the results obtained.

6.5.1.3 *Level of completeness.* It should be clearly noted that all the results cited in the previous section refer to mathematical coordinate determinations for a few well selected and clearly identifiable points. These results do not necessarily reflect the planimetric mapping potential of the sensors, since the results do not account for the level of mapping detail which can be interpreted from the imagery. For example, variations in scene contrast in Landsat imagery or changes in radar illumination

Table 6.12 Root mean square (rms) accuracies obtained with an analytical plotter and SMART program using Seasat and SIR-A data. X denotes the along-track co-ordinate (Leberl et al., 1985)

	Radar data and systems					
	Seasat (Los Angeles)			SIR-A (Greece)		
Number of ground control points	16	4	2	31	4	3
Number of orientation points	10	0	8	6	0	6
r.m.s. X(m)	58	73	72	62	89	86
r.m.s. Y(m)	134	174	169	113	132	119

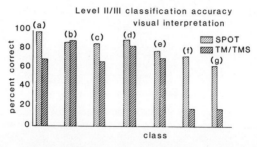

Figure 6.40 Landsat TM and SPOT–percent completeness for selected land cover classes. (a) Roads, highways and railways; (b) forested wetland; (c) commercial property; (d) single family housing; (e) rivers, streams and ponds; (f) barren transitional areas; (g) urban grassland. (Adapted from Welch, 1985.)

conditions can affect the interpretability of the features being mapped from each source of data. Consequently, several authors have assessed the level of mapping completeness which can be achieved from the three sources of data described.

The identification of planimetric features from photography produced by the ESA Metric Camera is reported in Baudoin (1985). The results indicate detectability of various features in areas of high and low contrast. For areas of high contrast, the threshold width for linear features such as roads, rivers and hedges which will ensure a 100% level of success is 18 m. For low-contrast targets, the values are two to three times greater. For areas of woodland and isolated buildings, it was not possible to state threshold values for identification, even in areas of high contrast. Generally, the level of detail which can be interpreted from the metric camera photography allows only about 50–60% of the detail required for mapping at 1:100 000 to be obtained. Similar results for the earlier Skylab mission are reported in Welch (1982).

The results of tests to assess the levels of completeness from Landsat MSS are reported in Welch and Pannell (1982). The tests, carried out using imagery of three test sites in China, indicate that overall only 41% and 37% of the detail contained on a 1:250 000 map can be obtained from visual interpretation of black-and-white Landsat band 5 and 7 imagery respectively.

More recent analysis of the relative performance of SPOT and Landsat TM imagery is reported in Welch (1985), and a selection of the results is illustrated in Fig. 6.40. The results indicate that accuracy levels of over 80% can be realized for selected land use/land cover types using SPOT imagery, whereas completeness levels of 65–70% are more typical of the results obtainable from Landsat TM imagery. It is suggested that the level of completeness is suitable for limited planimetric mapping at scales from 1:250 000 to 1:100 000.

The difficulties associated with SAR imagery from Seasat are summarized by Petrie (1985). The two main difficulties experienced were, firstly, deficiencies in image quality, and secondly, detection difficulties associated with the orientation of features relative to the direction of the radar signal. In the first case, the effect of background clutter considerably reduces the interpretability of the imagery. In the second case, the detection of linear features in particular seems to be highly dependent on their orientation relative to the SAR signal. The consequence of this is to produce rather inconsistent sets of results where features may or may not appear, depending on their orientation. It is essential that methods of improving the interpretability of such imagery are developed, and some progress in this field has been reported by Rye and Wright (1985), using segmentation techniques. However, it would seem that, until some improvements occur in this area, SAR imagery from space is unlikely to be a significant source of information for planimetric mapping, and may be of greater use in mosaic form for planning purposes (Van Roessel and DeGodoy, 1974).

6.5.1.4 *Contouring and digital terrain modelling (DTM)*. Ackermann and Stark (1985) concluded that, on the basis of an assessment of photography of Bavaria, the height accuracy from the Metric Camera was about ± 20 m, and this would be appropriate for contouring with an 80–100 m contour interval. Using a substantial number of known control heights, the results were reduced to ± 15 m and 50–60 m respectively.

A more rigorous investigation into the accuracy of DTMs produced automatically from the Metric Camera using digital image correlation techniques is reported in Boochs and Decker (1986). By comparing known and observed heights, and filtering the dataset to eliminate gross errors, height errors of between ± 15 m and ± 31 m were obtained, the variations being caused by changes in image contrast, object geometry and object types within the test sites. Engels *et al.* (1985) suggested that the DTM produced from this data may be suitable for the production of orthophotographs at a scale of 1:50 000.

The potential of Landsat imagery (both MSS and TM) for height determination is very limited, given the very poor geometry of the lateral overlap between successive passes. Generally, the accuracies achieved have been at best ± 60 m for Landsat MSS (Dowman and Mohammed, 1981), and similar results have been obtained with Landsat TM imagery (Cooper *et al.*, 1986).

More promising results using simulated SPOT imagery have been reported by Cooper *et al.* (1986) and Swann *et al.* (1986). Using an edge-based hierarchical matching algorithm, height errors of between ± 7 m and ± 10 m were obtained. Further discussion of the potential applications of DTMs and SPOT imagery can be found in Baudoin (1984).

The heighting accuracy of spaceborne SAR imagery is also very poor. Stereo radar imagery obtained from the 1984 SIR-B mission has been evaluated by several researchers. The results reported include both manual extraction of heights on a modified analytical plotter (Leberl *et al.*, 1986), and automated height determination using correlation techniques (Simard *et al.*, 1986). The results using the latter technique have proved inconsistent, and at present do not appear suitable for the operational determination of heights. The work carried out by the former researchers is also disappointing: differences between the radar-generated DTM and the map-derived

Table 6.13 Suitability of various satellite sensors for topographic mapping

	Photographic sensors		Scanning sensors		Synthetic-aperture radar
	Metric Camera	Large Format Camera	Landsat Thematic Mapper	SPOT	Seasat SIR-A/B
Plan accuracy	***	***	**	***	*
Height accuracy	—	—	—	**	—
Map compilation	**	**	*	**	*
Map revision	***	***	*	****	*
Map intelligence	***	***	***	****	**

—	Unsuitable
*	Suitable for 1:250 000 mapping
**	Suitable for 1:100 000 mapping at a relaxed specification (70% detail)
***	Suitable for 1:100 000 mapping
****	Suitable for 1:50 000 mapping.

DTM are about ± 60 m. Clearly the results are not currently of use on an operational basis, although improvements in the imaging geometry of future satellite SAR systems, such as Radarsat, may enable greater levels of accuracy to be obtained (Derenyi and Stuart, 1984).

6.5.1.5 *Suitability of satellite sensors for topographic mapping.* It is evident from the results provided in the previous three sections that, at present, the potential of satellite-based remote-sensing systems for topographic mapping is restricted to scales of between 1:100 000 to 1:250 000. Table 6.13 summarizes the suitability of the various sensors which have been discussed. For the future, the more likely application of the satellite data will be for map revision and map intelligence operations. In the latter case, Landsat TM imagery is currently being used by the Canadian government for detecting change on its 1:50 000 series in the more isolated northern regions. Mapping of the areas where a sufficient level of land cover change has occurred is then performed using conventional aerial photogrammetric techniques.

6.5.2 *Thematic mapping for engineering projects*
The results reported in the previous section have highlighted some of the reasons why, at present, the operational use of satellite remote sensing for small-scale topographic mapping is limited. Remote sensing data has, however, been used more successfully to produce thematic maps for engineering projects, since for such applications users are satisfied with a relatively lower level of positional accuracy and completeness in comparison with the requirements for topographic mapping. At larger scales also, remote sensing has a role to play, and the following sections indicate some of the potential uses of such imagery during the reconnaissance/planning, feasibility, design, construction and maintenance and operation stages of a civil engineering project (Fig. 6.41).

6.5.2.1 *Reconnaissance/planning stage.* The first stage of any major civil engineering project generally involves some form of preliminary reconnaissance study of the region

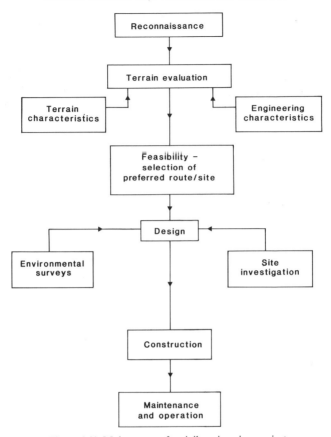

Figure 6.41 Main stages of a civil engineering project.

or project area. The primary aim of this study is to collect together all available data concerning the physical characteristics of the terrain in order to assess its likely influence on the overall design of the engineering works.

Traditionally, the engineer has carried out this reconnaissance stage by examining existing maps of the region; both topographic maps and specialist thematic maps (e.g. geological, geomorphological and pedological) are of greatest use at this initial stage of the project. In addition, examination of engineering reports produced for previous projects in the region may also provide valuable reference material. The main problems associated with this approach are, firstly, that in many regions existing maps may be at a very small scale (< 1:500 000), significantly out of date, and in some instances may not exist at any scale. Secondly, whilst engineering reports may give valuable data about a specific site, they may not be appropriate to or representative of regional terrain conditions. Consequently, many engineering projects now involve an examination of satellite and aerial imagery at this early stage.

The use of Landsat MSS data for projects of this type has been reported by Beaumont (1982). In this case, use of Landsat data for water resources planning in North-East Somalia is reported. Visual interpretation of Landsat data enabled a series of

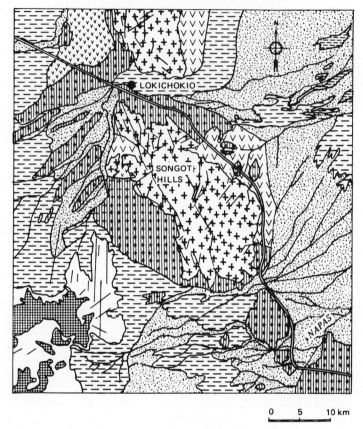

Figure 6.42 Geological map of Northern Kenya as interpreted from 1:250 000 Scale Landsat MSS false-colour composite image (Dick, 1982).

transparent overlays to be produced, illustrating drainage, surface water, groundwater potential and land capability over the region. A similar approach in relation to agricultural development planning in Southern Sudan is reported in Beaumont (1985).

The use of remote sensing for terrain evaluation purposes is also of considerable importance. Terrain evaluation is a form of thematic mapping in which the terrain is classified in a hierarchical manner into units having common landscape patterns and engineering characteristics. The former element of the process can be carried out very cost-effectively using satellite imagery in conjunction with aerial photography (Overseas Unit, 1982).

The use of Landsat MSS imagery for transport planning in Northern Kenya and Southern Sudan is reported in Dick (1982). Geomorphological and geological mapping was carried out for this project by visual interpretation of single-band and false-colour composite images. Figure 6.42 illustrates some of the results which were obtained.

Interpretation of Landsat data may also be useful for the provision of additional supplementary information on hydrological phenomena such as flooding (Deutsch and

Ruggles, 1978) and water quality monitoring (Alfoldi and Munday, 1978). A more general review of hydrological applications can be found in Blyth (1985).

Landsat has also been used for estimating urban populations over large areas (Forster, 1983), and for monitoring change, using multitemporal imagery (for example the expansion of urban areas, changes in soil erosion, variations in river channels, the impact of desertification and so on).

6.5.2.2 *Feasibility stage.* The aim of the feasibility stage of a project is firstly, to select potential routes/or sites, and secondly, to select from these the best from the available options. Beaven and Lawrance (1982) identified the following engineering requirements for the feasibility study where remote sensing could be of assistance:

(i) Determination of foundation conditions
(ii) Location of drainage patterns, calculation of catchment areas and positioning of culverts
(iii) Identification of spoil areas and possible borrow pits
(iv) Location of possible bridge sites
(v) Identification of major hazard areas, for example, poorly drained soils, spring lines, unstable ground
(vi) Location of possible sources of constructional materials.

In developing regions, the location of low-cost construction materials such as calcrete may be important (Beaven and Lawrance, 1982; Lawrance and Toole, 1984). Calcrete is a calcareous material which generally accumulates in large bare depressions in arid regions. It is a widespread source of low-cost road construction material, and consequently its location and distribution can have a considerable effect on the positioning of a new road alignment. Supervised classification techniques have been used to identify sources of calcrete, and Fig. 6.43 illustrates one such classification. The change in colour of the overlying sand deposits from their normal red or brownish hue to a neutral grey has been found to be correlated with calcrete deposits. Figure 6.44 illustrates the discrimination of these regions by the use of principal components analysis.

A comprehensive review of the role of remote sensing for highway feasibility studies can be found in Lawrance and Beaven (1985).

6.5.2.3 *Design.* Remote sensing has a particularly valuable role to play at the design stage of a project. At this stage a detailed site investigation of the proposed site or route will be required, together with supplementary environmental information which may influence the design of the project being considered. A critical stage of the site investigation stage is the desk study. The desk study provides the engineer with an overview of the project area, and is normally carried out using aerial photography. Not only can valuable information be obtained during this interpretation about possible engineering problems, but it can also be very important for the planning of the subsequent ground investigations. A general review of the potential of aerial photography in this respect is provided by Dumbleton (1983) and Matthews (1985), and a selected summary of the literature which has examined specific geotechnical problems is presented in Table 6.14.

Of increasing importance in recent years has been the role of remote sensing in providing environmental engineering data for input into the design process. A

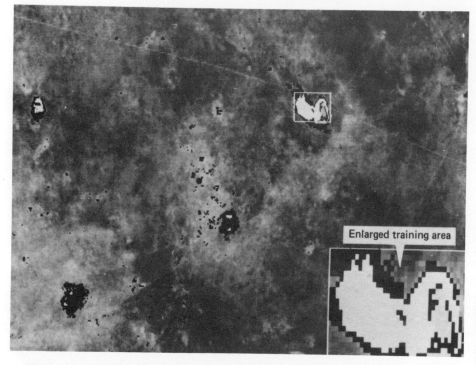

Figure 6.43 Location of calcrete deposits in Botswana–supervised classification and training area (Beaven and Lawrance, 1982).

comprehensive review of remote sensing in this context is provided by Mason and Amos (1985). Information which may be extracted from remote sensing imagery (often thermal IR linescan) includes data on energy conservation (heat loss), power-station cooling-water patterns, groundwater and spring detection, pipeline and field drainage patterns and hydraulic leakage from earth dams.

6.5.2.4 *Construction stage.* The role of remote sensing is of much less importance during construction than at the preceding planning and design stages. Nevertheless, aerial photography can be used to provide a valuable historial record of construction activities and may also be used to plan construction activities. Singhroy (1987) provides an interesting case history concerning the use of large-scale colour IR aerial photography and aerial video during pipeline construction in Canada. Both types of data were used to plan construction activities, determine site conditions and assess environmental effects before, during and after construction activity.

6.5.2.5 *Post-construction maintenance.* Post-construction applications of remote sensing are often related to the need to monitor the impact of the civil engineering project on the environment or to assist in the maintenance of the projcet. Examples which are relevant in this context include aerial traffic investigations using helicopters and airborne video cameras (Mountain and Garner, 1981), or the location of road surfaces suceptible to icing during winter months (Stove *et al.*, 1987). On a regional scale,

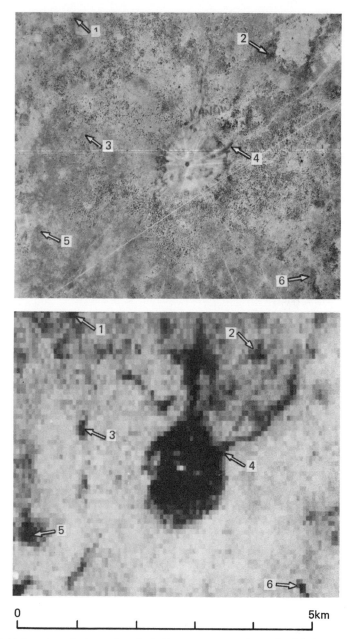

Figure 6.44 Location of calcrete deposits in Botswana–comparison of aerial photograph (top) and Landsat MSS image (bottom). Although the aerial photograph has a high spatial resolution, it is unable to distinguish areas of grey sand, where calcrete occurs, from the surrounding reddish brown sands. The Landsat image (after principal components analysis) enhances the difference and shows them as variations in dark tones (Beaven and Lawrance, 1982).

Table 6.14 Remote sensing and site investigations—selected references

	Aerial photography	MSS	Thermal IR linescanning
Landslides/ slope instability	Matthews and Clayton (1987)	Liu (1985)	Chandler (1975)
Solution features	Norman and Watson (1975)	Coker et al. (1969)	Edmonds et al. (1986)
Derelict/ contaminated land	Bullard (1983)	Coulson and Bridges (1984)	Coulson and Bridges (1984)
Surficial/ subsurface materials surveys	Caiger (1970)	Lynn (1984)	Singhroy and Barnett (1984)

satellite remote sensing may be used to evaluate the environmental impact of a project. For example, multitemporal Landsat MSS and TM imagery has been used to evaluate the historical development of irrigation practices in Morocco.

In conclusion, Table 6.15 illustrates the recommended use of remote-sensing techniques at each stage of a civil engineering project.

6.6 Conclusions and future trends

The primary objective of this chapter has been to discuss some of the surveying and civil engineering projects which may benefit from the use of airborne and spaceborne remote sensing. With a few exceptions, the use of remote sensing in these fields is currently still at the research and development stage. Before the technology is used routinely on an operational basis, it will require further advances, particularly with regard to the spatial resolution of spaceborne sensors.

For topographic mapping applications, the interpretability of the currently available satellite imagery is a major obstacle to its use even for very small-scale mapping. Similarly the accuracies attainable for height determination, even from SPOT imagery, limit its application to small-scale map revision. The sensors which are widely believed to offer the greatest benefits, particularly in the UK, are the SAR systems. Again, however, at present the discrimination of features on the imagery is very poor and generally inadequate for topographic mapping. It would seem therefore that aerial photography and the photogrammetric techniques discussed in Chapter 7 will remain the pre-eminent solution for routine topographic mapping operations.

In the creation of thematic maps for engineering projects, the development of more sophisticated processing and analysis techniques based on the use of geographical information systems (GIS) (Chapter 11), will be of considerable importance for the future. By using such systems, the engineer will eventually be able to examine a wide range of vector map data, and attribute textual data and raster-based remote sensing data. With the ability to interrogate this database and generate maps which consider the interrelationships between several different variables, the surveyor and civil engineer will have a very powerful capability at his disposal.

Table 6.15 Recommended use of remote-sensing techniques during a civil engineering project

Project phase	Photograph/image scale	Satellite						Aircraft			
		Photography	Landsat MSS	Landsat TM	SPOT	MOMS	Side-Looking Radar	Photography	MSS	Thermal Scanners	SLAR
Reconnaissance/planning	1:1 000 000 to 1:50 000	***	****	***	****	***	***	**	—	—	—
Feasibility	1:100 000 to 1:20 000	***	**	***	***	***	***	**	—	—	—
Design	1:20 000 to 1:2000	—	—	*	*	—	—	****	***	***	***
Construction	1:2000 to 1:500	—	—	—	—	—	—	****	*	*	*
Post-construction/maintenance	1:2000 to 1:500	—	—	—	—	—	—	****	*	*	*

**** Very useful   *** Very useful (but coverage limited)   ** Useful   * Of limited use   — inadequate.

References

Ackermann, F. and Stark, E. (1985) Digital elevation model from Spacelab metric camera photography. *ESA Spec. Publ.* **SP-209**, 9–12.

Alfoldi, T.T. and Munday, J.C. (1978) Water quality analysis by digital chromaticity mapping of Landsat data. *Can. J. Remote Sensing* **4**, 2.

Ali, A.E. (1986) Optical and digital Seasat SAR processing techniques: a comparison of accuracy results. *Int. Arch. Photogramm. and Remote Sensing* **26**(3/1), 28–39.

Amin, A.M. (1986) Geometrical analysis and rectification of thermal infrared video frame scanner imagery and its potential applications to topographic mapping. Ph.D. Thesis, University of Glasgow, 257 pp.

Annett, R.J. (1985) Radarsat: low earth orbit platform. *Proc. Conf. on Advanced Technology for Global Environmental Monitoring*, CERMA and Remote Sensing Society, 111–123.

Bagot, K.H. (1985) Digital processing of remote sensing data. In *Remote Sensing in Civil Engineering*, eds. T.J.M. Kennie and M.C. Matthews, Surrey University Press [Blackie], Glasgow and London, 87–105.

Baudoin, A. (1984) Applications of SPOT and DTM to cartography. *Proc. 18th Int. Symp. on Remote Sensing of Environment, Paris*, **1**, 61–75.

Baudoin, A. (1985) Identification of planimetric features from the Metric camera. *ESA Special Publication* **SP-209**, Metric Camera Workshop, 69–75.

Beaumont, T.E. (1977) Techniques for the interpretation of remote sensing imagery for highway engineering purposes. *TRRL Rept* **753**, Transport and Road Research Laboratory, Crowthorne, Berks., 24 pp.

Beaumont, T.E. (1982) Land capability studies from Landsat satellite data for rural road planning in North East Somalia. In *Proc. OECD Symp. on Terrain Evaluation and Remote Sensing for Highway Engineering in Developing Countries. TRRL Rept* **SR 690**, Transport and Road Research Laboratory, Crowthorne, Berks., 86–95.

Beaumont, T.E. (1985) Interpretation of Landsat satellite imagery for regional planning studies. In *Remote Sensing in Civil Engineering*, eds. T.J.M. Kennie and M.C. Matthews, Surrey University Press [Blackie], Glasgow and London, 164–201.

Beaven, P.J. and Lawrence, C.J. (1982) Terrain evaluation for planning and design. *TRRL Suppl. Rept* **725**, Transport and Road Laboratory, Crowthorne, Berks., 24 pp.

Bernstein, R. (1978) *Digital Image Processing for Remote Sensing*. IEEE Press, New York, 473 pp.

Blyth, K.B. (1985) Remote sensing and water resources engineering. In *Remote Sensing in Civil Engineering*, eds. T.J.M. Kennie and M.C. Matthews, Surrey University Press [Blackie], Glasgow and London, 289–334.

Braun, H. and Velten, E. (1986) The first ESA Rémote Sensing satellite: status and outlook. *Proc. ISPRS Comm. 1 Symp. Progress in Imaging Sensors, Int. Arch. of Photogramm. and Remote Sensing* **26**, 1, 329–334.

Bullard, R.K. (1983) Abandoned land in Thurrock: an application of remote sensing. *Working Paper* **8**, North East London Polytechnic, London.

Burnside, C.D. (1982) *Mapping from Aerial Photographs*. Granada, London, 303 pp.

Burnside, C.D. (1985) The future prospect of data acquisition by photographic and other airborne systems for large scale mapping. *Photogramm. Record* **11**(65), 495–506.

Caiger, P.B. (1970) Aerial photographic interpretation of road construction materials in S. Africa with special reference to its potential to influence route location in underdeveloped territories. *Photogrammetria* **25**, 151.

Chandler, P.B. (1975) Remote detection of transient thermal anomalies associated with the Portuguese Bend landslide. *Bull. Assoc. Eng. Geol.* **12**(3), 227–232.

Chevrel, M., Courtois, M. and Wells, G. (1981) The SPOT satellite remote sensing mission. *Photogramm. Eng. and Remote Sensing* **47**, 1163–1171.

Coker, A.E., Marshall, R. and Thompson, N.S. (1969) Application of computer processed multispectral data to the discrimination of land collapse (sinkhole) prone areas in Florida. *Proc. 6th Int. Symp. on Remote Sensing of the Environment*, Michigan, Vol. 1, 65–69.

Cooper, P.R., Friedmann, D.E. and Wood, S.A. (1986) The automatic generation of terrain models from satellite images by stereo. *Acta Astranautica* **14**, 10 pp.

Coulson, M.G. and Bridges, E.M. (1984) The remote sensing of contaminated land. *Int. J. Remote Sensing* **5**(4), 659–669.

Curran, P.J. (1985) *Principles of Remote Sensing.* Longman, London, 282 pp.
Derenyi, E.E. and Stuart, A.J. (1984) Geometric aspects of spaceborne stereo imaging radar. *Int. Arch. Photogramm. and Remote Sensing* **XXV**(A7), VII, 153–162.
Deutch, M. and Ruggles, F.H. (1978) Hydrological applications of Landsat imagery used in study of the 1973 Indus river flood, Pakistan. *Water Resources Bull.* **14**(2), 261–274.
Dick, O.B. (1982) Remote sensing for transport planning and highway planning. *TRRL Suppl. Rept* **690**, Transport and Road Research Laboratory, Crowthorne, Berks., 172 pp.
Doyle, F.J. (1982) Satellite systems for cartography. *Int. Arch. of Photogramm.* **24**(1), 180–18.
Dowman, I.J. and Mohamed, M.A. (1981) Photogrammetric applications of Landsat MSS imagery. *Int. J. Remote Sensing* **2**, 105–113.
Dowman, I.J. and Morris, A.H. (1982) The use of synthetic aperture radar for mapping. *Photogramm. Record* **10**(60), 687–696.
Dumbleton, M.J. (1983) Air photographs for investigating natural changes, past use and present condition of engineering sites. *TRRL Rept* **LR 1085**, Transport and Road Research Laboratory, Crowthorne, Berks.
Edmonds, C.N., Kennie, T.J.M. and Rosenbaum, M. (1986) The application of airborne remote sensing to the detection of solution features in limestone. *Proc. 22nd Ann. Meeting of the Engineering Group of the Geological Society, Plymouth Polytechnic,* 109–120.
Elachi, (1980) Spaceborne imaging radar—geologic and oceanographic applications. *Science* **216**, 1073–1082.
Elachi, C. (1982) Shuttle imaging radar experiences. *Science* **218**, 996–1004.
Engels, H., Muller, W. and Gonecny, G. (1985) Application of Spacelab Metric Camera imagery for map production. *Int. Arch. Photogramm. and Remote Sensing* **26**(4), 170–182.
Everitt, J.H. and Nixon, P.R. (1985) False colour video imagery: a potential remote sensing tool for range management. *Photogramm. Eng. and Remote Sensing* **51**(6), 675–679.
Fearns, D.C. (1984) Microcomputer systems for satellite image processing. *Earth Orient. Appl. Space Technol.* **4**(4), 247–254.
Forstner, G. (1983) Some urban measurements from Landsat data. *Photogramm. Eng. and Remote Sensing* **49**(12), 1693–1707.
Fusco, L., Frei, U. and Hsu, A. (1985) Thematic mapper: operational activities and sensor performance at ESA/Earthnet. *Photogramm. Eng. and Remote Sensing* **51**(9), 1299–1314.
Goetz, A.F.H., Vane, G., Solomon, J.E. and Rock, B.N. (1985) Imaging spectrometry for earth remote sensing. *Science* **228**, 1147–1153.
Graham, R.W. and Read, R. (1975) Small format aerial photography from microlight platforms. *J. Photogr. Sci.* **32**, 100–109.
Heath, W. (1980) Inexpensive aerial photography for highway engineering and traffic studies. *TRRL Suppl. Rept* **632**, Transport and Road Research Laboratory, Crowthorne, Berks., 24 pp.
Hodgson, R.M., Cady, F.M. and Pairman, D. (1981) A solid state airborne sensing system for remote sensing. *Photogramm. Eng. and Remote Sensing* **47**(2), 177–182.
Isong, M. (1987) An investigation of the geometric fidelity of spaceborne and airborne thematic mapper imagery and its potential application to topographic mapping. *M. App. Sci. Dissertation*, University of Glasgow, 242 pp.
Jensen, H., Graham, L.C., Porcello, J. and Leith, E.N. (1977) Side looking airborne radar. *Scientific American* **237**(6), 84–94.
Kennie, T.J.M., Dale, C.D. and Stove, G.C. (1986) A preliminary assessment of an airborne thermal video frame scanner. *Int. Arch. Photogramm. and Remote Sensing* **26**(7/2), 285–290.
Lawrance, C.J. and Beaven, P.J. (1985) Remote sensing for highway engineering projects in overseas countries. In *Remote Sensing in Civil Engineering,* eds. T.J.M. Kennie and M.C. Matthews, Surrey University Press [Blackie], Glasgow and London, 240–268.
Lawrance, C.J. and Toole, T. (1984) The location, selection and use of calcrete for bituminous road construction in Botswana. *TRRL Rept* **LR 1122**, Transport and Road Research Laboratory, Crowthorne, Berks.
Leberl, F.W., Domik, G. and Kobrick, M. (1985) Mapping with aircraft and satellite radar images. *Photogramm. Record* **11**(66), 647–666.
Leberl, F.W., Domik, G., Raggam, J. and Kobrick, M. (1986) Radar stereomapping techniques and application to SIR-B images of Mt. Shasta. *IEEE Trans. on Geoscience and Remote Sensing* **GE-24**(4), 473–480.
Liu, J.K. (1985) Remote sensing for identifying landslides and for landslide prediction—cases in

Taiwan. *Int. Conf. on Remote Sensing Data Acquisition, Management and Applications*, 223–232.

Lynn, D.W. (1984) Surface material mapping in the English fenlands using airborne multispectral scanner data. *Int. J. Remote Sensing* **5**(4), 699–713.

Matthews, M.C. (1985) Interpretation of aerial photography. In *Remote Sensing in Civil Engineering*, eds. T.J.M. Kennie and M.C. Matthews, Surrey University Press [Blackie], Glasgow and London, 204–239.

Matthews, M.C. and Clayton, C.R.I. (1987) The use of oblique aerial photography to assess the extent and sequence of landslipping at Stag Hill, Guildford, *Proc. 20th Ann. Meeting of the Engineering Group of the Geological Society*, Geological Society, London.

Mason, P.A. and Amos, E.L. (1985) Environmental engineering applications of thermal infrared imagery. In *Remote Sensing in Civil Engineering*, eds. T.J.M. Kennie and M.C. Matthews, Surrey University Press [Blackie], Glasgow and London, 269–288.

Meier, H.-K. (1985) Image quality improvement by forward motion compensation. Paper presented at 40th Photogrammetric Week, Stuttgart, Institute of Photogrammetry, University of Stuttgart, 20 pp.

Meisner, D.E. and Lindstrom, O.M. (1985) Design and operation of a color infrared video system. *Photogramm. Eng. and Remote Sensing* **51**(5), 555–560.

Menuguette, A.A.C. (1985) Evaluation of metric camera photography for mapping and coordinate determination. *Photogramm. Record* **11**(66), 699–710.

Mohamed, M.A. (1977) Photogrammetric analysis and rectification of Landsat MSS and Seasat SAR imageries. Ph.D. Thesis, University of London, 180 pp.

Mountain, L.J. and Garner, J.B. (1981) Semi-automatic analysis of small format photography for traffic control studies of complex intersections. *Photogramm. Record* **10**, 331–342.

NOAA (1978–1986) *Landsat Data User's Notes*, Issues 1–36. National Oceanographic and Atmospheric Administration, Washington DC.

Norman, J.W. and Watson, I. (1975) Detection of subsidence conditions by photogeology. *Eng. Geology* **9**, 359–381.

Overseas Unit (1982) Terrain evaluation and remote sensing for highway engineering in developing countries. *TRRL Suppl. Rept* **690**, Transport and Road Research Laboratory, Crowthorne, Berks., 172 pp.

Petrie, G. (1978) Geometric aspects and cartographic applications of side looking radar. *Proc. 5th Ann. Conf. Remote Sensing Society*, University of Durham, 21 pp.

Petrie, G. (1983) The philosophy of digital and analytical photogrammetric systems. *Proc. 39th Photogrammetric Week, Stuttgart*, Institute for Photogrammetry, University of Stuttgart, 27 pp.

Petrie, G. (1985) Remote sensing for topographic mapping. In *Remote Sensing in Civil Engineering*, eds. T.J.M. Kennie and M.C. Matthews, Surrey University Press [Blackie], Glasgow and London, 119–161.

Petrie, G. (1987) General photogrammetry: an introduction to basic concepts, instrumentation and procedures. *ASPRS Int. Conf. and Workshop on Analytical Instrumentation, Phoenix, Arizona*, 36 pp.

Rudd, R.D. (1974) *Remote Sensing: A Better View*. Duxbury Press, N. Scituate, Mass., 135 pp.

Rye, A.J. and Wright, A. (1985) Segmentation—the extraction of spatial information from radar images. *Int. Conf. on Advance Tech. for Monitoring and Processing Global Environmental Data*, London, 305–317.

Simard, R., Plourde, F. and Toutin, T. (1986) Digital elevation modelling with stereo SIR-B image data. *Int. Arch. Photogramm. and Remote Sensing* **26**(7/2), 161–166.

Singhroy, V. (1988) Case studies on the application of remote sensing data to geotechnical investigations in Ontario. In *Geotechnical Engineering Applications of Remote Sensing and Remote Data Transmission, ASTM Spec. Tech. Publ.* **967**, American Society for Testing and Materials, Philadelphia, 9–45.

Singhroy, V. and Barnett, P. (1984) Locating subsurface mineral aggregate deposits from airborne imagery: a case study in southern Ontario. *Int. Symp. on Remote Sensing for Exploration Geology, Colorado Springs*, 523–539.

Stove, G.C., Kennie, T.J.M. and Harrison, L. (1987) Airborne thermal mapping for winter highway maintenance using the Barr and Stroud IR18 thermal video frame scanner. *Int. J. Remote Sensing* **8**(7), 1077–1084.

Swann, R., McDonald, J.S., Westwell Roper, A., Wood, S. and Laing, W. (1986) The automated extraction of digital terrain models from satellite imagery. *Proc. 20th Int. Symp. on Remote Sensing of the Environment, Nairobi*, 16 pp.

Tait, D.A. (1987) *Small Computers in Image Processing and Mapping*. Remote Sensing Society, London, 103 pp.

Taranik, J.V. (1978) Principles of computer processing of Landsat data for geological applications. *USGS Open File Report* **78-117**, 50 pp.

Tomlins, G.F. (1983) Some considerations in the design of low cost remotely piloted aircraft for civil remote sensing applications. *Canadian Surveyor* **37**, 157–167.

Traynor, C.P. and Jeans, T.G. (1982) The development of a satellite borne earth imaging system for UOSAT. *Radio and Electronic Engineer* **52**(8/9), 398–402.

Trevett, J.W. (Ed.) (1985) *The European SAR 580: Final Report*. ESA/Joint Research Centre (JRC), Ispra Publication.

Trevett, J.W. (1986) *Imaging Radar for Resources Surveys*. Chapman and Hall, London, 313 pp.

Van Roessel, J.N. and de Godoy, R.C. (1974) SLAR mosaics for Project RADAM, *Photogramm. Eng.* **40**(6), 583–595.

Welch, R. (1982) Image quality requirements for mapping from satellite data. In *Proc. ISPRS Commission 1, Symp. on Primary Data Acquisition*, 50–54.

Welch, R. and Panell, C.W. (1982) Comparative resolution of Landsat 3 MSS and RBV image data of China. *Photogramm. Record* **10**(59) 575–586.

Welch, R. (1985) Cartographic potential of SPOT image data. *Photogramm. Eng. and Remote Sensing* **51**(8), 1085–1090.

Welch, R., Jordan, T.R. and Ehlers, M. (1985) Comparative evaluations of the geodetic accuracy and cartographic potential of Landsat 4 and 5 thematic mapper data. *Photogramm. Eng. and Remote Sensing* **51**(11), 1799–1812.

Wolf, P.R. (1983) *Elements of Photogrammetry*. 2nd edn., McGraw-Hill, 628 pp.

Wong, K.W. (1975) Geometric and cartographic accuracy of ERTS-1 imagery. *Photogramm. Eng.* **41**(5), 621–635.

7 Analogue, analytical and digital photogrammetric systems applied to aerial mapping

G. PETRIE

7.1 Introduction: the role of aerial photogrammetric methods in engineering surveys

It is very difficult to conceive of the planning without the aid of photogrammetry of any civil engineering projects covering considerable areas of terrain—those concerned with the construction of roads, railways, dams, reservoirs, airfields, pipelines, large housing developments and opencast mining. Certainly there is no other method of survey which will generate the required data about the terrain surface in the form of accurate plans, maps, photomaps, elevation models, profiles and contours in a timely and economic manner as is required for the planning of all large civil engineering projects nowadays.

However, photogrammetric methods also have their limitations. For example, they have little or no part to play in the work of setting out roads, buildings, bridges, tunnels and other structures in the field and of controlling the construction of these structures. Nor does it make any sense to use aerial photogrammetric methods for the survey of an individual building site—the cost and complexity of taking suitable aerial photography, establishing ground control points, carrying out stereoplotting in a photogrammetric instrument and verifying and completing the plotted detail in the field would make the whole process uneconomic as compared with field survey methods. But as soon as the area to be surveyed or mapped becomes larger than can be covered by one or two pairs of aerial photographs, the situation changes drastically. There is no way in which ground survey methods, whether employing (i) mechanical, optical or electronic methods of measuring distance, (ii) optical or electronic methods of measuring angles, or (iii) some combination of these methods, can compete with photogrammetric methods of survey and mapping. This is especially the case in rough or rugged terrain where there is no possibility of ground survey procedures, however automated, producing detailed data about the terrain surface within a reasonable time frame.

In addition, photogrammetry has for many years offered the possibility of capturing or generating terrain data in computer-compatible form for the purposes of digital mapping, terrain modelling and computer-aided design. Furthermore, the advent of automated or semi-automated methods of collecting highly accurate terrain elevation data in pre-programmed patterns using stereoplotting instruments under computer control gives a method of data collection for extensive areas for which there is no comparable rival or alternative.

Yet another unique possibility offered by photogrammetry is the generation of photomaps, photomosaics and orthophotographs which offer an attractive alternative to the traditional form of line map. These have been adopted widely in North America and Western Europe (though not in the UK), especially in those areas, such as deserts and semi-arid land, swamps and wetlands, woodlands and forests, wilderness areas and mountainous terrain, which are very difficult to represent using the lines and symbols of the traditional type of topographic map. Thus photogrammetry can generate a unique product for those areas which are especially sensitive from the environmental point of view and often of considerable economic or scientific importance. But the photomap approach has also been adopted still more widely as the basis for state-wide or national series. Such series exist in Sweden, Belgium and certain states in West Germany at medium (1:5000 to 1:20 000) scales, and in the USA, Australia, South Africa, etc. for the small-scale coverage of large areas which are largely undeveloped and contain little cultural detail. The popularity of such series, where they are available, is a tribute to the information which they convey to engineers, planners, field scientists and administrators requiring data about the land and its topography, cover and use.

7.2 Stereoplotting instruments

Although there are a large number of alternative methods available to photogrammetrists—for instance the radial triangulation and single photo methods of plotting planimetric detail, and the use of a stereoscope and parallax bar for the determination of heights—it is safe to say that 99% of aerial photogrammetric work is carried out using stereoplotting instruments. This may appear surprising to non-photogrammetrists, since these alternative methods often require comparatively simple and inexpensive equipment, which would be attractive to many users who may not have the financial resources to invest in a stereoplotting instrument. However, these alternative methods almost always involve approximations in their solution, resulting in a less accurate and therefore less acceptable result. Furthermore, they almost always involve very laborious and time-consuming procedures which are, in the end, often uneconomic to implement. Thus the stereoplotting instrument is used in almost all types of mapping work.

The approach taken in all stereoplotting instruments involves the use of a pair of overlapping photographs to form a reduced-scale three-dimensional model of the terrain within the instrument and to measure this exact model of the terrain very accurately to produce the required map or coordinate information instead of measuring the terrain directly using field surveying instruments. The type of model of the terrain which may be formed in the instrument can be either optical or mechanical in nature. Using the language of an engineer, one can say that the photogrammetric procedures carried out using a stereoplotting instrument based on such a model constitute an analogue process where the object (the terrain) being measured is simulated by another physical system (the optical or mechanical model) which is measured instead. This is the origin of the descriptive term 'analogue', where the *analogue* type of stereoplotting instrument is differentiated from an *analytical* type of photogrammetric instrument in which the solution is purely mathematical and the model is wholly numerical.

7.3 Analogue photogrammetric instrumentation

Analogue stereoplotting instruments have been built in profusion by many manufacturers. However, over the last five years, the manufacture of these instruments has declined to a marked extent. This appears to be due partly to the drop in overall demand, since most survey and mapping organizations in developed countries already have a very large installed base of such instruments, while those in developing countries which have a large and unsatisfied demand for such equipment cannot afford to buy it. However, the drop in the production of analogue stereoplotting instruments is also due to the growth in the use of analytical instruments, especially in the more advanced countries. In spite of this changing situation, the analogue type of instrument is still in very widespread use, often updated by the addition of digital components such as encoders, micro- and minicomputers controlling interactive graphics screens and terminals, and microprocessor-controlled plotting tables driven by stepping motors. Thus, at present, analogue stereoplotting instrumentation still executes the vast majority of maps and surveys carried out for engineering purposes using aerial photogrammetry.

The use of *optical models* (in so-called optical projection instruments) has been very common in the past in North America. However, the alternative mechanical projection approach resulting in the use of *mechanical models* has been adopted by most of the leading European manufacturers of photogrammetric equipment. While the optical projection instruments are simpler and less expensive to purchase, to operate and to maintain, in general the mechanical projection instruments are more flexible and accurate, even if they are also more complex in their design and more expensive to purchase. Thus mechanical projection instruments are still in widespread use in Europe and have also secured a good market in North America.

7.3.1 Optical projection instruments

Basically, an instrument of this type comprises two identical projectors with the same geometric characteristics (format size, focal length, angular coverage) as the camera which took the photography (Fig. 7.1). Positive copies (film or glass diapositives) of the

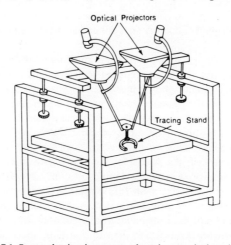

Figure 7.1 Stereoplotting instrument based on optical projection.

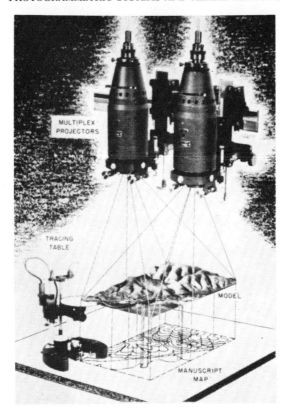

Figure 7.2 Measurement of an optical model to plot a contoured map.

pair of overlapping photographs which cover the area of interest are placed in the two projectors and are illuminated from above so that the images are projected downwards into the space below the projectors. Through the procedure of *relative orientation* by which the photogrammetrist rotates and shifts the two projectors in an ordered manner, the two photographs are set in the same relationship to one another as they occupied in the air at the time of exposure. When this has been done successfully, the two bundles of corresponding rays from each projector will meet in the space below the projectors and so form an optical model of the terrain. This optical model can easily be viewed stereoscopically (in three dimensions) by the user, and measurements can readily be carried out on this model.

The actual measurements of this correct stereomodel of the terrain are carried out accurately using a free-moving measuring device which has a platen with a white surface and a small measuring mark at its centre (Fig. 7.2). This device can be used to plot a map from the stereomodel, the observer following the edge of the road, field boundary, forest or other feature precisely with the measuring mark, the correct position being given by a plotting pencil located directly below the mark. If the ground rises or falls, the measuring mark is continuously adjusted by the observer to the correct height in the model as the plotting of the feature continues. This means that continuous

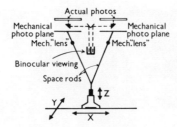

Figure 7.3 Stereoplotting instrument based on mechanical projection.

measurements of the heights of each ground feature are being made and these relate directly to the actual height present in the terrain.

Alternatively, the measuring mark can be set to a predetermined height and the observer can keep the mark continuously in contact with the optical model of the terrain at that height, so producing a plot of that particular *contour* on the map sheet. Each required contour can be measured in the same manner, the measuring mark being set to the corresponding height in the model in each case.

The prerequisite for these operations of plotting out the map detail and contours is that the orientation of the optical model be such that it is in the correct relationship with the terrain system—an operation called *absolute orientation* by photogrammetrists. This entails the scaling of the model so that the positions of certain ground control points, when plotted from the model, fit the corresponding positions of these points already plotted on the map at the desired scale. In addition, the model is rotated as a whole so that the correct heights will be measured in the model. These operations must be carried out prior to the commencement of the detailed measurements required to produce the map.

7.3.2 *Mechanical projection instruments*

The basic concept of mechanical projection is to duplicate mechanically all the optical components of the optical projection type of instruments. In the latter, there were photographs, projection lenses, projected rays and an optical model. In the equivalent mechanical projection instruments (Fig. 7.3), each photograph is duplicated mechanically (as the *mechanical photo plane*); the projected rays are duplicated mechanically by their equivalents (known as space-rods); the projection lenses are duplicated mechanically as the centre of a free-moving gimbal system; and the specific point in the optical model where a pair of intersecting optical rays would meet is the position where the corresponding pair of space-rods meet and form the *mechanical model point*.

Thus the basic measuring system in this alternative type of instrument is now wholly mechanical instead of being purely optical as in the previous type of instrument. Thus the terms 'mechanical projection instrument' and 'mechanical model' apply to it. However, it must be realized that these terms apply only to the projection or measuring system. It is still necessary for the observer to view the photographs optically and stereoscopically if three-dimensional measurements are to take place. This is achieved by a mechanical linkage which ensures an exact correspondence between what is being measured (using the mechanical projection system) and what is being viewed in three dimensions by the operator using the instrument's optical system. The latter is in principle a form of mirror stereoscope, by which the basic requirement, that the

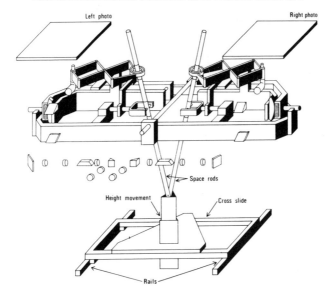

Figure 7.4 Mechanical projection instrument with cross-slide and rails.

observer's left eye see the left photo only and his right eye, the right photo only, is actually achieved. In this way, the observer or operator of the instrument can view and measure the model of the terrain stereoscopically, that is, in three dimensions, in much the same manner as was done in the optical projection type instrument.

The actual measurement of the mechanical model may be carried out by the operator using a free-moving measuring device of the same basic design as that employed in the optical projection instrument. Again, this may have a pencil attached to it to allow direct plotting of the detail required for a topographic map including the contours. As before, since the mechanical model being measured is a correctly scaled representation of the terrain, the detail plotted and the heights measured in this model relate directly to the actual positions and heights present in the terrain.

An alternative to the free-moving measuring device moving over the map sheet is to mount the intersection point (mechanical model point) of the space rods in a *three-dimensional coordinate measuring system* which comprises two rails and a cross-slide to allow the planimetric (X/Y) movement to take place, plus a telescopic movement in the vertical (Z) direction to allow heights to be measured (Fig. 7.4). This whole system can be moved freely by hand by the operator or it can be driven mechanically using handwheels and lead screws to any point in the model which has to be measured (Fig. 7.5). By coupling drive shafts and gears to the lead screws, a coordinatograph can be attached to the stereoplotting instrument, so permitting the final map to be plotted at a scale different to that of the actual model in the instrument.

7.3.3 *Digital data acquisition using analogue stereoplotting instruments*
Besides the graphical plotting discussed in the previous section, it is also possible to record all the measurements made in a stereoplotting instrument in a numerical digital form for subsequent processing in a computer-based mapping or information system or

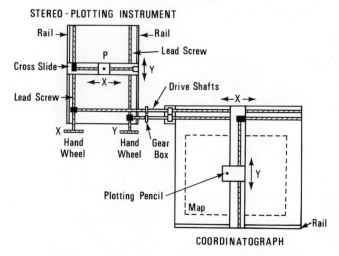

Figure 7.5 Coordinatograph attached to stereoplotting machine.

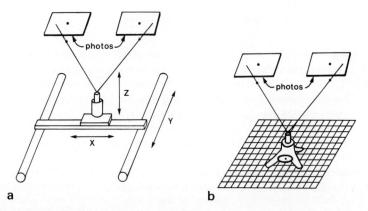

Figure 7.6 (a) Linear encoders attached to cross-slide system. (b) Tablet digitizer mounted on base surface.

in a computer-aided design (CAD) system. This approach is one which is being implemented to an ever-increasing extent, and many existing analogue stereoplotting instruments have been converted for the purpose.

All the three-dimensional (X, Y and Z) measurements made in a stereoplotting instrument may be *digitized* by mounting the measuring mark on a suitable cross-slide system and encoding each of the axes individually using linear or rotary encoders (Fig. 7.6). Alternatively, a tablet digitizer may be mounted on the base surface of the instrument to give positional (X, Y) information while the terrain heights (Z) are obtained in digital form by adding a linear or rotary encoder to the height measuring device. The former arrangement is preferred, since the use of cross-slides can easily allow measuring resolutions or accuracies of 5–10 μm, whereas the tablet digitizer is limited to 25–50 μm at best.

PHOTOGRAMMETRIC SYSTEMS TO AERIAL MAPPING 245

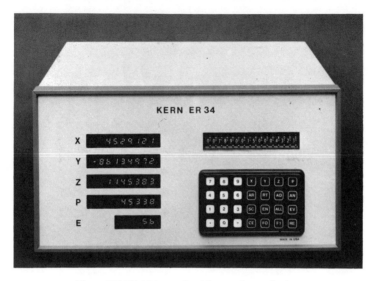

Figure 7.7 Digitizing unit with coordinate display.

A variety of *digitizing units* can be attached to the encoders mounted on an analogue type of stereoplotting instrument. These provide coordinate display, allow the selection of different measuring modes (such as point or stream digitizing) and allow headers, feature codes and descriptions to be entered into the unit. Trying to sort out and evaluate their differing capabilities can be quite bewildering to the non-specialist.

Hardware-based, firmware-based and software-based units are all used (Petrie, 1981). Hard-wired and firmware-based units are essentially stand-alone electronic boxes (Fig. 7.7) which decode and count the signals coming from the linear or rotary encoders mounted on the stereoplotting instrument and convert them to X, Y and Z model coordinate values which can be viewed on an LED or LCD display and recorded usually on magnetic tape, cassettes or cartridges. Software-based units (Petrie and Adam, 1980) are, as the name suggests, based on the use of an on-line computer which not only performs the data recording but can give prompts and assistance to the operator during the setting up of the stereomodel, as well as carrying out checks on the measured coordinate data. Some of these software-based units, such as the Kern PC-PRO, utilize a dedicated microcomputer such as an IBM-PC or clone attached on-line to the instrument; others use a large time-sharing computer, as in the system shown in Fig. 7.8, based on a large DEC PDP-11 minicomputer.

Essentially, all of these systems are carrying out the 'blind' digitizing technique also used widely for the digitizing of the detail and contours shown on existing maps. Checking must then be carried out using hard-copy plots generated at the end of a digitizing session. However, in a few cases, such as in the Kern DC-2B system (Fig. 7.9) or the Wild RAP system, a computer-driven plotting/drafting table has been attached on-line which can plot out all the terrain detail and the elevation or contour data measured by the photogrammetrist for his inspection and possible correction during the actual measuring operations.

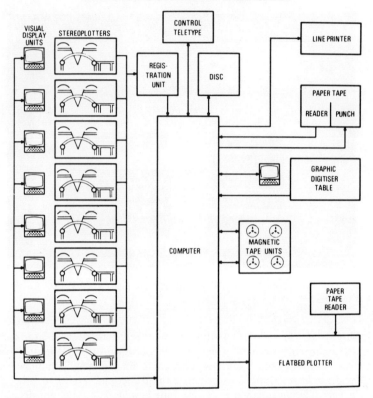

Figure 7.8 Stereoplotting instruments connected to time-sharing minicomputer.

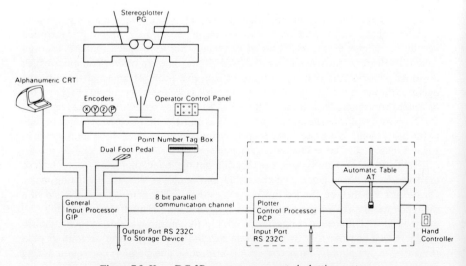

Figure 7.9 Kern DC-2B computer-supported plotting system.

Figure 7.10 Zeiss Oberkochen Planimat with intergraph graphics display and interactive tablet digitizer (in foreground) and superimposition unit (mounted on the left side of the stereoplotting instrument).

7.3.4 Interactive graphics workstations attached to analogue stereoplotting instruments

The direct attachment of graphics work-stations to analogue stereoplotting instruments to allow direct checking and interactive editing of the measured data is a current trend. Some photogrammetric manufacturers have developed their own integrated systems, such as Kern's MAPS 200 and 300 units based on the DEC PDP-11 series of computers, and Imlac or Tektronix graphics terminals which are attached to the firm's well-known PG-2 stereoplotting instruments (Newby and Walker, 1986).

However, many of these integrated systems have been supplied by Intergraph, working in collaboration with leading European manufacturers of stereoplotting instruments such as Zeiss Oberkochen and Wild. The Zeiss Oberkochen/Intergraph system is especially interesting in that the plotted contours are displayed on a graphics display mounted on the side of the stereoplotting instrument (Fig. 7.10). The graphical image of plotted or measured information displayed on the screen can also be transformed in accordance with the exterior orientation of the left photo and the terrain elevation and displayed on an auxiliary screen mounted on the left side of the stereoplotting instrument. It is then projected via a semi-reflecting mirror into the left optical channel of the stereoplotting instrument so that the operator may see exactly what has been measured or plotted superimposed on the stereomodel. He may then correct any errors or omissions as required. As will be seen later, if an analytical plotter is available, this *superimposition* technique can be extended further to generate stereosuperimposition of plotted detail such as contours on the stereomodel.

7.4 Analogue instrumentation for orthophotograph production

An alternative approach to photogrammetric mapping has become important in recent years: the production of orthophotographs and orthophotomaps as a supplement or an alternative to the traditional type of line map. The orthophotograph is a photographic image of the terrain from which all the tilt and relief displacements inherent in a single aerial or space photograph have been removed, i.e. it has the same geometric characteristics as the line map. The main way in which this is produced is again via the use of an analogue stereo-plotting instrument of the same basic design as has been described above (Petrie, 1977).

7.4.1 Optical projection instruments

In devices used to produce orthophotographs, the overlapping photographs are set up in the stereoplotting instrument (thus removing the effects of tilt) and the optical model of the terrain is again formed and oriented and can be viewed and measured stereoscopically in 3-D. However, the measuring mark used to locate the point to be plotted in the stereoplotting instrument when used to construct maps is replaced by a narrow slit, through which only one of the projected images is admitted (using filters) to be recorded on a sheet of photographic film (Fig. 7.11a). The plotting procedure is also different. Instead of continuously following individual features—roads, field boundaries, streams—as is done using a measuring mark, the slit is made to traverse the stereomodel systematically in a series of parallel profiles (Fig. 7.11b), forming a raster pattern. Along each profile, the slit is continually raised or lowered to remain in contact with the model as the ground rises or falls. All the detail present in the stereomodel is then recorded orthogonally on the film located on the datum surface below the model. When the whole model has been scanned, the film is developed as a negative orthophotograph from which positive film or paper copies may be made.

More common nowadays is an arrangement by which an additional *rectifying projector* (or orthoprojector) is attached on-line to a standard analogue stereoplotting instrument (Fig. 7.12). This projector exactly duplicates one of the projectors of the instrument, i.e. it contains the same photograph (diapositive) and is equipped with rotations which allow it to be tilted to the same values as are present in the equivalent projector of the stereoplotting instrument. Systematic scanning of the stereomodel in a raster pattern takes place as before, but exposure of the orthophotograph is made using the additional rectifying projector in which all the X, Y, Z movements made using the measuring device in the stereoplotting instrument are duplicated by the scanning slit in the additional rectifying projector.

However, a still more important point is that this arrangement of an additional rectifying projector allows the production of orthophotographs in an *off-line* mode. This involves the use of a storage device on which the measured profiles can be stored either by scribing them directly on a template or by recording them digitally on magnetic tape (Fig. 7.13). This record is later used to control the operation of the additional rectifying projector in producing the orthophotograph.

The off-line operation offers considerable advantages both economically and from the point of view of the quality of the resulting orthophotograph:

(i) Erroneous measurements can be eliminated and the profile remeasured, whereas in the on-line type of operation, if the operator makes a mistake during the scanning/measuring operation, the orthophotograph has already been exposed

PHOTOGRAMMETRIC SYSTEMS AND AERIAL MAPPING 249

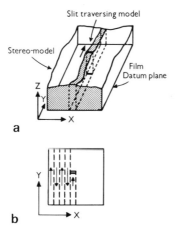

Figure 7.11 (*a*) Slit traversing optical stereomodel to produce an orthophotograph. (*b*) Raster scan pattern followed by the scanning slit.

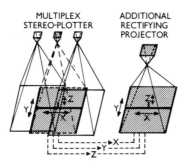

Figure 7.12 Additional rectifying projector attached to a stereoplotting machine.

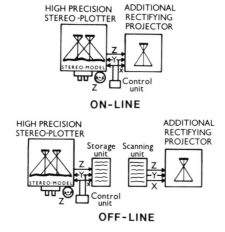

Figure 7.13 On-line and off-line modes of operation of an additional rectifying projector.

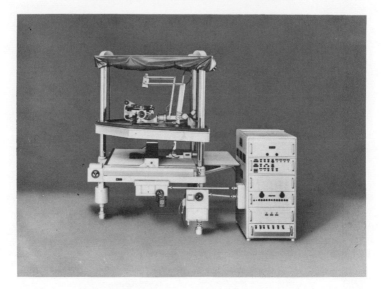

Figure 7.14 Zeiss Oberkochen GZ-1 orthoprojector.

and the affected area appears as a mismatch or anomaly in the photographic image which is readily discerned by the user.

(ii) During an on-line operation, the orthophotoprojector is idle 50% of the time—while diapositives are being changed, during orientation of the stereomodel, etc. With an off-line operation, the orthoprojector can be kept continuously in operation using the measured profiles from more than one stereoplotting instrument. For example, in the Swedish National Survey, a single orthoprojector can deal with the profiles measured in three or four high-precision stereoplotters. This is possible because the orthoprojector can be run at a higher speed off-line than can be achieved on-line when the operator is actually making the profile measurements, particularly in difficult terrain.

A well-known example of an additional rectifying projector of the type discussed above is the Zeiss Oberkochen GZ-1 Orthoprojector (Fig. 7.14) which has been used extensively for orthophotograph production in both the on-line and off-line modes.

7.4.2 *Non-projection methods of orthophotograph production*
All of the devices discussed in section 7.4.1 are of the *optical projection* type. Despite their widespread use, they are burdened with the same defect as stereoplotting instruments based on this principle—a lack of flexibility in handling different types of aerial photography. If an aerial camera lens of different focal length or distortion characteristics has been employed to take the photography, it is necessary to change the projector to one with a corresponding focal length or matching distortion pattern. It is extremely expensive to keep a range of projectors which will handle all the types of aerial photography likely to be encountered.

With stereoplotting instruments, this shortcoming has been overcome by the development of the instruments based on *mechanical projection*. As previously

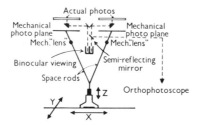

Figure 7.15 Optical transfer with the image of the right-hand photograph in a mechanical projection instrument tapped off or transferred for rectification.

discussed in section 7.3.2, a mechanical model is formed in which the pair of corresponding optical rays for the point being measured is replaced by a pair of metal rods or rulers. However, although such instruments can be and are used as the measuring stereoplotter in an on-line or off-line system using an additional rectifying projector, the use of mechanical projection does not lend itself to the direct on-line production of an orthophotograph within the instrument itself, since there is no optical projection.

To meet these difficulties and to allow the direct (on-line) use of analogue mechanical projection and analytical stereoplotters for orthophotograph production, two other methods of producing orthophotographs have been developed. These are *optical transfer* and *electronic transfer*.

7.4.2.1 *Optical transfer instruments.* As already described in section 7.3.2, in mechanical projection instruments the rays which occurred at the times of exposure of the overlapping pair of photographs, are re-projected as mechanical space-rods which act as the measuring elements of the instrument. Coupled to the rods are microscopes which view or scan the actual photographs and transmit the images via separate optical trains to the observer's eye. At some point in this train, the image can be tapped off or transferred optically via a semi-reflecting mirror (Fig. 7.15) to produce the orthophotograph, hence the term optical transfer.

However, there is quite a different set of problems to solve in rectifying this image (removing relief and tilt displacement and bringing it to a constant scale) using optical transfer as compared with those inherent in the optical projection method. First, since there is no optical re-projection of the rays with a variation in the projected distance corresponding to variations in terrain height, the transferred image has to be continuously changed in scale. Second, because of the frontal viewing of the photograph instead of projected viewing of a stereomodel, the effects of tilt are also unrectified both in scale and convergence.

In the optical transfer device, which is attached on-line to the stereoplotter, the scale rectification is performed using a pair of prisms which shorten or lengthen the image path (Fig. 7.16) while the image rotation to correct for convergence is produced by a Dove prism, both actions being carried out under the control of a small analogue computer or a microprocessor. The image is recorded on a sheet of film wrapped around a drum which can be rotated and shifted in accordance with the scanning movements. To implement this solution, the stereomodel is scanned in the stereoplotting instrument in a similar raster pattern to that used in optical projection devices.

ENGINEERING SURVEYING TECHNOLOGY

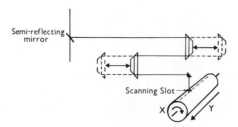

Figure 7.16 Scale rectification by variation in the optical path length; image rotation to correct for convergence.

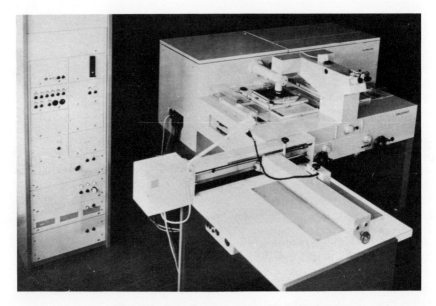

Figure 7.17 Zeiss Jena Topocart with Orthophot.

Drop-line profiles (section 7.4.3) can be recorded, either graphically on the stereoplotter's coordinatograph, or digitally on magnetic media.

A well-known example of this type of optical transfer device is the Orthophot (Fig. 7.17) produced by Zeiss Jena as an attachment to its Topocart stereoplotting instrument. Other similar units have been produced by Wild (the PPO-8) and Galileo (the Orthosimplex).

7.4.2.2 Electronic transfer instruments. This is yet another method of orthophotograph production, in principle similar to that of optical transfer. It involves the scanning of a pair of overlapping aerial photographs in an analogue stereoplotting instrument followed by their rectification (scaling and shaping) by digital or electronic analogue means and finally the printing of the orthophotograph using a cathode ray tube (CRT). While in principle it is similar to optical transfer, all the optical components and procedures of the latter method are replaced by their electronic equivalents. The model

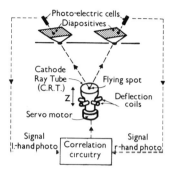

Figure 7.18 Correlator as applied in an optical projection instrument.

is invariably applied in conjunction with automated method of scanning and measuring a stereomodel.

The device which is used for these tasks is the so-called *correlator*, the best known being the Stereomat device originally conceived by G.L. Hobrough in Canada in 1959 and applied later to a variety of instruments, notably the Wild B8 and A2000 Stereomats. The design of the correlator is shown in Fig. 7.18 as applied to an optical projection type of plotter. A CRT replaces the measuring mark of the stereoplotter, and on its face is generated a spot of light which can be made to scan across the surface of the tube. This flying spot is then projected upwards through the objective lenses of the two projectors of the stereoplotting instrument and passes through the two diapositives. The amount of light reaching each of the monitoring photoelectric cells (which measure light intensity) depends on the density of the image detail present on the diapositive. Since the spot scans each of the two corresponding images, separate electrical signals are generated for each photograph. These signals are compared in the correlator to give measurements of parallax. The parallax signals in the y-direction are used to control small motors which can tilt the projectors to carry out a relative orientation and thus form a stereomodel. As is well known, the parallax signals in the x-direction give a measurement of height and can be used to operate the Z-servo motor so that the CRT is always set to the correct height in the stereomodel during scanning or profiling for production of the orthophotographs.

As applied in the Wild B8-Stereomat, the Wild B8 stereoplotting instrument is used as the basis of the system, and uses a mechanical projection system, so a separate CRT has to be provided for the scanning of each photograph (Fig. 7.19a). However, the actual process of correlation is still carried out as described above. For production of the orthophotograph, profiling is carried out in the same way as with optical projection and optical transfer devices. A series of parallel profile scans is made in the Y-direction with a step-over of a predetermined amount in the X-direction until the whole stereomodel is scanned (Fig. 7.19b). The orthophotograph is printed by taking the signal from one of the photoelectric cells and applying it to control the light output on the face of a third (printing) CRT (Fig. 7.20). The printing CRT generates the same scan pattern as that on the scanning CRT, but the pattern can be modified slightly so as to provide the corrections necessary for scale, terrain slope and tilt. The printing CRT in the B8-Stereomat is mounted on a cross-slide system in a light-tight box located behind the B8 (Fig. 7.21). The CRT image is projected through a lens to be imaged on a sheet of

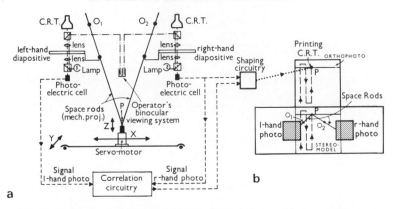

Figure 7.19 (*a*) Correlator as applied in a mechanical projection instrument; (*b*) raster scan pattern in both the mechanical projection instrument and the orthophoto printing unit located behind the instrument.

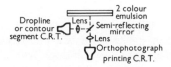

Figure 7.20 Printing CRT for orthophotograph exposure.

Figure 7.21 Wild B8 Stereomat printing unit.

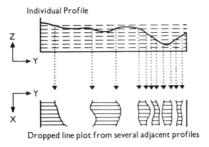

Figure 7.22 Drop line plot.

film. A fourth CRT image is mounted alongside the third to produce a drop-line plot or, more exactly, a *contour segment plot*. The film which is placed in the light-tight box actually has two emulsions: one is blue-sensitive and is used to record the orthophotograph image; the other is red-sensitive, to record the drop-line or contour segment. The two superimposed images are separated by conventional photomechanical processing.

Besides the Wild B-8 Stereomat, a number of similar systems have been produced by other manufacturers, such as the Itek EC-5 correlator attached to the Zeiss Oberkochen Planimat stereoplotter and the Zeiss Jena Topomat system which is essentially a correlator-equipped, automated version of the Topocart/Orthophot combination mentioned in section 7.4.2.1.

7.4.3 *Heights and contours obtained during the production of orthophotographs*
It is also possible to acquire height information during the systematic scanning of the stereomodel for production of the orthophotograph, so eliminating the need to measure the heights and contours in an additional measuring operation. Thus, during each scan, the operator or the correlator is keeping the measuring mark or the centre of the slit continuously at the height of the stereomodel (Fig. 7.11). This *height profile information* can be recorded in any one of several different ways:

(i) Graphically as an inked or scribed trace
(ii) Graphically as a 'drop-line plot'
(iii) Digitally on magnetic tape, cartridge or cassette.

The drop-line plot records the X-Y position where the profile cuts the required contour levels. The name derives from the form of the record, which usually comprises a series of lines and spaces, the end of each being the contour position along a profile. Taking a series of adjacent profiles, the ends of the corresponding lines and spaces can be joined to form the appropriate contours (Fig. 7.22).

7.5 Analytical photogrammetric instrumentation

As already mentioned, the alternative and increasingly important type of photogrammetric instrumentation is that based on the use of analytical photogrammetric procedures in which the optical or mechanical models of the analogue approach are replaced by purely numerical models based on analytical/mathematical solutions which nowadays are invariably implemented in suitably programmed computers. From the instrumental point of view, there are several alternative ways in which these

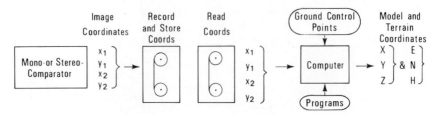

Figure 7.23 Comparator and off-line solution.

solutions may be implemented in practice. These may be classified into three main groups:

(i) The use of a comparator as the measuring device and the execution of the analytical photogrammetric solution as an off-line process carried out later in a computer
(ii) The use of a comparator but with the numerical computational solution executed as an on-line process and
(iii) The construction and operation of an analytical plotter in which the computer is totally integrated into the design of the instrument and the numerical solution is always an on-line process, executed in realtime and resulting in a continuously oriented stereomodel.

(i) *Comparator and off-line solution*
The basic arrangement is shown in Fig. 7.23. The measuring device may be a monocomparator in which the individual photographs comprising a stereopair are measured separately. The x, y image coordinates of each point on the first photograph are measured in the instrument, encoded using digitizers and then recorded in computer-compatible form. When the corresponding points on the second photograph of the stereopair have been measured and recorded, the two sets of measured coordinate data can then be merged together to implement the analytical photogrammetric solution. It will be seen that if a stereopair is being measured and a stereomodel formed, the use of a monocomparator as the measuring device always results in an off-line computational solution since the measurement of the points whose position and height have to be determined takes place separately and sequentially on the individual photographs of the stereopair.

Alternatively, the measuring instrument can be a stereocomparator, in which stereoviewing is possible and the measurement of both photographs takes place simultaneously. The image coordinates—in this case, x_1, y_1 for the left photograph and x_2, y_2 for the right photograph—are again stored and recorded. Later, the numerical computational solution is implemented to determine the terrain coordinates of all the points measured in the stereocomparator.

(ii) *Comparator and on-line solution*
In this arrangement, the comparator is attached on-line to a computer (Fig. 7.24). The basic measuring process and computational procedure is the same as that described for the off-line solution, but obviously this arrangement allows the model to be formed and the terrain coordinates to be calculated as soon as the minimum number of points needed to implement the analytical solution has been measured. Once this has been

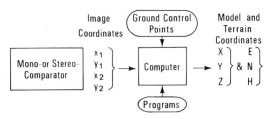

Figure 7.24 Comparator and on-line solution.

achieved, then the measurement of any other additional points in the comparator will of course result in the computation of their corresponding terrain coordinates.

While a monocomparator may be attached directly to the computer to implement this solution, there is little advantage in doing so if a stereomodel has to be measured, since the plates would have to be changed before the second set of measurements can be obtained and the computational solution take place. A development of this on-line approach is the digital monoplotter, in which a monocomparator, or even a lower-cost tablet digitizer, is attached on-line to the computer in which a file of existing digital terrain model (DTM) coordinate data is held. The orientation of the photograph is executed analytically by space resection and the continuous plotting of detail on the photograph may then be carried out monocularly, for instance for map revision purposes.

If a stereomodel as distinct from a single photograph has to be measured using a comparator in conjunction with an on-line computational solution, it is almost invariably implemented using a stereocomparator where the simultaneous measurement of both photographs comprising the stereopair can take place. Quite a number of stereocomparator on-line computer combinations exist in practice—in many cases, the attachment of the computer to the stereocomparator coordinate output device and the writing of the necessary software has been carried out by the users themselves.

A direct development of this approach is the so-called image space plotter, in which the computed terrain coordinates are plotted out graphically as points or lines as soon as they have been determined. Using the language of the engineer, this procedure may be characterized or described as being an on-line, open-loop solution. In contrast to the third approach of the analytical plotter, there is no oriented stereomodel, nor is there a need to implement the formation and solution of the mathematical model in real time.

(iii) Analytical plotter
The basic arrangement of an analytical plotter is shown as a block diagram in Fig. 7.25. In this type of instrument, the computer will always be attached on-line to the measuring elements of the instrument, which, in principle, will be similar to a stereocomparator. However, the analytical plotter will also feature a closed-loop system in which the computer provides a real-time solution of the analytical photogrammetric equations (i.e. it forms the necessary mathematical and numerical model), and issues control signals to motors and driving elements which also move the plates to the required positions in real time. The result is that an oriented stereomodel is formed and continuously maintained, and the photogrammetrist can measure the planimetric detail and carry out contouring in exactly the same manner as in an analogue type of stereoplotting instrument.

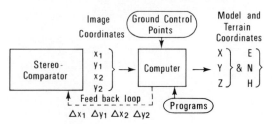

Figure 7.25 Analytical plotter.

As will be seen later, two different computational solutions are possible, based on the use of image or object coordinates respectively. The decision as to which of these solutions is implemented has a profound effect on the design of the analytical plotter and its eventual capabilities (Helava, 1980). In turn, these affect, in a fundamental manner, the provision of the software required to implement the analytical plotter concept. This includes:

(i) The *real-time program* which computes image or object point coordinates from the measured input data
(ii) The *orientation programs* which carry out numerically and mathematically the same or similar inner, relative and absolute orientation procedures to those which are implemented mechanically or optically in analogue instruments
(iii) The *application programs* which perform the digital data acquisition and processing for any one of several operations: mapping; terrain modelling; aerial triangulation; the control of automatic image correlation; the processing of measurements of deflection, deformation or dimensions in non-topographic photogrammetry; etc.

7.5.1 *Algorithms for use with analytical photogrammetric instrumentation*
From the algorithmic point of view, there are two basic approaches to implementing analytical photogrammetric procedures—the first involving the measurement and input of image coordinates as the raw data to the computational process, and the second based on the use of object (model or terrain) coordinates in a corresponding but different computational process (Helava, 1980).

7.5.1.1 *Image coordinates primary.* The first approach is one in which the inputs are the x, y coordinates of points measured on the photographs, i.e. the comparator coordinates, and the outputs are the X, Y and Z model coordinates of these points. In turn, these model coordinates of each point can be transformed into the corresponding easting (E), northing (N) and height (H) coordinates of the terrain system. This approach is that followed when the points have been measured in a comparator, whether a monocomparator in which each photograph is measured individually or a stereocomparator in which both photographs constituting a stereopair are measured simultaneously. It is also the approach implemented in certain types of analytical plotter.

The detailed mathematical basis of this approach will be discussed in the next chapter. In practice, one or other of two alternative solutions may be adopted within this general approach. In the first, the coordinates of each camera station and the tilts present on each photograph are determined using the method of *space resection*.

The measured x, y image coordinates and the known E, N and H coordinate values of the ground control points required for their computation are linked together by the so-called *collinearity equations*. Once the projection centre coordinates and tilts of each photograph have been determined using these equations, the terrain coordinates of all other points measured on the stereopair may be computed using the *space intersection technique*, again based on the collinearity equations.

The alternative, mathematical solution involves the so-called *coplanarity approach*. The corresponding x, y image coordinates measured for common points in the stereopair of photographs are used to generate parallaxes from which the tilts of each photograph may be calculated: i.e. an analytical relative orientation is achieved. Once these tilt values have been determined, it is then possible to compute the values of the X, Y and Z model coordinates from the measured image coordinates (x, y) of any point, again using specific forms of the collinearity equations. The computed X, Y and Z model coordinates are then converted by a simple three-dimensional linear conformal or affine transformation into the corresponding E, N and H terrain coordinate values.

7.5.1.2 *Object coordinates primary.* The second approach is the inverse of the first in which the inputs to the analytical photogrammetric solution are the X, Y and Z model coordinates of the measured points. From these, the corresponding image coordinates (x, y) of each point may be computed for each of the component photographs of the stereopair. Simultaneously the X, Y and Z model coordinates of each measured point may also be transformed into the corresponding terrain coordinate values (E, N and H). The use of object coordinates (model or terrain coordinates) as the primary input to the solution is perhaps less obvious to the non-expert, but in fact it is the more straightforward and easier solution to implement in analytical plotters, especially if there is a requirement to produce contours or to generate a digital terrain model in which the measured points have to be located at predetermined positions in the terrain coordinate system under computer control. Thus the majority of analytical plotters follow the 'object coordinates primary' approach.

In mathematical terms, this approach is once again implemented using the collinearity equations with the terms rearranged so that the input values to the computational process are the measured or given object coordinate values (i.e. X, Y, Z or E, N, H coordinates) and the output values are the corresponding x, y image or plate coordinates. These latter values are the positions to which the plates (photographs) are being driven to by the motors and feedback control mechanisms of the analytical plotter to maintain an oriented, parallax-free stereomodel in which the photogrammetrist can carry out the stereoplotting of line detail or the measurement of individual elevation points or of continuous contours or height profiles.

7.5.2 *Comparators, monoplotters and image space plotters*

These instruments comprise those which fall into groups (i) and (ii) as defined in the introductory section 7.5—they are the analytically-based instruments which have no oriented stereomodel. Thus measurement is carried out on a point-by-point basis, with either off-line or on-line computational processing of the measured image coordinate data.

The most basic measuring instrument used in photogrammetry is one which can measure x, y image coordinates on a photograph. While in principle this could be carried out using a simple tablet digitizer, the resolution and measuring accuracy

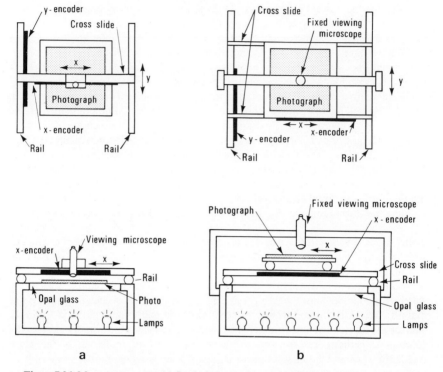

Figure 7.26 Monocomparator: (*a*) fixed plate; moving optics, (*b*) fixed optics; moving plate.

(0.05–0.2 mm) of such a device is too low for most photogrammetric purposes. The photogrammetric instrument which is mostly used for image coordinate measurement is the *comparator*. This is a mechanically-based instrument with a measuring resolution and accuracy in the range of 0.001–0.02 mm (1–20 μm), which is an order of magnitude better than that of the tablet digitizer.

7.5.2.1 *Monoscopic instruments.* The construction of a simple monocomparator is shown in Fig. 7.26. It is very similar in principle and construction to the rectangular coordinatographs (Fig. 7.5) used as output plotting devices attached to stereoplotting instruments. The photograph, which will normally be a film diapositive, is fixed in position over a light box. The observation system, which may be either a monocular or binocular monoscopic microscope or a closed circuit television (CCTV) camera and monitor, is mounted on a trolley or carriage which can move in the *x*-direction along a cross-slide (Fig. 7.26*a*). The scanning in the *y*-direction is achieved by the cross-slide itself being mounted on two rails to allow such a movement to take place. The actual pointing to a specific mark or object on the photograph is made under high magnification using a measuring mark such as a dot, cross or circle controlled by slow-motion screws akin to those of a theodolite. The *x*- and *y*-coordinate values of the position may be read off linear glass or metal scales, but nowadays are almost always measured by electronic linear or rotary encoders which output the coordinate values in digital form.

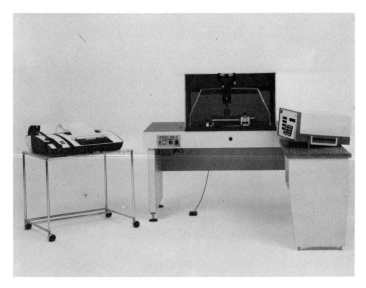

Figure 7.27 Kern MK-2 Monocomparator.

Alternative designs of monocomparator may have a fixed position of the optical viewing system, beneath which the photograph moves on a cross-slide system (Fig. 7.26b), the position of any specific point again being measured using digital encoders. However, the end product of this alternative arrangement is the same as the former in terms of the x, y comparator coordinates of a point. In the most accurate form of monocomparator, as encountered in the Zeiss Jena Ascorecord, the measuring resolution (least count) is 0.1 μm and the accuracy (standard deviation) is $\pm 0.5\,\mu$m. Such an instrument is used primarily on tasks such as the calibration of grid plates which require the very highest accuracies, rather than the routine measurement of points on aerial photographs. Typical monocomparators used for normal accurate photogrammetric measurement are the Kern MK-2 (Fig. 7.27) and the Zeiss Oberkochen PK-1 and PEK-1 instruments with measuring resolutions of 1 μm and accuracies of ± 2 to $\pm 3\,\mu$m over the standard aerial photographic format of 23 × 23 cm. At the less accurate end of the measuring range (least count 5 μm or 10 μm; accuracy $\pm 15\,\mu$m) is the inexpensive PI-1A monocomparator of Surveying and Scientific Instruments (Fig. 7.28) which has been used extensively in the UK for measurements of both aerial and close-range terrestrial photography (Mayes, 1985).

As already mentioned in section 7.5, the monocomparator can be attached on-line to a small computer and plotter to carry out digital monoplotting, for example for map revision purposes (Makarovic, 1973; Radwan and Makarovic, 1980). A very early example was the Bendix LR-2 Digital Portable Line Drawing Rectifier (Forrest, 1972) which in fact featured a tablet digitizer, a special purpose-built digital computer and a flatbed coordinatograph. A modern example is the Kern DMP-1 Digital Monoplotter (Fig. 7.29), introduced in 1987. This follows the first of the two mechanical and optical arrangements for a monocomparator described above, but with the enlargement of the measuring range to 50 by 40 cm. This allows either two standard-format (23 × 23 cm)

Figure 7.28 Surveying & Scientific Instruments PI-1A monocomparator attached to a BBC microcomputer.

Figure 7.29 Kern DMP-1 Digital Monoplotter.

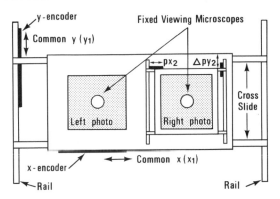

Figure 7.30 Stereocomparator.

film diapositives to be accommodated, or a single enlarged film transparency to be placed over a backlit surface. The resolution or least count is $5\mu m$, and the comparator coordinates are passed direct to an IBM-PC/AT microcomputer which carries out the required analytical photogrammetric solution.

With such an instrument, a number of useful mapping tasks may be undertaken, especially map revision carried out in conjunction with existing digital map and terrain model data (Besenicar, 1978; Masry and McLaren, 1978). The analytical space resection technique establishes both the coordinates of the camera station and the tilts present on the photograph. Thereafter, the measured comparator coordinates of any measured point can be transformed into the corresponding terrain (E, N and H) coordinates of the point where the projected line or ray meets the terrain surface defined by the digital terrain model. The attachment of a drum of flatbed plotter ensures the continuous plotting of the additional data measured along the successive points defining the line feature or area required to be added to the existing map data. A good description of the procedure is given in Burnside *et al.* (1983), where the measuring instrument is the SSI PI-1A monocomparator mentioned above.

7.5.2.2 *Stereoscopic instruments.* In a stereocomparator, two overlapping photographs are inserted into the instrument and the measurement of the corresponding points on both photographs takes place simultaneously under stereoscopic viewing. The construction of such an intrument is shown in Fig. 7.30. The two photographs are mounted on a common stage which sits on a cross-slide and rail system, allowing the operator to translate both photographs together from one point to another, either manually or under the action of motors controlled by handwheels, switches or a joystick. When the point to be measured is reached, any residual x- or y-parallax may be eliminated using small displacements px_2 and py_2 of the second photo stage relative to the first. The final pointing may be made using measuring marks which fuse together to present a single 3-D floating mark to the operator under stereoviewing. All four axes (x, y, px and py) will be equipped with coordinate measuring scales, which nowadays will take the form of linear or rotary encoders with a least count of 1 to 2 μm, giving a digital output which can be recorded on magnetic media in an off-line solution or input directly to a computer in an on-line solution.

In recent years, several of the mainstream photogrammetric manufacturers (Wild,

Figure 7.31 Zeiss Jena Dicometer.

Kern, Galileo) have ceased to offer such an instrument. Thus the availability of stereocomparators has been limited largely to Zeiss Oberkochen with its PSK series of instruments and Zeiss Jena with its Stecometer and Dicometer models (Fig. 7.31). The reason for this relative paucity of choice is not clear, but obviously the considerably lower cost of the monocomparator has its attractions, notwithstanding the extra effort of point marking and transfer to ensure the accurate identification of corresponding points before measurement takes place.

Both the monocomparator and the stereocomparator find their principal application in the measurement of individual points—ground control points and pass points—in the process of providing additional control points through analytical aerial triangulation. However, in the same way that the monocomparator can be used in a limited mapping role as a digital monoplotter when connected to an on-line computer, so can a stereocomparator-type instrument. Such an arrangement is not usually implemented with an expensive high-precision stereocomparator because the mapping tasks on which it will normally be applied require only moderate accuracy. Thus a specifically designed optical measuring instrument built to lower standards of mechanical accuracy, though with the same layout, is normally used for the purpose. The result is the so-called *image space plotter*, a concept originally put forward by Forrest in 1970.

The general layout of this type of instrument is shown in Fig. 7.32. The image coordinates primary approach is followed, so the handwheels drive the x- and y-movements of the left photo under the control of the operator. The footdisc is used to impart the px motion for height (Z) measurement. All three coordinate values (x_1, y_1, px_2) are encoded digitally and sent to the computer. However, no correction values or movements are calculated or imparted by the computer to ensure an oriented,

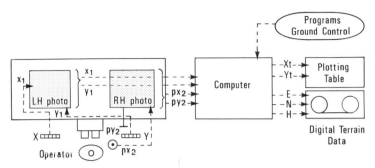

Figure 7.32 Image space plotter.

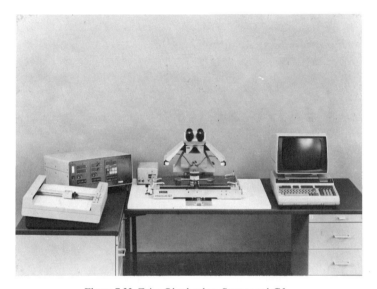

Figure 7.33 Zeiss Oberkochen Stereocord G3.

parallax-free model. The removal or measurement of y-parallax values for this purpose can only be implemented manually by the operator for an individual point. In practice, this means that the measurement of a stereomodel is rather cumbersome and can only be conducted on a point-by-point basis.

On the other hand, the lack of control devices (such as stepping motors) and feedback mechanisms, and the lighter, less accurate construction, mean that the instrument is much less expensive than an analytical plotter with a fully oriented model, and the required computations may be somewhat simpler. On the output side, the computer will generate the X, Y and Z model and E, N and H terrain coordinates which can be recorded on magnetic media and can also be used to plot the positions of each measured point on an attached plotting table. The individually plotted points can be connected up by the pen of the plotter to form the boundaries of a property, the outline of a building, the edge of a road, the perimeter of a wood or whatever other feature has to be mapped. However, because of the lack of a continuously oriented stereomodel, it is impracticable to carry out the direct measurement of contours. Also, it is impossible to drive an image space plotter to a series of predetermined positions under computer control as is required in some forms of terrain modelling, since there are no feedback or control elements available for this purpose. As an alternative, individual elevation points may be selected and measured manually, and contours can then be interpolated and plotted later as a subsequent off-line operation if a suitable terrain modelling program is available.

Obviously this type of instrument is considerably less expensive than any other type of analytically-based stereoplotting instrument, but it is also much less capable in terms of its possible output and more limited in its accuracy. Instruments which implement this approach include the Zeiss Oberkochen Stereocord G3 (Fig. 7.33) (Schwebel, 1984), and the earlier models of the APPS (Analytical Point Positioning System) constructed for the US Army Engineers Topographic Laboratories (ETL).

7.5.3 *Analytical stereoplotting instruments*

Three main types of analytical plotters may be distinguished; one implements the image coordinates primary approach discussed above in section 7.5.1.1, while the other two are based on the alternative object coordinates primary approach (section 7.5.1.2):

(i) The first group comprises those analytical plotters based on the image coordinates primary approach which maintain a continuously oriented or parallax-free model
(ii) The second group comprise the classical 'helava' type analytical plotters which follow the object coordinates primary approach and again maintain a continuously oriented and parallax-free model
(iii) The third group comprise those analytical plotters equipped with correlators for the automatic measurement of height. The majority of such instruments constructed to date implement the object coordinate primary approach. As will be seen later, they can be subdivided into those which utilize area correlation techniques and those which carry out correlation along epipolar lines.

7.5.3.1 *Analytical plotters with image coordinates primary.*

The general layout of this type of instrument is given in Fig. 7.34. The measuring part of the instrument is, in all major respects, akin to that of a stereocomparator. The two handwheels, which are operated manually by the photogrammetrist, physically move the left plate to each point being measured or plotted. In this way, the x- and y-coordinates of the left-hand image (x_1 and y_1) are generated. The footwheel, again operated by the photogrammetrist, imparts the required height value (H) by displaying the right photograph in the x-direction, i.e. it imparts the corresponding x-parallax (px_2) to the right-hand image. The final manual input is the y-parallax movement (py_2) which can be input by the operator during the initial orientation phase of the operation when the instrument is being set up before plotting of the maps or measurement of individual points.

All four possible manual inputs are sent to the computer as digital image coordinate values (x_1, y_1, px_2, py_2) using encoders. Once the orientation phase has been completed, the small additional corrections Δpx_2 and Δpy_2 required to maintain a fully oriented model are generated continuously and in real time by the computer for every position being measured by the operator. The actual physical shifts of the right-hand plate corresponding to these corrections are carried out using stepping motors.

At the same time, the computer is continuously calculating the model (X, Y, Z) and

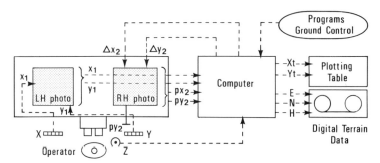

Figure 7.34 Analytical plotter with image coordinates primary.

Figure 7.35 Galileo Digicart 20 analytical plotter.

terrain (E, N, H) coordinates of each measured point, which can be transferred and recorded on a tape or disc. In addition, the corresponding positions (X_t, Y_t) of the plotting device on an attached drawing table or coordinatograph can be continuously computed and plotted in the form of points, symbols and lines, the required movements being made by digital stepping motors operating under the control of the analytical plotter's computer.

Instruments which implement this particular solution include the Galileo Digital Stereocartograph (Inghilleri, 1980), and Digicart 20 (Fig. 7.35), the Autometrics APPS IV (Greve, 1980), the AP 190 produced by Carto Instruments A/S and Cartographic Engineering (Carson, 1985, 1987) and the HDF Maco analytical plotters. The ordinary stereoplotting of planimetric detail and contours can be implemented in these instruments without difficulty. However, a limitation is that, since the terrain coordinates are not a primary input (they are output values), it is impossible to drive to

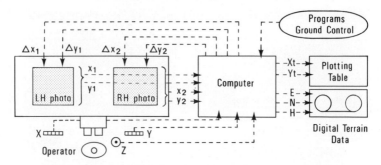

Figure 7.36 Analytical Plotter with object coordinates primary.

specified positions under computer control. This can be important in some applications, for example the measurement of elevations at specific points as required for digital terrain modelling.

7.5.3.2 *Analytical plotters with object coordinates primary.* This is the classic type of instrument favoured by Helava (1957/8), the originator of the analytical plotter concept (Helava, 1980). The basic layout of this type of instrument is shown in the block diagram in Fig. 7.36. Again the measuring part of the instrument corresponds closely to that of a stereocomparator. However, it will also be seen that the movements of the X and Y handwheels and the Z footdisc made by the operator are encoded and go directly to the control computer. In this way, the object coordinates X, Y and Z are input to the computer. This is a major difference to the arrangement used in the previous type of analytical plotter, where the handwheels directly drive the left-hand plate of the instrument. In the 'object coordinates primary' type of instrument, the computer outputs are, in the first place, the four image coordinates x_1, y_1 (for the left photograph) and x_2, y_2 (for the right photograph). These values are applied continuously to the stepping motors of each plate and ensure that the correct positions corresponding to the input coordinates are being viewed and measured by the operator, and that simultaneously the stereo-model is being maintained in an oriented, parallax-free condition. In addition, the control computer will also calculate and send in real time the required table coordinate positions (X_t, Y_t) to the automated drawing table or to a graphics terminal to ensure continuous plotting or display of the measured detail or contours. Alternatively or simultaneously the model (X, Y, Z) or terrain (E, N, H) coordinate data can be recorded on magnetic media.

It will be seen that this particular design approach leads to the most capable and flexible form of analytical plotter. Although it requires more control mechanisms, motors, and drives, it is easier to implement contouring on this type and, unlike the previous type of analytical plotter, it is easy to ensure that the instrument can be positioned at prespecified or predetermined points under computer control. As a result of these considerations, most of the larger photogrammetric manufacturers have adopted this approach. Thus the Wild Aviolyt AC-1, BC-1 and BC-2 instruments; the Zeiss Oberkochen Planicomp C 100, 110, 120 and 130 and P1, P2 and P3 series; the Kern DSR-1 and DSR-11 instruments; and the OMI AP/C3, AP/C4 and AP 5 analytical plotters all implement this solution, as do a number of instruments constructed by smaller manufacturers.

Figure 7.37 Yzerman APY Analytical Plotter.

All of these instruments address the conventional aerial photogrammetric marketplace in that they can accommodate standard 23 × 23 cm format aerial photography and can carry out aerial triangulation for the provision of additional control points to allow the stereoplotting of individual models for topographic map compilation. In addition, most manufacturers can provide software for the acquisition of digital terrain model (DTM) data. This allows the instrument to be driven to pre-defined points in the terrain coordinate system (E, N, H), including, in some cases, the implementation of progressive sampling by which the density of the sampling on a grid basis is varied according to the local roughness of the terrain surface (Ebner and Reinhardt, 1984; Reinhardt, 1984, and Chapter 10).

Those analytical plotters based on the 'object coordinates primary' approach which are produced by smaller manufacturers are often lower-cost instruments which are aimed at very specific markets or applications. Thus, for example, the APY instrument (Fig. 7.37) is designed specifically for map revision purposes including the optical reprojection of the image of an existing map into the viewing system (Yzerman, 1984). Other instruments, such as the Adam Technology MPS-2, are designed to accommodate 35 mm or 70 mm photography primarily for non-topographic applications, though obviously such an instrument can accommodate 70 mm aerial photography if required. Further details of these instruments can be found in section 8.5.2.2.

The integration with *graphics workstations* which has occurred with analogue (optical and mechanical projection) stereoplotters is also taking place with analytical plotters. Thus, for example, the Kern DSR-1 and DSR-11 analytical plotters can be supplied with MAPS 200 or 300 workstations already mentioned (Fig. 7.38) and the Zeiss Oberkochen Planicomp series with Intergraph Interact dual-screen workstations. In spite of the current enormous expense of such units, there has been sufficient demand for them for Intergraph to announce its entry into the field on its own account with the recent introduction of its Intermap Analytic instrument (Fig. 7.39) with complete integration of the analytical plotter and the Interpro desktop graphics

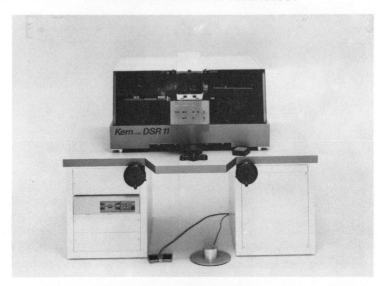

Figure 7.38 Kern DSR-11 analytical plotter.

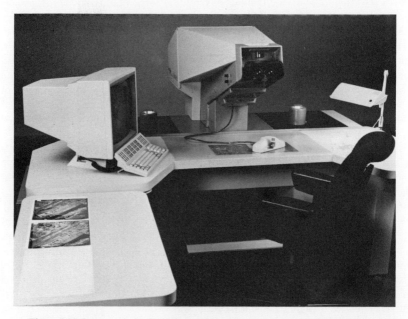

Figure 7.39 Intergraph Analytic Stereoplotter (courtesy Intergraph [GB] Ltd.).

workstation into a single unit. A similar development and integration has taken place in the new Wild System 9-AP analytical plotter which utilizes a Sun graphics workstation as the graphics component of the instrument.

Associated with the introduction of both the latter instruments has been the development of stereosuperimposition. In the simpler type of superimposition applied in analogue stereoplotting instruments (section 7.3.4), a monoscopic image of the plotted map detail is injected into one of the optical channels of the instrument and viewed and compared by the operator against the detail present in the stereomodel. Thus essentially a planimetric 2-D image of the plotted detail is being compared with the 3-D image of the stereomodel. Inevitably, there will not be a complete registration of the plotted detail against the same detail as seen in 3-D in the stereomodel. However, for checking purposes to ensure the completeness of the plotted detail, these mismatches are not too serious, and so the method of monoscopic superimposition has been useful.

However, it is also possible to carry out the whole checking operation three-dimensionally by stereo-superimposition of the plotted detail on the stereomodel (Beerenwinkel et al., 1986). This means that images of the same plotted detail have to be injected into each of the two optical channels, each displaced to fit exactly the detail present in the respective photograph. Thus the orthographic detail of the plot has to be reprojected back into the perspective geometry of each of the photographs of the stereopair. This operation has to be done in real time to ensure correct parallax-free viewing and must be carried out in perfect synchronization with the X, Y movements of the analytical plotter, taking into account the height of the terrain and the absolute orientation of the stereomodel. Needless to say, stereosuperimposition adds considerable capability to an analytical plotter but, at the present time, only at a considerable additional cost—the Wild S9-AP instrument is one of the most expensive analytical plotters presently available on the market.

7.5.4 Comparison of analogue and analytical plotters
Various advantages result from utilizing an analytical plotter instead of the analogue type of stereoplotting instrument (Ebner, 1981; McKay et al., 1985).

(i) The very high measuring accuracy (1 to 3 μm in the most accurate instruments), allied to the possibility of correcting for systematic errors such as lens distortion and film deformation by numerical methods means that a *more accurate determination of position and height* can be achieved.

(ii) It is possible to form and measure stereomodels using photography taken with cameras with focal length values, format sizes and tilts which are *beyond the range of normal analogue stereoplotting instruments*. Examples are the very long focal lengths ($f = 30$ cm to 1.20 m) and enlarged format sizes (23 × 46 cm) used in spaceborne cameras; the direct plotting of vertical or high or low oblique photography taken with non-metric reconnaissance cameras, including panoramic cameras; and the measurement and plotting of metric or non-metric terrestrial photography for industrial purposes.

(iii) It is also possible to use analytical plotters to measure images, plot maps and form terrain models from the non-photographic *remotely sensed imagery* discussed in Chapter 6. Examples include the measurement of stereo-models formed from overlapping side-looking radar images and from the stereo-imagery produced by the SPOT pushbroom scanner.

(iv) The analytical plotter offers the possibility of *automating* substantial parts of the photogrammetric procedure so that certain tasks can be carried out more rapidly and efficiently. In turn, productivity is increased and can be achieved with much less effort on the part of the photogrammetrist—provided of course the requisite procedures and solutions can be devised and then implemented efficiently in the form of appropriate software.

While photogrammetrists have been aware of these advantages for some time, the extremely high cost of the hardware and software was such as to inhibit most manufacturers and users from attempting to implement the analytical plotter approach. The result was that between 1962 (when the first instruments appeared), and 1975, the use of analytically-based systems and methodology was largely confined to a few military mapping agencies in the United States, where the need for the provision of terrain data in a timely manner from photography and other imagery was such as to justify the huge expenditure required to develop, acquire, operate and maintain these systems. Furthermore, the manufacture and integration of these analytical systems was confined almost entirely to a single partnership between OMI (Italy), which provided the optical and mechanical components, and Bendix (USA) which contributed the electronic and computational components of the systems (Konecny, 1980).

However, in 1976, a great change took place, with the simultaneous introduction of analytical plotters by several of the mainstream photogrammetric instrument manufacturers (Zeiss Oberkochen, Kern, Matra) at the ISP Congress held at Helsinki. Previously these manufacturers had built only analogue stereoplotting instruments. The main reason for this dramatic change was the development of small, fast, reliable and capable minicomputers such as the DEC PDP11, Data General Nova and Hewlett Packard HP-1000 series. These were produced at a cost which allowed the construction of analytical plotters at a price permitting them to compete effectively with the largest and most accurate types of analogue instrument, if not with the smaller topographic type of stereoplotters. Since then, the situation has steadily changed still further in favour of the analytical plotter with the dramatic improvement in the price:performance ratio of the small computers and the control components needed for the construction of analytical plotters. The current situation (in 1988) is that the production of analogue stereoplotting instruments has declined to the point where only three models of instrument are available from photogrammetric manufacturers in the Western countries—though others are still made in Eastern Europe, most notably by Zeiss Jena (East Germany), where the development of computer technology has been less rapid.

7.5.5 *Analytically-controlled orthophotoprinters*

Besides these different types of analytical plotters, another class or type of analytical photogrammetric device is the digitally-controlled orthophotoprinter. Essentially these work in much the same manner as the optical transfer devices already described in section 7.4.2.1. However, instead of operating as an on-line device in which the required scale variation and the image rotation of the tapped-off image are implemented by optical elements controlled by analogue electronic control circuitry, these same functions are carried out off-line using similar optical components but this time under the control of a small minicomputer or microprocessor. The concept was first realized in the OMI OP/C instrument in 1968–70. This was an on-line device attached to an OMI AP/C analytical plotter with the analytical solution executed in the control computer of the AP/C. However, more importantly, this concept has since been implemented as a stand-alone computer-controlled orthophotoprinter capable of being operated off-line. Examples are the Wild OR-1 Avioplan controlled by a Data General Nova minicomputer, the Zeiss Oberkochen Orthocomp Z-2 (Faust, 1980; Meier, 1984) and the OMI OP/C-2, both of the latter being controlled by a

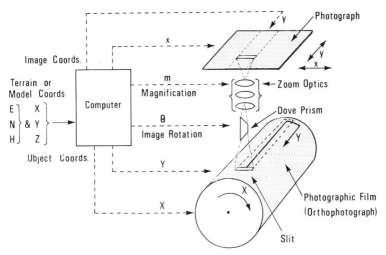

Figure 7.40 Principle and construction of an analytically controlled orthophotoprinter.

microprocessor and firmware which execute the analytical solution for the production of an orthophotograph.

The principle and construction of an analytically-controlled orthophotoprinter is shown in Fig. 7.40. On the optical side, a duplicate of one of the diapositives of the stereopair provides the input image which is rectified by the zoom optics (giving the required optical magnification and scale change) and the dove prism (which imparts the required image rotation for correction of convergence) both acting under computer control. The transformed orthographic image is recorded on photographic film mounted on a revolving drum in much the same manner as the optical transfer device shown in Fig. 7.16. On the algorithmic/computational side, the solution is based on the input of model (X, Y, Z) or terrain (E, N, H) coordinates which are usually provided by prior profile scanning in an analogue stereoplotting instrument or analytical plotter or, more commonly nowadays, from an existing regular grid digital terrain model (DTM). The computer calculates and outputs (i) the corresponding x, y image coordinates derived via the collinearity equations, and (ii) the parameters to control the zoom magnification and the slit rotation of the orthophotoprinter.

In essence, the computational solution can be visualized as being that for half an analytical plotter. These analytical orthophotoprinter devices such as the Zeiss Oberkochen Z-2 Orthocomp (Fig. 7.41) can be seen to extend the orthophotograph concept still further since, like the analytical plotter, they do not suffer from the focal length and tilt restrictions of a purely analogue device.

7.5.6 *Analytical plotters equipped with correlators*

Just as analogue stereoplotting instruments have been adapted for automated photogrammetric operations, so have analytical plotters. From the point of view of the algorithmic or computational approach followed in this type of instrument, usually it is the same as or similar to that described above for the manually operated Helava-type instruments. Thus the inputs are the model $(X, Y$ and $Z)$ or terrain $(E, N$ and $H)$

Figure 7.41 Zeiss Oberkochen Z-2 Orthocomp.

coordinates, while the outputs are the image coordinates (x_1, y_1, x_2, y_2). However, the major difference lies in the use of correlators to automate the measuring process, with the control computer ensuring the correct positioning of the correlator which then automatically measures the height and outputs the elevation (H) value and also, if required, an orthophotograph. Thus this type of analytical instrument carries out much the same function as correlator-equipped analogue instruments such as the Wild B-8 Stereomat and Zeiss Jena Topomat discussed earlier. Indeed, the development of the Bunker-Ramo UNAMACE, an early analytically-based, correlator-equipped system optimized for the automated production of orthophotographs and terrain profiles, proceeded in parallel with that of the B8-Stereomat. However, the UNAMACE was designed specifically for use by a military mapping agency, the US Defense Mapping Agency (DMA), and was not accessible to non-military users.

Current types of analytical plotters equipped with correlators may be differentiated into two main types—those employing area correlators and utilizing correlation along epipolar lines.

7.5.6.1 *Area correlation.* The analytical instruments employing area correlation techniques are typified by the Gestalt Photo Mapper (GPM) manufactured in Canada (Fig. 7.42). This operates on the basis of a systematic patch-by-patch correlation of the stereomodel, instead of the series of close parallel profile scans carried out in a raster mode or pattern used in the B8 Stereomat and UNAMACE instruments. The output from a single stereomodel using the current GPM-2 or IV instrument is an orthophotograph and a dense matrix of some 500 000 to 750 000 terrain elevation points. The elevation matrix constitutes an excellent basis for the generation of an accurate contour map or perspective block diagram of the terrain (Fig. 7.43), if required. In addition, it is possible to output a stereomate image to allow the

PHOTOGRAMMETRIC SYSTEMS AND AERIAL MAPPING 275

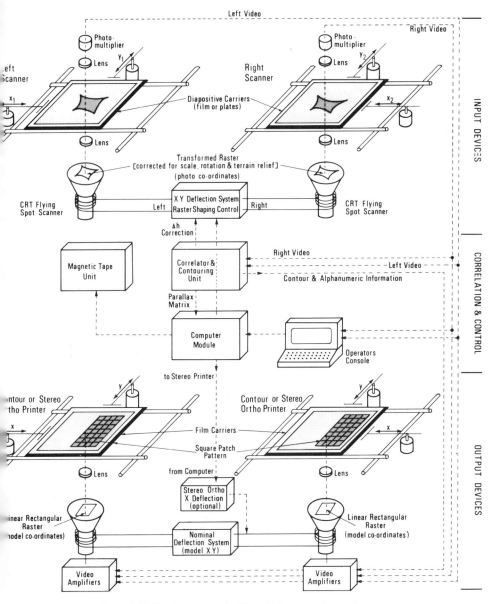

Figure 7.42 Northway Gestalt GPM-2–diagram of overall system.

stereoviewing of the terrain in combination with the planimetrically correct orthophotograph.

Unfortunately, the cost of buying and running such automated systems is extremely high, and can normally only be justified where there is a very large programme of mapping or terrain modelling to be carried out (as is the case with national mapping

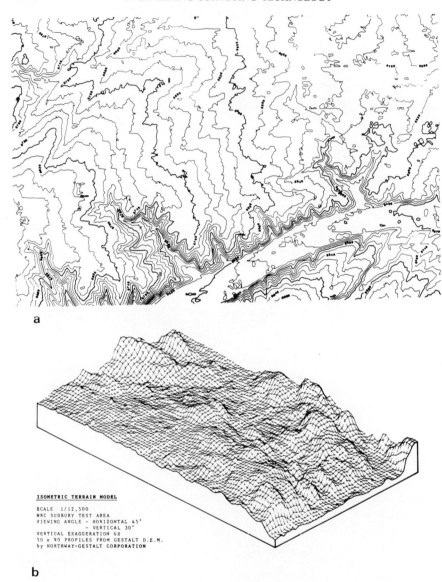

Figure 7.43 Examples of output from Northway-Gestalt GPM-2 automated system–(a) contour plot of elevation data; (b) perspective (isometric) view of NRC Sudbury test area, Canada.

agencies) or where the overriding requirement is for rapid measurement and cost is a subsidiary factor (as is the case in military mapping agencies). Terrain elevation data has to be acquired on a vast scale by the latter for use in aircraft simulators and in the terrain following and avoidance systems used in low-flying aircraft and cruise missiles.

Besides these large government mapping organizations, a few commercial mapping firms, such as Teledyne Geotronics (USA) and Northway (Canada), offer an automated

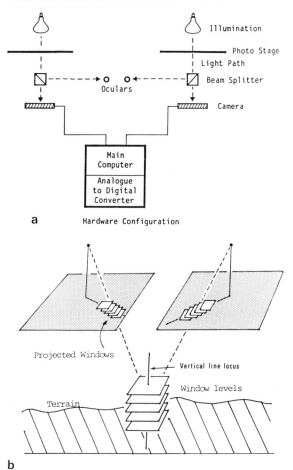

Figure 7.44 Kern VLL (vertical line locus) concept.

photogrammetric terrain data acquisition service based on the use of GPM instruments. Unfortunately, these are all located in North America. Although early examples of the GPM instruments were installed in bureaux in Europe in the 1970s, the demand for their services was too low for them to be operated economically, and the instruments were returned to North America.

A very recent development has been the introduction of less expensive correlator units, for which the input data are acquired using small cameras equipped with CCD areal arrays to image the corresponding patches of each photograph and output the images as matrices of digital values. These CCD cameras may be added to existing analytical plotters such as the Zeiss Oberkochen Planicomp and Kern DSR instruments. In the latter case (Cogan and Hunter, 1984), this allows the implementation of the *VLL (Vertical Line Locus) concept* (Fig. 7.44) in which a number of windows are defined in the model space at different Z-values along the vertical line locus of a point. Next the corresponding image points of the corners of each window are

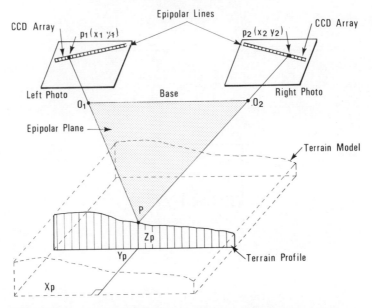

Figure 7.45 Automated analytical plotter based on epipolar line correlation.

determined for each photograph and the densities of all the points lying within the patch are obtained as digital values via the CCD cameras. These are then matched for each model window using computational methods and a best estimate made of the elevation value of the terrain point.

In functional terms, this method of correlation is optimized for the acquisition of terrain elevation data through the addition of relatively inexpensive components to an existing standard analytical plotter. This contrasts with the purpose-built GPM-2 instrument which is optimized for the correlation procedure and carries out the direct production of orthophotographs as well as terrain elevation data, albeit at a considerably greater cost.

7.5.6.2 *Epipolar line correlation.* The basic concept was introduced simultaneously but independently by Helava and Chapelle (1972) and by Masry (1972) in papers given at the 1972 ISP Congress in Ottawa (Fig. 7.45). The idea is to first establish the positions of corresponding sets of epipolar lines on each photograph of a stereo-pair. The scanning of the image densities along a pair of corresponding epipolar lines then takes place using, for example, a CCD linear array or camera mounted below each photograph. Correlation is then greatly simplified, since the search and matching of the density values of common points merely has to be carried out along specific lines instead of over areas, as in the area correlation technique. This technique was first implemented in the OMI/Bendix AS-11-BX instrument developed for the US Defense Mapping Agency in the mid-1970s, and then in the Rastar device designed by Hobrough and attached to an OMI AP/C-3 analytical plotter at the University of Hanover in 1978.

It is surprising that so far the use of these analytically-based, correlator-equipped systems has been confined largely to military mapping organizations and to a few university or other research laboratories. The cost of the large systems such as the GPM or AS-11-BX is certainly part of the story, but so far even the lower-cost systems such as the Kern VLL Correlator have not found much application outside the research environment.

7.6 Digital photogrammetric instrumentation

At the time of writing, the photogrammetric community is eagerly awaiting the results of present research and development into all-digital photogrammetric systems. Many of the principal problems and issues associated with this development and some potential solutions are outlined by Petrie (1984). The present analytical photogrammetric instrumentation is digital only in the sense that digital computers are used to implement and solve the analytical photogrammetric procedures. Invariably the actual imagery used in the analytical instrumentation takes the form of hard-copy film transparencies derived from aerial photography exposed by frame cameras. Furthermore, the instrumentation itself is still based on the use of the highest-quality mechanical and optical components in the comparator-based mensuration and viewing parts of each analytical instrument. These are supplemented by stepping motors, drive shafts, encoders, electronic coordinate counters and displays and many other mechanical and electrical components, which are essential parts of the present analytical plotter technology.

However, it is certainly the view of many photogrammetrists that, in time, the present analogue and analytical photogrammetric technology will largely disappear. The possibilities and prospects offered by digital image processing techniques are such that this vision appears certain to be realized. The time-scale for this development is, however, highly uncertain since, at the present time, many formidable obstacles have to be overcome.

7.6.1 *Digital image data acquisition and processing*

One of the most serious problem is the acquisition of digital image data of the terrain which are comparable in terms of resolution, coverage and geometric stability with the current hard-copy images produced by aerial photographic cameras. The requirements in terms of digital image data can be seen in Table 7.1, which sets out the number of pixels to be imaged, recorded and processed for a single metric frame camera image for pixel sizes of 10, 20 and 50 μm respectively.

The figure of 530M pixels for a single high-resolution image recorded in one band may be compared with the 25M pixels comprising a single Landsat TM scene. At the present time, the highest-resolution solid-state frame camera using an areal array of

Table 7.1 Digital image data produced by digitizing a metric frame camera photograph (23 × 23 cm format)

Pixel size (μm)	Pixels per line for 23 cm	Pixels per image (23 × 23 cm)	If image is quantized at 8 bits per pixel (bits)
10	23 000	529 000 000	4 232 000 000
20	11 500	132 250 000	1 058 000 000
50	4 600	21 160 000	169 280 000

CCDs has a resolution of 1000 × 1000 pixels, giving 1M pixels, so there is a very long way to go before an all-digital frame camera gives the same resolution and format size as the present highly-developed photogrammetric cameras.

As a result, digital imaging technology is currently concentrated on the use of *scanners*, either using optical-mechanical scanning, as for example with the Landsat TM scanner, or linear arrays, as deployed in the MOMS and SPOT pushbroom scanners described in Chapter 6. With these scanners, an image of the terrain is built up sequentially by a series of individual line scans as the instrument is borne forward by the movement of the spacecraft. While this procedure produces a very acceptable image in the very stable conditions existing in the near-vacuum of space, experience with airborne line scanners shows the difficulties which occur when scanner images are obtained from aircraft operating in the more turbulent conditions existing in the Earth's atmosphere. These include gaps occurring between lines, double imaging and changes in orientation between lines, so that, in geometric terms, the situation is very different to that of the airborne frame camera with its stable geometry, fixed orientaton and simultaneous exposure of the whole photograph at the moment of exposure.

On the processing side, only the largest and fastest computers existing at present are in sight of being able to process stereopairs of frame photographs (each composed of 500M pixels) in real time in the manner carried out routinely with analogue and analytical photogrammetric instrumentation on hard-copy images. Nevertheless, in spite of these difficulties, much progress is being made. Digital image data are currently being acquired through the scanning of the standard format aerial photographic transparencies off-line in a high-resolution micro densitometer utilizing a mobile x/y stage plate as employed in a monocomparator (Kibblewhite, 1981). This produces a digital image of the required resolution. The alternative is to use CCD cameras to image and convert to digital values only that part of the hard-copy photograph which has to be viewed and measured at a specific moment of time. This relieves the computational and processing requirements greatly, though it is still necessary to provide accurate motor-driven mechanical stages to achieve this.

7.6.2 *Digital analytical instruments*

A discussion of many of the individual components required to implement an all-digital photogrammetric system is given in Petrie (1984). The design of a video-based stereocomparator in which no hard-copy image exists is given in Fig. 7.46. Many alternative methods of achieving stereoviewing of digital images are also set out in this paper.

The layout of a prototype video/digital stereo-plotting instrument—the Matra Video Correlator Compiler (CVCC)—for use in generating three-dimensional (X, Y, Z) coordinates from stereopairs of aerial photographs is shown in Fig. 7.47. The stereoviewing is implemented by using two video screens (CRT) which are viewed simultaneously through a semi-reflecting mirror. The required image separation is produced by insertion of horizontally and vertically polarized sheets in front of the two screens, the operator wearing spectacles with corresponding filters to produce the 3-D stereoscopic model of the terrain. Obviously by a combination of these various developments the route is opening towards an all-digital stereoplotter based on the mathematical solutions and algorithmic approaches already in use in analytical plotters—though, as mentioned above, the digital image processing required to maintain the continuous orientation and formation of the stereo model is a considerable task.

PHOTOGRAMMETRIC SYSTEMS AND AERIAL MAPPING 281

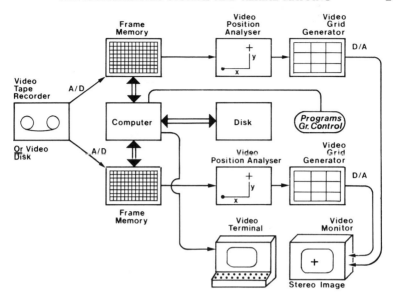

Figure 7.46 Video-based stereocomparator.

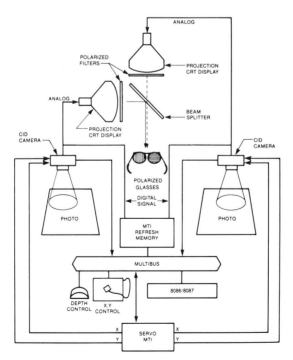

Figure 7.47 Matra Video Correlator Compiler (VCC).

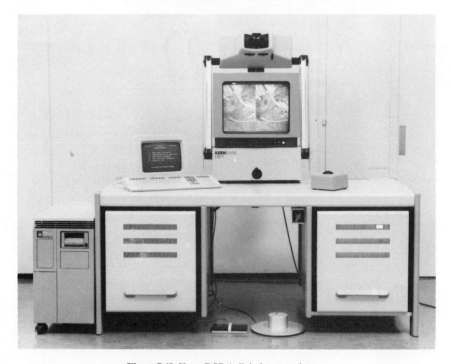

Figure 7.48 Kern DSP-1 digital stereo-plotter.

Finally in 1988, Kern produced an all-digital stereoplotter, the DSP-1. This utilizes digitized aerial photographs and employs digital processing techniques for image movement, measurement and displays based on the use of large amounts of fast-access RAM. A high resolution monitor (Fig. 7.48) with split-screen facilities is used to view the resulting stereo-models. While the main processor is a DEC MicroVAX, an additional Motorola 68020 acts as display processor and Inmos Transputers are used for the high-speed parallel processing needed for stereocorrelation in automatic height determination.

7.6.3 Digital production of orthophotographs

By far the greatest progress in implementing an all-digital photogrammetric system has been in the production of orthophotographs. The general principle of carrying out this procedure by differential rectification is given in the papers by Keating and Boston (1979) and by Konecny (1979), and is illustrated in Fig. 7.49. The input photographic image is first digitized using a scanning micro-densitometer. Standard analytical photogrammetric procedures determine the position of the perspective centre and the orientation of the photograph. The rectification of the input image is undertaken using the collinearity equations in conjunction with a digital terrain model (DTM) of the area being mapped. The rectified image of the orthophotograph is output on a raster-based film plotter. The whole procedure has been successfully implemented as the DIDAK package by Professor Wiesel and his colleagues at the University of Karlsruhe,

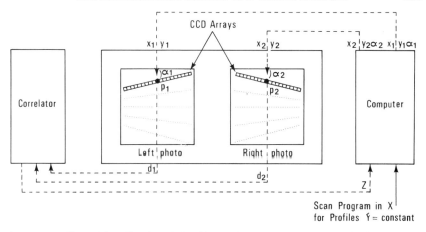

Figure 7.49 Differential rectification of digital images for orthophotograph production based on the use of a digital terrain model (DTM).

reported on in a series of papers by Wiesel, Schweinfurth, Peterle and Kuhn published in 1985. This system is now available commercially from Kern as the KDOSS (Kern Digital Orthophoto Software System).

7.7 Application of aerial photogrammetric instruments to engineering surveys

The main application to engineering of the analogue and analytical photogrammetric instrumentation, discussed above in sections 7.3–7.5, lies in the production of maps and orthophotographs, usually at large scales, and of accurate terrain elevation data in the form of spot heights and contours. The topographic maps may be produced either in purely graphical form or in digital form as the geographically-referenced input to a computer-aided design (CAD) system, or to a land or geographic information (LIS or GIS) system. If orthophotographs are required as the product on which the engineering planning and design will be based, these will be delivered as hard-copy photographic images, perhaps enhanced by a grid, symbols and lettering to become an orthophoto-map. This will be supplemented by the measured elevation profile data collected during the production of the orthophotograph. Alternatively, a digital terrain model (DTM) consisting of a dense matrix of heights may be generated photogrammetrically, and may be supplemented by graphic products such as contour plots and perspective block diagrams of the terrain directly from the DTM data.

7.7.1 *Mapping from aerial photographs*

The scale of the final map, the positional and elevation accuracies which can be achieved using photogrammetric methods, and the possible contour interval, are dependent on various interrelated factors, but chiefly:

(i) The scale and resolution of the aerial photography
(ii) The flying height at which the photography was taken
(iii) The base:height ratio (i.e. the geometry) of the overlapping photographs
(iv) The accuracy of the stereoplotting equipment used for the measurements.

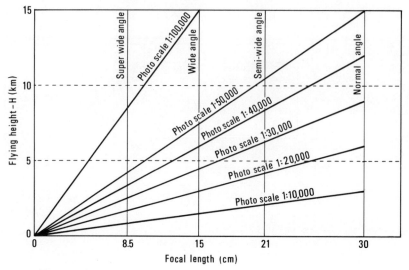

Figure 7.50 Photo scales related to flying height and camera focal length.

Table 7.2 Relationship between scale and resolution of aerial photography and the resulting mapping scale and contour interval.

Photographic scale	Ground resolution @ 40 lp/mm	Flying height (m)		Mapping scale	Enlargement Factor (photo:map)	Contour interval
1:3000	0.075 m	450	→	1:500	6 ×	0.5 m
1:5000	0.125 m	750	→	1:1000	5 ×	1 m
1:10 000	0.25 m	1 500	→	1:2500	4 ×	2 m
1:25 000	0.625 m	3 750	→	1:10 000	2.5 ×	5 m
1:50 000	1.25 m	7 500	→	1:50 000	1 ×	10 m
1:80 000	2.0 m	12 000	→	1:100 000	0.8 ×	20 m

Addressing first the matter of the *photographic scales* which may be encountered in terrain mapping and modelling, these are directly related to the flying height H of the aircraft and to the focal length f of the aerial camera used for the photographic mission. The possible range of scales over the range of flying height H from 0 to 15 km over which civilian survey aircraft can operate has been plotted in Fig. 7.50 for those focal lengths f commonly used in aerial survey work: $f = 30$ cm (normal-angle); $f = 21$ cm (semi-wide-angle); $f = 15$ cm (wide-angle) and $f = 8.5$ cm (super-wide-angle). These correspond to angular coverages of 56°, 75°, 93° and 120° respectively (Fig. 6.5), when used with the standard format size of 23 × 23 cm.

It will be seen that the use of these cameras gives rise to a wide range of photographic scales, from 1:3000 to 1:5000 at the lower end, to 1:100 000 scale at the upper end of the spectrum. The actual choice of photographic scale will usually depend on the required scale of mapping and the desired contour interval. Table 7.2 summarizes the relationship between these parameteres for the case of standard (23 × 23 cm) format photography taken with a wide-angle ($f = 15$ cm) camera and a 60% forward overlap over the scale range 1:3000 to 1:80 000.

These figures represent a rough yardstick of current practice by commercial air survey companies employing high-precision stereoplotting instruments on mapping contracts. It will be noticed that the ratio between photographic scale and map scale declines markedly as one goes from large-scale mapping for engineering purposes (where enlargement factors of 6 are normal) to small-scale topographic mapping (where slight enlargements or even reductions are common). This is due to the fact that, even for small-scale topographic maps at 1:50 000 to 1:100 000 scales, it is necessary to detect, interpret and map features of rather small dimensions such as individual buildings, secondary roads, rivers and streams. If the scale and ground resolution of the photographic imagery becomes too small, then the completeness of the map will suffer, leading to a large expense incurred through the necessity for extensive field completion work by surveyors.

Besides the plotting of maps from aerial photographs in the traditional form of a line map, the data can of course be generated in digital form for use in computer-based mapping systems. This is often done to exploit the advantages (productivity, flexibility and economy in map production) of the in-house digital mapping systems installed in most surveying and mapping organizations. It may also be possible to supply the digital map information in a form which can be exploited by an engineering firm or organization which has a suitable computer-aided design (CAD) system or a planning organization in possession of a geographical information system (GIS). However, this may require the transformation of the digital data from one format to another, using for example the Standard Interchange Format (SIF), which may require considerable additional processing, including restructuring of the data. An excellent account of the many issues which may arise in this context is given by Walker (1987).

7.7.2 Terrain elevation data from aerial photographs

The volume of terrain elevation data which is generated by the photogrammetric scanning and measurement of stereomodels is very high, particularly if it is obtained using correlator-equipped stereoplotting instruments which can measure huge amounts of elevation in a short time. Figure 7.51 shows the total amount of elevation data generated during the scanning of a single stereomodel for production of an orthophotograph. The number of points which are measured and recorded can amount to several tens of thousands, rising to hundreds of thousands of points, depending on the interval or spacing between measured profiles and the sampling interval along each individual profile. In the case of the GPM-2 instrument based on the use of area correlation, some 500 000 to 750 000 elevation points per stereomodel are obtained routinely from a single stereomodel (Blais et al., 1985).

Turning next to the *accuracy* with which the spot heights or elevations for terrain modelling may be measured, as noted above, this depends firstly on the relationship of the distance between successive photographs (the air base B) and the flying height (H) in the so-called base:height ($B:H$) ratio, which summarizes the geometrical characteristics of the photographs. Secondly, the heighting accuracy will also depend on the accuracy of the specific stereoplotting instruments on which the measurements are carried out. Lastly, and perhaps most importantly, it depends on the flying height H at which the photography was taken.

The first of these three factors, the base:height ($B:H$) ratio, is dependent on the focal length f and angular coverage of the taking camera, giving values of 0.3 with normal-angle ($f = 30$ cm and $60°$ coverage); 0.45 with semi-wide-angle ($f = 21$ cm and $75°$

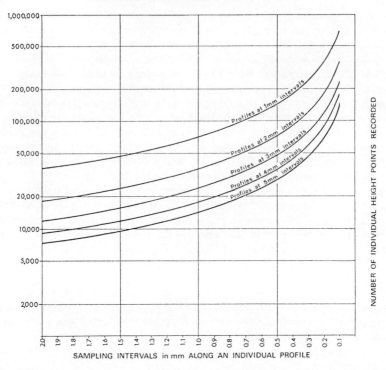

Figure 7.51 Total amount of elevation data generated from the production of a single orthophotograph.

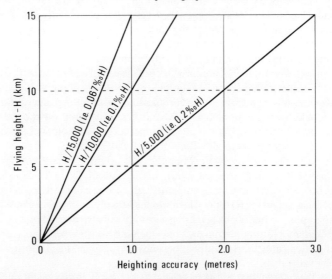

Figure 7.52 Heighting accuracy from flying heights in the range 0–15 km using stereoplotting instruments with heighting accuracies of 1:5000, 1:10 000 and 1:15 000 of the flying height (H) respectively.

coverage); 0.6 with wide-angle ($f = 15$ cm and $90°$ coverage) and 1.0 with super-wide-angle ($f = 8.5$ cm and $120°$ coverage) photographs. In general terms, the larger the value of the base:height ratio, the more accurate the measurements of terrain elevation.

For a particular base:height ratio, the heighting accuracy will be related directly to the flying height H. Taking the usual range of stereoplotting equipment available to air survey companies and national mapping agencies, the expected accuracy of spot heights using wide-angle photography will lie in the range 1/5000 to 1/15 000 of the flying height H, depending on the type of instrument used (Fig. 7.52). From this it will be seen that the expected heighting accuracies of terrain elevations expressed as root mean square errors (rmse) will lie in the range ± 0.1 to 3.0 m over the possible range of flying heights—up to a maximum of $H = 15$ km in the case of high-flying jet photographic aircraft. This accommodates the accuracy requirements of almost all terrain modelling carried out by surveyors and civil engineers.

7.8 Conclusion

From this chapter, it is clear that a great variety of photogrammetric instrumentation exists which, when used with aerial photography of a suitable scale and configuration, is capable of supplying the engineer with the maps and other topographic information which are the essential prerequisite to the planning and design of most large construction projects. This situation is unlikely to change in the foreseeable future, though the photogrammetric technology itself is in a period of rapid change, with analogue instrumentation currently being supplanted by analytical instrumentation, and digital instrumentation almost certain to replace it in the future.

References and bibliography

Beerenwinkel, R., Bonjour, J.-D., Hersch, R.D. and Kolbl, O. (1986) Real-time stereo image injection for photogrammetric plotting. *Presented Paper, ISPRS Comm. IV Symp., Edinburgh*, 11 pp.

Besenicar, J. (1978) Digital map revision. *New Technology for Mapping. Proc. ISP Comm. IV Symp., Ottawa*, 221–232.

Blais, J.A.R., Chapman, A.M. and Kok, A.L. (1985) Digital terrain modelling and applications. *Tech. Pap., ACSM-ASPRS Fall Conv.*, 646–651.

Burnside, C.D., Walker, A.S., Hampton, J.N. and Soffe, G. (1983) A digital single photograph technique for archaeological mapping and its application to map revision. *Photogramm. Record* **11**(61), 59–68.

Carson, W.W. (1985) The development of an inexpensive image plotter. *Photogramm. Record* **11**(65), 525–541.

Carson, W.W. (1987) Development of an inexpensive analytical plotter. *Photogramm. Record* **12**(69), 303–306.

Cogan, L. and Hunter, D. (1984) Kern DSR1/DSR11 DTM collection and the Kern Correlator. *Presented Paper, XV ISPRS Congr., Rio de Janeiro*, 11 pp.

Ebner, H.J. (1981) The analytical plotter and numerical photogrammetry. *Photogramm. Record* **10**(58), 409–420.

Ebner, H.J. and Reinhardt, W. (1984) Progressive sampling and DEM interpolation by finite elements. *Bildmessung und Luftbildwesen* **52**(3), 3–9.

Faust, H.W. (1980) Orthocomp Z2, the analytical orthoprojector from Carl Zeiss. *Bildmessung und Luftbildwesen* 1980/4.

Forrest, R.B. (1972) A digital portable line-drawing rectifier. *Bendix Tech. J.* **5**, No. 1.

Greve, C.W. (1980) The Analytical Photogrammetric Processing System—IV (APPS-IV). *Proc. ASP Analytical Plotter Symp. and Workshop*, 79–85.

Helava, U.V. (1957/8) New principle for photogrammetric plotters. *Photogrammetria* **14**(2), 89–96.

Helava, U.V. (1957/8) Mathematical methods in the design of photogrammetric plotters. *Photogrammetria* **14**, 2.
Helava, U.V. (1980) The concepts of the analytical plotter. *Proc. ASP Analytical Symp. and Workshop*, 12–29.
Helava, U.V. and Chapelle, A. (1972) Epipolar scan correlation. *Bendix Tech. J.* **5**(1), 19–23.
Hobrough, G.L. (1959) Automatic stereo-plotting. *Photogramm. Eng.* **25**(5), 763–769.
Inghilleri, G. (1980) Theory of the DS Analytical Systems. *Proc. ASP Analytical Plotter Symp. and Workshop*, 101–111.
Keating, T. and Boston, D. (1979) Digital orthophoto production using scanning microdensitometers. *Photogramm. Eng. and Remote Sensing* **45**, 735–740.
Kibblewhite, E.J. (1981) Automated measurement of astronomical photographs. *Photogramm. Record* **10**(58), 427–433.
Konecny, G. (1980) How the analytical plotter works and differs from analog plotter. *Proc. ASP Analytical Plotter Symp. and Workshop*, 30–75.
Konecny, G. (1979) Methods and possibilities for digital differential rectification. *Photogramm. Eng. and Remote Sensing* **45**(6), 727–734.
Kuhn, H. (1985) Perspective terrain presentations for planning using digital image processing. *Photogrammetria* **40**, 137–153.
Makarovic, B. (1973) Digital mono-plotters. *ITC Journal*, 1973–4, 583–600.
Masry, S.E. (1972) The analytical plotter as a stereo-microdensitometer. *Presented Paper, Comm. II, XII ISP Cong., Ottawa*, 17 pp.
[Also published as 'Digital correlation principles', *Photogramm. Eng.* (1974), 303–305.]
Masry, S.E. and McLaren, R. (1978) Digital map revision. *New Technology for Mapping. Proceedings, ISP Comm. IV Symp.*, 253–266.
Mayes, M.H. (1985) On line data processing for the PI-1A monocomparator using the BBC microcomputer. *Photogramm. Record* **11**(65), 515–523.
McKay, W.M., Dowman, I.J. and Farrow, J.E. (1985) Experiences with analytical plotters. *Photogramm. Record* **11**(66), 729–743.
Meier, H.-K. (1984) Analytical orthophotography: instrument development design and application. *Photogramm. Record* **11**(64), 371–382.
Newby, P.R.T. and Walker, A.S. (1986) The use of photogrammetry for direct digital data capture at Ordnance Survey. *Int. Arch. Photogramm. and Remote Sensing* **26**(4), 228–238.
Peterle, J. (1985) A concept for topographic map updating using digital orthophotos. *Photogrammetria* **40**, 87–94.
Petrie, G. (1977) Orthophotomaps. *Trans. Inst. Brit. Geogr.* **2**(1), 49–70.
Petrie, G. (1981) Hardware aspects of digital mapping. *Photogramm. Eng. and Remote Sensing* **47**(3), 307–320.
Petrie, G. (1984) The philosophy of digital and analytical photogrammetric systems. *Proc. 39th Photogrammetric Week*, Spec. Publ. No. 9, Institute for Photogrammetry, University of Stuttgart, 53–68.
Petrie, G. and Adam, M.O. (1980) The design and development of a software-based photogrammetric digitizing system. *Photogramm. Record* **10**(55), 39–61.
Radwan, M.M. and Makarovic, B. (1980) Digital mono-plotting system—improvements and tests. *ITC Journal*, 1980–3, 511–534.
Reinhardt, W. (1984) A program for progressive sampling for the Zeiss Planicomp. *Proc. 39th Photogrammetric Week*, Special Publication No. 9, Institute for Photogrammetry, University of Stuttgart, 83–90.
Schwebel, R. (1984) The extended performance range of the G3 Stereocord. *Proc. 39th Photogrammetric Week*, Special Publication No. 9, Institute for Photogrammetry, University of Stuttgart, 27–38.
Schweinfurth, G. (1985) Orthophotomaps from digital orthophotos. *Photogrammetria* **40**, 77–85.
Walker, A.S. (1987) Input of photogrammetric data to geographical information systems. *Photogramm. Record* **12**(70), 459–471.
Weisel, J. (1985) Digital image processing for orthophoto generation. *Photogrammetria* **40**, 69–76.
Yzerman, H., (1984) The APY System for analytical photogrammetry. *Photogramm. Record* **11**(64), 407–414.

8 Close-range photogrammetry

D.M. STIRLING

8.1 Introduction

Photogrammetry can be described as the science of obtaining reliable measurements by means of photographs. The 'father of photogrammetry' is generally accepted to be the Frenchman Aimé Laussedat, who pioneered terrestrial photogrammetry for architectural recording around 1860. Towards the end of the 19th century, terrestrial photogrammetric techniques were applied to alpine and glacier mapping. Around 1900, a number of important developments took place in instrument design, including the production of the first stereocomparator (7.5, 8.5) in 1901, and the first analogue stereoplotting instruments for use with terrestrial photographs in 1908. World War I, with the resulting rapid developement of aircraft, caused a shift in emphasis to aerial photogrammetry, using air photographs for reconnaissance and mapping. The first analogue stereoplotting instrument (7.3, 8.5) for aerial mapping was developed in 1921, and the use of terrestrial photogrammetry became restricted to specialist applications.

The terminology used to describe different types of photogrammetry is rather confused, but the following definitions are regularly used today. *Terrestrial photogrammetry* is often used to describe all applications of photogrammetry which do not use aerial photographs. *Non-topographic photogrammetry* is a common term used to cover all non-mapping applications. The phrase *close-range photogrammetry* is usually used to describe the situation where the camera-to-object distance is less than 300 m, and the term 'terrestrial photogrammetry' is now used where the taking distance is greater than this. This is the definition of close-range photogrammetry which has been used for selecting the contents of this chapter.

8.2 Principles of photogrammetry

Figure 8.1 illustrates the situation that exists when a camera photographs an object. The camera produces a *central perspective projection* of the object on the negative where the centre of the camera lens, O, is the *perspective centre* for the projection. Therefore a point A on the object is imaged at a' on the negative, and an object point B is imaged at b'. The angle θ subtended at O by A and B is recreated inside the camera by θ', the angle subtended at O by a' and b'. Therefore a camera may be regarded as a form of theodolite which instantaneously records an infinite number of angles between an infinite number of points on the object photographed. Thus taking a photograph is a remarkably efficient way of recording information. The mathematics of photogrammetry, in effect, allow any desired angle to be recreated from measurements between two

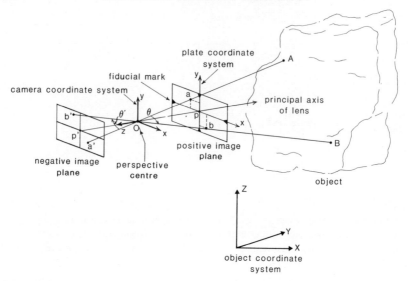

Figure 8.1 Photogrammetric coordinate systems and the geometry of a central perspective projection.

images on the photograph. By measuring the positions of a series of image points on a photograph, the resulting series of angles produces what is known as a *bundle of rays*.

8.2.1 *Coordinate systems*

Figure 8.1 also illustrates the main coordinate systems used in photogrammetry. These may be divided into two main groups:

(i) Those which are defined within the *image space*, i.e. lying on the negative or imaging side of the perspective or projection centre *O*
(ii) Those which are defined within the *object space*, i.e. lying in the object side of the perspective or projection centre *O*.

(i) *Image coordinate systems.* Three main coordinate systems are defined within the image space. These are the plate coordinate system, the comparator system and the camera coordinate system.

The two-dimensional photograph has a rectangular *x*, *y* coordinate system, generally referred to as the *photo* or *plate coordinate* system, which has its origin in the principal point of the photograph. This system can be based on the negative image plane, which reflects the situation in the camera when the photograph is exposed. Alternatively, it can utilize the positive image plane which reflects the situation in which the photographs are normally viewed and measured. Thus from Fig. 8.1, point *A* on the object will be imaged at point *a* in the positive image plane which will have plate coordinates x_a, y_a with respect to the principal point. Similarly, *B* will be imaged at point *b* which will have plate coordinates x_b, y_b.

It is often not practical to measure plate coordinates directly, so an additional two-dimensional system of rectangular coordinates may be required for the image plane. This is the *comparator coordinate system*, with an arbitrary origin which will often be the zero points of the scales or encoders used as the measuring elements of the comparator. Obviously this is a convenient system, since it is based on the measuring

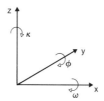

Figure 8.2 The three rotations around the camera coordinate system axes.

axes of the instrument, and again produces x, y image coordinates, albeit with a different origin and orientation of the coordinate axes as compared with those used in the plate coordinate system.

The third coordinate system to be considered is the *camera coordinate system* with its origin at the perspective centre O. This is another three-dimensional right-angled system with its x- and y-axes parallel to the x- and y-axes of the plate coordinate system. The z-axis is in the direction from the perspective centre towards the negative plane. Therefore an image point with plate coordinates x_a, y_a has camera coordinates of x_a, y_a, z_a. The term f, for the focal length of the camera, which can be variable (section 8.4), is often substituted for z_a.

(ii) *Object coordinate systems.* Two main coordinate systems are defined within the object space. These are the model coordinate system and the terrain coordinate system.

The former is a three-dimensional, right-angled coordinate system within which the points present in the analogue or mathematical/numerical model formed from the stereopair of overlapping photographs may be defined. It is therefore known as the *model coordinate system*, within which the X_M, Y_M, Z_M model coordinates of a point may be measured or defined in an analogue or analytical stereoplotting instrument.

The other main coordinate system used in the object space is the *terrain coordinate system*, which is again a three-dimensional right-angled coordinate system used to define points within a local, national or global reference system. This can be any convenient system, such as the Ordnance Survey National Grid in which the coordinates of a point are expressed in terms of its easting (E), northing (N) and height (H) values. Alternatively its position can be expressed in terms of its geographic coordinates—longitude (λ), latitude (ϕ) and height (H)—or even in the geocentric system with its origin at the centre of the Earth.*

8.2.2 Rotation systems

A camera is free to move in space along the three axes of the camera coordinate system. It can also rotate around these three axes. Thus a camera is said to have six degrees of freedom, three translations and three rotations. Figure 8.2 illustrates the conventional system used for describing these rotations. The x camera axis is taken as the primary axis of rotation, and the rotation around this axis is known as the ω-rotation, which is positive if it is clockwise when viewed in a positive direction along the axis of rotation. The secondary axis of rotation is the y camera axis, and the rotation around this axis is known as ϕ. The z camera axis is the tertiary axis and the rotation around this axis is known as K.

A similar series of rotations of the stereomodel as a whole can take place within the

---

*In close-range applications, this coordinate system is normally referred to as the object coordinate system (X, Y, Z) as shown in Fig. 8.1, and this notation is used in the remainder of the chapter.

model coordinate system. These are often referred to as *common* rotations, and are Ω around the x model axis, Φ around the y model axis and K around the z model axis.

8.3 Mathematical models used in photogrammetry

The mathematics of photogrammetry consists of a series of transformations between the various coordinate systems detailed in the preceding section. These transformations are often described as *orientations*, a term originally adopted to describe the operating procedures used in an analogue stereoplotting instrument (8.5.1).

8.3.1 *Inner orientation*

Inner orientation describes the procedures used to obtain coordinate values in the plate coordinate system. This involves transforming measurements in the comparator coordinate system into the plate coordinate system.

The relationship between the xy, plate coordinates, and the comparator coordinates, $x_c y_c$, is shown in Fig. 8.3 and can be expressed as:

$$x_c = a_0 + x \cdot \cos \alpha - y \cdot \sin \alpha$$
$$y_c = b_0 + x \cdot \sin \alpha + y \cdot \cos \alpha \tag{8.1}$$

If the photograph has suffered from deformation (8.4.5) such as is caused by film shrinkage, then an affine transformation may be used:

$$x_c = a_0 + \lambda_x \cdot x \cdot \cos \alpha - \lambda_y \cdot y \cdot \sin(\alpha + \beta)$$
$$y_c = b_0 + \lambda_x \cdot x \cdot \sin \alpha + \lambda_y \cdot y \cdot \cos(\alpha + \beta) \tag{8.2}$$

where λ_x and λ_y are the scale factors along the x and y plate axes respectively and

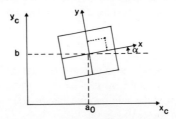

Figure 8.3 The principle of a monocomparator and the relationship between comparator coordinates $(X_c\ Y_c)$ and plate coordinates (xy).

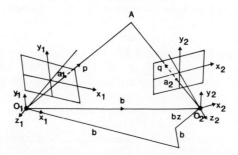

Figure 8.4 The principle of the coplanarity condition for relative orientation.

β is the non-orthogonality between the x- and y-axes. Additional terms can be added to allow for the effects of the unflatness of the film or plate (8.4.5). Corrections for lens distortion (8.4.1) can also be included.

8.3.2 Relative orientation

Relative orientation describes the procedure for forming a stereomodel from measurements on two overlapping photographs, i.e. a stereopair. This utilizes the coplanarity condition. An object point A (Fig. 8.4) is imaged on two photographs at points at a_1 and a_2 respectively. The light ray from A through the perspective centre O_1 produces the spatial vector p. Similarly, the ray through O_2 produces vector q, and b denotes vector $O_1 O_2$, known as the *base*. The coplanarity condition is:

$$b \cdot p \times q = 0 \qquad (8.3)$$

i.e. the three vectors b, p and q lie in the same plane. If O_1 is taken as the origin of the model coordinate system with axes parallel to the O_1 camera axes, then equation (8.3) can be written as

$$\begin{vmatrix} 1 & x_1 & m_{11}x_2 + m_{21}y_2 + m_{31}z_2 \\ by/bx & y_1 & m_{12}x_2 + m_{22}y_2 + m_{32}z_2 \\ bz/bx & z_1 & m_{13}x_2 + m_{23}y_2 + m_{33}z_2 \end{vmatrix} = 0 \qquad (8.4)$$

where m_{11} to m_{33} are the nine elements of a three-dimensional orthogonal rotation matrix resulting from the sequential rotations ω, ϕ and κ of photograph 2 with respect to photograph 1.

8.3.3 Absolute orientation

Model coordinates, $X_m Y_m Z_m$, produced after a relative orientation are normally transformed into terrain coordinates, E, N, H, using a three-dimensional similarity (linear conformal) transformation:

$$E = X_{O_m} + \lambda(r_{11} X_m + r_{12} Y_m + r_{13} Z_m)$$
$$N = Y_{O_m} + \lambda(r_{21} X_m + r_{22} Y_m + r_{23} Z_m) \qquad (8.5)$$
$$H = Z_{O_m} + \lambda(r_{31} X_m + r_{32} Y_m + r_{33} Z_m)$$

where X_{O_m}, Y_{O_m}, Z_{O_m} are the coordinates of the origin of the model coordinate system, λ is the scale factor between the two systems, and r_{11} to r_{33} are the nine elements of a three-dimensional orthogonal rotation matrix resulting from the sequential rotations Ω, Φ and K.

This procedure is normally known as an *absolute orientation*.

8.3.4 Collinearity equations

An alternative to carrying out a relative orientation followed by an absolute orientation is to use a different mathematical model, namely:

$$x_a = -f \frac{m_{11}(X_A - X_O) + m_{12}(Y_A - Y_O) + m_{13}(Z_A - Z_O)}{m_{31}(X_A - X_O) + m_{32}(Y_A - Y_O) + m_{33}(Z_A - Z_O)}$$

and $\qquad (8.6)$

$$y_a = -f \frac{m_{21}(X_A - X_O) + m_{22}(Y_A - Y_O) + m_{23}(Z_A - Z_O)}{m_{31}(X_A - X_O) + m_{32}(Y_A - Y_O) + m_{33}(Z_A - Z_O)}$$

where m_{11} to m_{33} are the nine elements of a three-dimensional orthogonal rotation matrix resulting from the sequential rotations ω, φ and κ of the camera coordinate system relative to the object coordinate system.

The two equations in (8.6) are known as the *collinearity equations*, and state that the object point A, the perspective centre O and the image point a lie on a straight line, i.e. they are collinear. This is a direct relationship between the plate and object coordinate systems, and does not involve any intermediate coordinate systems such as the model coordinate system required using coplanarity.

Both the collinearity and coplanarity conditions and equations can be used in analytical close-range photogrammetry. However, the coplanarity solution applies only to a *stereopair*, whereas two collinearity equations can be written for a single *photograph*. Therefore, any number of photographs can be used by adding more collinearity equations to the solution. This approach is far more flexible than coplanarity, and its uses in close-range photogrammetry are detailed in the following sections.

8.3.5 *Use of the collinearity equations*

Two special applications of the collinearity equations refer specifically to problems connected with cameras, and so are discussed in 8.4.4. More general uses are detailed below.

8.3.5.1 *Space resection.*

Unless the camera is set up over a predetermined point and pointed in a known direction, a procedure which is very difficult to carry out precisely, it is normal for the position and orientation of the camera to be computed from the measurement of images of points whose object point coordinates have been determined by some other method. These points are known as *control points*. This procedure is known as a *space resection* and is similar in principle to the standard resection technique carried out with a theodolite but is three-dimensional. For each image point, two collinearity equations can be formed as observation equations, one for the x plate measurement x_a, and one for the y plate measurement y_a. There are six unknowns in a space resection—the coordinates of the perspective centre, $X_O Y_O Z_O$, and the three rotations, ω, φ and κ—often referred to as the *camera parameters*. Therefore at least three non-collinear control points with known object point coordinates, $X_A Y_A Z_A$, must be imaged on the photograph for a solution to be possible. As in all aspects of surveying, it is usual to have redundant observations, so four or more control points should be used, and these control points should be well distributed throughout the volume of the object and the area of the image.

A space resection is often referred to as a single-stage or *bundle orientation* to distinguish it from the sequential use of relative and absolute orientations discussed previously.

8.3.5.2 *Spatial intersection (reprojection).*

If the six camera parameters are known, from a space resection or otherwise, it is then possible to find the object coordinates of unknown points which are imaged on the photographs. From the equations in (8.6) there are now three unknowns—the object coordinates of the unknown point, $X_A Y_A Z_A$. However, the image point will yield only two observation equations, one each for x_a and y_a. Therefore, at least two photographs upon which the unknown point is imaged are required to enable all three coordinates of the point to be found. Usually

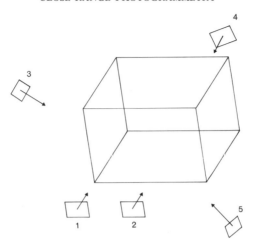

Figure 8.5 The geometry of a stereopair (photographs 1 and 2) compared with a multi-station approach (photographs 1, 3, 4 and 5).

only two photographs are used, but more photographs (Fig. 8.5) will give a greater redundancy (see 8.3.5.3). Only one photograph is required if one of the three coordinates of the object point is known. This is possible in special cases when the object photographed is completely flat or, say, the Z coordinate of the point is known (see 8.6.1.2 and 8.6.1.6 for examples).

8.3.5.3 *Phototriangulation and bundle adjustments.* When the number of photographs required to cover an object increases beyond two, the provision of three or more coordinated control points on the object by direct measurement becomes very time-consuming. In this case, a technique known as *phototriangulation* may be employed. The term 'aerial triangulation', which refers to this technique when used in aerial mapping, may be more familiar to some readers.

The coplanarity condition (section 8.3.2) is used in many aerial triangulation software packages. In this method, the coplanarity condition is used to carry out relative orientations to form adjacent stereomodels, which are then joined together using similarity transformations. Cooper (1979) describes a close-range application where this form of aerial triangulation package was used. However, many of these packages adjust plan (X, Y) and height (Z) separately, and in close-range photogrammetry with photographs taken from many positions and orientations, it becomes difficult or even impossible to select a reference plane on which to base these two adjustment procedures. The collinearity solution, on the other hand, adjusts X, Y and Z simultaneously, and thus the problem of selecting a suitable plane does not arise. Therefore this approach is more suited to many close-range applications.

Phototriangulation methods which utilize the collinearity solution for coordinate determination are referred to as *bundle adjustments*. For each photograph taken of the object, a bundle of rays can be recreated from measurements of the plate coodinates of the image points. The triangulation method links these individual ray bundles together into a *block* and calculates the object coordinates of the imaged points and the camera parameters for each photograph. This is similar to a triangulation network carried out using ground survey methods where the 'bundles' of rays measured at stations with theodolites are linked together and adjusted. For each measured image point, two observation equations can be derived from the collinearity equations (8.6). If the coordinates of the object point are known, then the observation equations will be the same as those for a space resection (8.3.5.1). If the object point coordinates are not known, then the two observation equations are modified to include the object coordinates among the unknowns, along with the six camera parameters. Most bundle triangulation packages are sufficiently flexible for partially known object points, e.g. levelled points where Z is known but not X and Y, to be included and the observation equations arranged accordingly. Provided that there are at least the same number of observations, i.e. plate measurements, as unknowns in the block, then a solution is possible. Because there are still six unknown camera parameters per photograph, at least three image points must be measured on each photograph, but usually four or more well-distributed points are measured to add redundancy to the adjustment. Object points whose coordinates are unknown must be measured on two or more photographs in the block. Also within the block there must be at least seven known object coordinates to provide a datum (X, Y, Z), scale (λ) and orientation (Ω, Φ, K) for the block as a complete unit. Additional known coordinates should be provided to give a suitable redundancy in the solution.

A bundle triangulation approach can therefore greatly reduce the number of control points which have to be coordinated on site, from at least three per photograph to a theoretical, but not recommended, minimum of three points for the whole object, thus greatly reducing the time spent on site carrying out a conventional control survey.

The other major use of bundle triangulation procedures is in applications which require both high precision and high reliability. In the conventional reprojection from a stereopair (8.3.5.2) the four observation equations for the three unknowns of the object point coordinates yield a redundancy of only one. It is a long-established principle in engineering surveying that coordinating points by theodolite intersection produces significantly higher precision and reliability if the point is observed from three or more stations than if it was observed from only two. This is true also for photogrammetry. Furthermore, using three or more photographs not only increases the redundancy, but enables the geometry of the spatial intersections to be improved. In many engineering applications, highly convergent photography is used to improve the results compared with the standard stereopair (Granshaw, 1980; Fraser and Brown, 1986). When three or more photographs (Fig. 8.5) are used to coordinate each point, the term *multi-station photogrammetry* is used. However, the increase in precision and reliability obtained using multi-station photography has to be offset against the reduced number of points that can be measured on the object. When only two photographs are used, they are usually viewed and measured stereoscopically. In this way, as long as the object has sufficient texture on it to enable the images to be fused, then any number of points on the object can be measured by fusing the two measuring marks together in the stereoimage. However, in multi-station work three or more photographs cannot be viewed and measured simultaneously, and so the problem of identifying exactly the same point on all the photographs, often taken from widely differing viewpoints, will degrade the accuracy of the solution. To overcome this restriction, it is normal to affix specific targets to the object, but only a finite number of points can be targeted.

In many uses of photogrammetry, and almost always in aerial mapping, the normal practice is to regard the conventionally surveyed control points as fixed and error-free and to fit the photogrammetric measurements to this control. However, in a number of high-precision close-range applications, the inherent accuracy of the photogrammetric measurements is degraded if this approach is taken. Plate coordinates can be measured to a few micrometres (μm) or better, which at a photograph scale of, say, 1:100, represents a few tenths of a millimetre on the object. It is extremely difficult, but not impossible (section 1.5), to obtain this level of accuracy using conventional techniques. Therefore, in a situation such as this, it is desirable to use the photogrammetric measurements to improve the survey results rather than to use the results of the survey to degrade the photogrammetry. There are two ways in which this can be carried out.

Firstly, the survey observations can be adjusted in a normal survey adjustment package, preferably one that adjusts all three coordinates simultaneously, which will yield the adjusted coordinates of the control points and their covariance matrix. These are then used as input into a bundle program adjustment that allows the control points to move as well as all other points and camera stations. In this way, a 'best fit' overall solution will be achieved with correct weighting being assigned to the survey results and the photogrammetric measurements.

The other solution to the problem is to enter all the survey and photogrammetric measurements together into one overall bundle adjustment program and adjust the combined networks as one homogeneous unit.

It should be noted that both approaches should give the same results from the same set of measurements.

In normal surveying adjustment procedures and in the bundle methods discussed above, it is standard practice to use absolute coordinates for the network based on a suitable datum. This can cause problems in deformation studies since it may be extremely difficult to select datum points close to the object being studied which can be regarded as stable. Selecting different points in the network to define the datum can produce completely different results (Cooper *et al.*, 1984) for the deformation as the formation of the normal equations for the least-squares adjustment will be different. Therefore, if only relative deformation of the object within itself is important, as opposed to absolute movement with respect to some fixed point, then the best solution is to have no datum points at all. This is a so-called '*free*' *adjustment*. Normal least-squares techniques for the adjustment in this case will fail, since the matrix of coefficients of the normal equations will be singular. One way to overcome this problem is to use a generalized matrix inversion routine, but this greatly increases the computing required for a solution (Cooper, 1985). It is preferable to employ the method of 'inner constraints' by introducing arbitrary constraints into the system. As long as the *same* constraints and starting values are used for the object point coordinates for the adjustment of each epoch's observations, then the results can be more readily used to estimate the deformation.

8.4 Cameras

Since the photograph is the major unit of data in photogrammetry, it is of vital importance that the camera selected for a particular project is capable of producing an image of sufficient quality for the required task. Figure 8.1 and equation (8.6) refer to plate and camera coordinate systems and, in the majority of cases in photogrammetry, it is necessary for the camera employed to define these coordinate systems on the photograph. An exception to this rule is discussed later in 8.4.4. Cameras which are capable of supplying this information are called *metric* cameras, and cameras which do not define these coordinate systems are known as *non-metric* cameras. Occasionally the term *semi-metric* camera may be encountered, but this really refers to a non-metric camera which has been modified to act as a low-precision metric camera.

Figure 8.1 shows that the angle subtended at the perspective centre O by A and B is θ. In an ideal projection, the angle θ' subtended at O by a and b will also be θ. However, no lens system is absolutely perfect, and the three points A, O and a will not be exactly collinear since, as it passes through the lens, the ray will deviate slightly. This deviation is known as *lens distortion*. If the distortion is small, it may be neglected in all but high-precision applications. However, if it is large, it must be allowed for in the solution, or the results will be virtually meaningless.

Before discussing cameras in more detail, it is necessary to define a number of commonly used terms for which reference should be made to Fig. 8.1. *Format* is the term used to describe the size of the image area. For nearly all modern air survey cameras this is 230×230 mm. Close-range cameras tend to use smaller formats because they have to be carried and positioned on sites relatively easily. The Zeiss Jena UMK 10/1318 (Fig. 8.6), for example, has a usable format of 165×120 mm while the Rolleiflex 3003 is a small 35 mm metric camera. The *principal axis* of the lens is defined by the ray which passes through the perspective centre O, and intersects the film plane normal to it at p, the *principal point* of the camera, which is the origin of the plate coordinate system. The distance Op is called the *principal distance* or, more commonly, the *focal length*, f, of the lens, and this is the z coordinate of the camera coordinate system.

8.4.1 *Metric cameras*

In a metric camera, all the terms defined above are determined by calibrating the camera after manufacture. In the focal plane of the camera there is usually a series of fiducial or index marks, normally four, positioned either in the corners or midway along the sides of the format, such that two lines drawn between the pairs of opposite fiducials will intersect at, or very close to, the principal point p. The intersection of these two lines produces the *fiducial centre*. Any discrepancy between the fiducial centre and the principal point can be determined during the calibration and noted on the manufacturer's calibration certificate. Subsequent measurements of plate coordinates with respect to the fiducial centre can be corrected to the principal point using this information. An alternative method of defining the principal point is to fit a glass plate in the focal plane of the camera with a series of crosses engraved on the glass which will be imaged on the photograph. This is called a *réseau*, and the central cross should be close to the principal point. Again, any slight discrepancy is determined during calibration.

The *focal length* f also needs to be known to a high degree of accuracy. In all air survey cameras, the lens is of the fixed-focus type and is set at a taking distance of infinity. In close-range photogrammetry, the object photographed is obviously not infinitely far away, and so cameras which are not focused at infinity are desirable. These can have either a fixed focus or a variable focus. The Wild P-32, for instance, is supplied with a fixed focus setting of 25 m. By reducing the aperture, the depth of field can be increased to between 3.3 m and infinity. Alternatively, the camera can be supplied with one of a number of fixed-focus settings down to 0.7 m. The larger Wild P-31 can be supplied with a series of spacer rings which can be used to vary the focal length by

Figure 8.6 Zeiss Jena UMK 10/1318 metric camera on orientation mount.

calibrated amounts. Greater flexibility is afforded by using a variable-focus camera, such as some versions of the Zeiss Jena UMK. The nominal $f = 100$ mm lens version of this camera has a variable focal length which allows focusing distances of between 1.4 m and infinity.

Metric cameras should have a minimal degree of lens distortion, and whatever distortion is present in the lens should be calibrated, so that even this small amount can be allowed for in high-precision applications. A fixed-focus camera has the lens design optimized for the set focal length of the camera. The Wild P-32, for instance, has a maximum radial distortion of only 4 μm. However, a variable-focus camera will have variable lens distortion, and the lens will have been designed to have minimum distortion at a particular focal length. This may result in more than one model of a particular camera design, such as for example in the case of the Zeiss Jena 100 mm UMK where the standard camera has minimum distortion with the lens set at infinity, whereas the close-range model has its minimum distortion for a setting of 2.3 m. When using such a camera at a setting which is markedly different from the designed optimum, then the lens distortion which is present on the image will be quite large. The calibration certificate should include distortion values for a series of focus settings throughout the available range of the camera.

The necessity for maintaining calibrated values for the principal point, focal length and lens distortion over a reasonable period of time requires the construction of the camera to very high standards of stability and, in the case of variable focus cameras, repeatability. Any mis-centring of the lens or change in the focal length over time will lead to errors in the measured results. Metric cameras are therefore built to the same high standards as all other pieces of surveying equipment and are, as a consequence, expensive.

The UK National Physical Laboratory has developed a revolutionary new camera which is virtually distortion-free. This camera, known as the Centrax (Burch and Forno, 1982), consists of a perfectly spherical lens with a liquid centre. The camera is of the fixed focus type but has a depth of field of 300 mm to infinity and a maximum distortion, on tests with the prototype, of only 0.1 μm. The main limitation to the use of this camera is that the object to be measured must be covered in luminous or point light source targets to produce a well-defined light ray for the lens to image. The main application of this camera will be in the laboratory measurement of engineering and scientific components. It is thought that a precision of ± 1 μm per metre in defining the size of the object will be possible using this system.

8.4.2 *Stereometric cameras*

When two metric cameras are rigidly attached on a bar with their principal axes parallel to one another and perpendicular to the bar, the arrangement is known as a stereometric camera. The advantage of a stereometric camera over two separate tripod-mounted cameras is that it guarantees that the resultant stereopair can be set up in an analogue stereoplotting instrument (see 8.5.1). The distance between the two perspective centres is called the base.

Individual metric cameras can also be mounted on a base bar to form a stereometric camera. Two cameras can be mounted simultaneously, or one camera moved along the bar, thus maintaining the parallelism of the principal axes.

Stereometric cameras are ideal for photographing fairly flat objects which are to be measured in an analogue stereoplotting instrument (8.5.1) and are widely used for architectural archival purposes where the base is positioned parallel to the main axis of the facade. Their main disadvantage is their sheer size and weight, especially when the base length is 1 m or more. When working on a confined site such as a chemical plant, it is very difficult to manoeuvre such a bulky piece of equipment among complex and cramped pipework. Such cameras are also very inflexible for many engineering applications where high precision is required, and are not suitable for multi-station procedures.

8.4.3 *Phototheodolites*

The first main use of photogrammetry was for topographic mapping of mountainous areas. The *phototheodolite* was developed specifically for this application. This is an instrument in which a metric camera is rigidly attached to some form of angle measuring device, which may be either a full theodolite as in the Wild P-30 or, as in the Zeiss Jena Photheo 19/1318, a full horizontal circle and a limited vertical circle. Using this type of instrument, the six camera parameters can be fixed either by setting up over a known point, or by resecting the camera station using the theodolite portion of the instrument. This involves measuring the ϕ rotation with the horizontal circle and setting ω and κ to zero by levelling the instrument. Therefore to provide control points on the object it is not necessary to climb high peaks to fix control points in the mountainous areas in which this technique was widely used. Aerial photogrammetric cameras have now virtually replaced the phototheodolite as an instrument for topographic mapping.

It is also possible to mount other metric cameras, such as the Wild P-32, on a theodolite, thus creating a system which is often, but mistakenly, described as a phototheodolite. This arrangement is used like a phototheodolite, but the attachment between camera and theodolite is not completely rigid, and after a period of use, the two units can become misaligned. Alignment should be checked frequently and adjusted when required.

Most other metric cameras are supplied, either as standard or as an optional accessory, with an orientation mount (Fig. 8.6) thus allowing the camera to be set up over a known point and values obtained for the ω, ϕ and κ rotations. Therefore this type of arrangement may also be considered as a phototheodolite. Stereometric cameras can have a similar mount with the added knowledge of the fixed base length providing a control for scale.

Using these phototheodolite 'look-alike' systems as true phototheodolites, i.e. fixing the camera parameters when taking the photographs and not using object control points, may appear to be a very attractive method, but should be used with caution. The temptation to reduce the amount of fieldwork required should be resisted in all but low-precision applications because of the uncertainties of the true camera parameters. As well as the possible misalignment of the camera axes mentioned above, which results in incorrect rotational values being used in the solution, there is the additional problem that the perspective centre may not be exactly above the centre of the instrument mount. Therefore there is an offset which can be variable and is often not known. If the camera-to-object distance is very short, then any uncertainties in the value of this offset can have serious consequences on the accuracy of the result. This offset value could, however, be included as an additional unknown in a bundle solution, which would then normally require some form of object control.

8.4.4 *Non-metric cameras*

There is always great pressure in photogrammetry to use non-metric cameras, since they are significantly cheaper than their metric counterparts. However, before succumbing to this pressure, it is very important to be aware of the limitations of these cameras. Non-metric cameras will not have a fiducial reference system. The corners of the format can be used, but the 'fiducial centre' defined by them may be significantly different from the principal point. The focal length is not known to a high precision, and is variable and not repeatable even when the same focus setting is selected. A film flattening device (8.4.5) is not normally fitted. Additionally, non-metric cameras have lenses designed for optimum picture quality, so lens geometry and hence distortion are of secondary importance, resulting in some lenses with very large distortions.

One way in which the low cost of these cameras can be utilized is to modify them to form so-called '*semi-metric*' *cameras*, which are actually true metric cameras, but of low precision. This normally involves fitting some fiducial reference system, usually a glass réseau, into the focal plane of the camera, and permanently fixing the focal length of the lens. The camera is then calibrated so that all the necessary information is known, and the camera is then used as a normal metric camera. However, as the basic camera does not have the inherent stability of a purpose-built metric camera, it should be recalibrated at comparatively short intervals.

If there are enough control points on the object, then non-metric cameras can be used directly by employing one of a number of analytical techniques. One of the most common of these is the *Direct Linear Transformation* (DLT) developed at the University of Illinois (Abdel-Aziz and Karara, 1971). In this non-rigorous solution, a fiducial reference system is not required, and eleven parameters are used to produce a direct linear relationship between object coordinates and photograph measurements. Up to five parameters for lens distortion can also be included. *Self-calibration* can also be used where the collinearity equations (Fig. 8.6) are expanded to include further unknowns for the displacement of the principal point, the focal length and lens distortion. In this system the camera is calibrated either for every individual photograph used or for a whole block of photographs depending on the project. All of these techniques require more object control points than when using metric cameras. In the basic eleven-parameter DLT solution, at least six control points are required, and this number rises to eight if the five additional lens distortion parameters are included. Therefore, although the camera may be cheaper, the amount of fieldwork required to provide the necessary control is significantly increased. A great danger with DLT is that, if additional measured points of detail are not close to one of the coordinated control points which were used to calculate the DLT parameters, then the coordinates of these additional points are significantly reduced in accuracy. Therefore, to obtain high accuracies with DLT, many well-distributed control points are required.

There are a number of special cases where it is impossible to use metric photography. One such case is where the object no longer exists and only non-metric photography of the object is available. E.H. Thompson (1962) was involved in the restoration of the eighteenth-century dome of Castle Howard, Yorkshire, destroyed by fire in 1940. Using various techniques and a series of old photographs of the building taken with non-metric cameras some time before the fire, it was possible to recover the dimensions of the dome so that it could be reconstructed to resemble the original as closely as possible.

Cheap cameras can be used in photogrammetry but they should be employed with care and should not be used at all if high precision is required.

8.4.5 *Photographic media*

If the high accuracy which close-range photogrammetry is capable of producing is to be exploited fully, as much attention must be paid to the photographic media being used as to any other aspect of the project. In traditional photogrammetry silver halide photographic emulsions are used (see 8.7 for some recent developments using other imaging techniques). The emulsion has to be supported on a base material which is either a flexible plastic film or a plane glass plate. In close-range work, the selection of base material is a question of balance between convenience, accuracy required and equipment available. Some cameras can accept only glass plates, some film only, and some both.

The dimensional stability of the base material is of prime importance and, where possible, glass plates should be used to maintain accuracy. The film bases used for aerial photography are almost as stable as glass, but are normally supplied on 230 mm wide rolls which require cutting to fit the smaller-format cameras usually used in close-range work. Commercially available films can be used; these are normally produced on less stable, cheaper base materials, but have the advantage of wide availability and can be processed by normal commercial laboratories. The Wild P-32, for instance, uses standard 120 mm roll film in its film cassette. If the camera is fitted with a réseau plate and an analytical solution is to be employed, then significant film distortion can be allowed for by transforming the measured images of the réseau marks back to their calibrated values. However, when doing this, great care should be taken to ensure that the mathematical model being employed is suitable for the disposition of the réseau marks in the camera (Fraser, 1982).

The unflatness of the base material may also be important and is much more difficult to model. To reduce unflatness to a minimum, the camera should incorporate some form of film flattening device. This can either be a vacuum back to hold the film flat, or a pressure plate system which presses the film forward on to a plane glass plate mounted in the focal plane of the camera. Even photographic glass plates are not perfectly flat, and humidity changes in the emulsion can cause plates to bow by as much as $35\,\mu$m (Gates et al., 1982), producing significant errors in the measurements of plate coordinates which can result in large and undetected systematic errors in the results. To overcome this problem in ultra-precision industrial applications, Geodetic Services Inc. of Florida has designed and built its own camera, the CRC-1. This camera uses full-size 230 mm roll aerial film, and has a vacuum back which flattens the film to within $1\,\mu$m in the negative image plane. In other metric cameras, however, the glass plates used will usually give better results than film.

The choice of emulsion may also be important. Normally black-and-white emulsions are used because they are cheaper and easier to process than colour. Colour can be used if interpretation of the image is important, as for instance in geotechnical studies (8.6.1.1), where subtle changes in the colour of the object provide important information about its nature. However, colour films, tend to be much less stable than black-and-white films because they are processed at significantly higher temperatures. Colour plates are made only to special order and are thus very expensive. If measurements of very fine detail are required, then a fine-grained film is necessary. A problem introduced here is that finer-grained films are 'slower' and require longer exposure times which may be difficult to achieve if the illumination of the object is poor or vibration is present.

8.5 Equipment for the measurement of photography

There are two types of computer which can be used to carry out the coordinate transformations in photogrammetry (8.3), namely mechanical analogue computers or electronic digital computers. In 1901 Pulfrich invented the stereocomparator, which enabled plate coordinates to be measured. The resultant transformations from the plate to object coordinates were carried out mathematically using tables of logarithms and trigonometric functions, a very tedious process. To speed up the measuring process, systems which measured model coordinates were developed, and the first analogue stereoplotting instrument was built in 1908. It is only in the last thirty years that analytical techniques have been revived with the advent of increasingly more powerful and cheaper digital computers.

8.5.1 Analogue stereoplotting instruments

The great majority of analogue stereoplotting instruments produced have been designed specifically for aerial mapping and have rather limited usefulness in close-

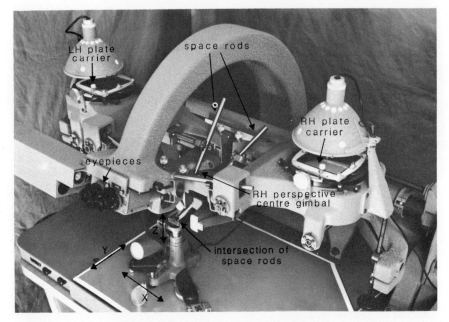

Figure 8.7 The Wild B-9 analogue stereoplotter.

range photogrammetric work. As discussed in Chapter 7, most analogue instruments use a system of mechanical projection (Fig. 8.7) where the optical ray that entered the lens of the camera at exposure is simulated by a metal space-rod with a gimbal or universal joint for the perspective centre. The two space-rods, one for each photograph, are joined together at the bottom and continually compute the intersection of the two rays as the measuring marks are moved over the photographs. The two projector assemblies can rotate around the gimbals to simulate the rotations that were present on the cameras at the moments of exposure. These movements are normally restricted to about 5° around each axis. This is a serious mechanical constraint which significantly limits the application of such instruments in close-range photogrammetry, since only photography with small rotations can be used. Another limitation is that the movement of the point of intersection of the space-rods in depth (Z) is limited, and so objects with large depth ranges in the direction towards the cameras may not be able to be accommodated on these instruments. The analogue stereoplotting instruments used also have very limited, or no, ability to handle lens or film distortions.

The setting of the values for inner, relative and absolute orientations (8.3) on an analogue stereoplotting instrument is usually carried out empirically by the operator using procedures developed for aerial mapping. These procedures are based on the assumption that the object photographed is either flat, or has depth variations which are very small compared with the taking distance. The solution is iterative, the number of iterations required increasing with the depth range of the object. For a typical aerial model it may be possible to set up the instrument in about thirty minutes. However, as the depth of the object increases, these empirical techniques begin to break down, and setting up the instrument becomes increasingly difficult or even impossible in certain circumstances. In such situations, it becomes necessary to use a digital computer to help the operator to set up the plotter by using the computer to calculate the rotations on the projectors so that the operator can set these manually (Shortis, 1981). It should be noted that these computer-aided orientation techniques still use the analogue devices in the plotter to carry out the photogrammetric solution, and so do not overcome the basic mechanical constraints of this type of instrument.

A number of analogue stereoplotting instruments, such as the Zeiss Jena Technocart and the Wild A-40 Autograph have been built specifically for terrestrial applications. These instruments have a much larger depth range than their aerial counterparts, so that objects with significant depth ranges can be accommodated. However, all these special instruments, apart from the Technocart, have no facility to introduce rotations to the projector assemblies, so that great care must be taken when acquiring the photography to ensure that the camera axes are exactly parallel. It is extremely difficult to do this with single cameras, and so these analogue stereoplotting instruments can only be used efficiently if a stereometric camera (8.4.2) is used for the photography.

8.5.2 *Analytical methods*

All the measuring instruments used in analytical photogrammetric work are two-dimensional coordinate measuring instruments, the transformations from two-dimensional instrument readings to three-dimensional object coordinates being carried out mathematically, normally in a digital computer. The two main categories of instrument are comparators and analytical plotters. Recently, some low-cost systems have been developed using simpler measuring systems. As long as the necessary formulae and information are included in the software, almost any type of photography can be accommodated in an analytical solution subject to the limitations discussed previously (8.3, 8.4).

8.5.2.1 *Comparators.*

The simplest form of instrument for photogrammetric measurement is the *monocomparator* (Fig. 7.26) which is basically an x-y measuring microscope. The photograph is placed on a stage plate and the microscope (or, in some instruments, television camera) is tracked across the photograph to the required image point. In other instruments, the viewing head is fixed and the stage plate is moved. This produces two-dimensional comparator coordinates which first have to be transformed into plate coordinates and then to object coordinates. Because only one photograph can be measured at a time, only well-defined or targeted points can be measured separately on two or more photographs, unless the object is known to be flat. The *stereocomparator* provides more flexibility by incorporating two stage plates and a binocular stereoviewing system (Fig. 7.30), so, untargeted points can be measured stereoscopically. On the Zeiss Jena Stecometer (Fig. 8.8), for instance, the x and y movements are split between the stage plates and the viewing optics. Differential movements of the plates, px and py, enable the measuring marks to be placed over the same feature on both photographs. For each measurement the operator has to manipulate the four handwheels controlling these movements. The four readings, x, y, px and py, can be used to compute x and y machine coordinates for each of the two photographs.

Because comparators are mechanically very simple, they can be very precise. The Stecometer, for example, has a precision of measurement of $\pm 3\,\mu$m. Monocomparators can be even more precise; the Zeiss Jena Ascorecord has a precision of $\pm 0.5\,\mu$m. For special applications, the UK National Physical Laboratory has fitted laser interferometers to give a precision of $\pm 0.02\,\mu$m (Oldfield, 1986).

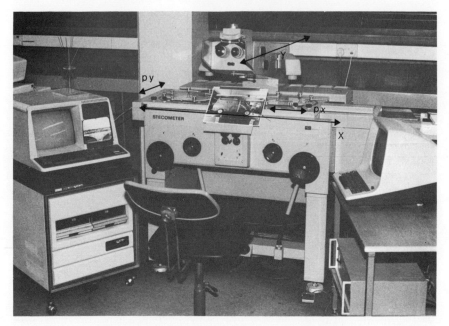

Figure 8.8 The Zeiss Jena Stecometer stereocomparator.

8.5.2.2 *Analytical plotters.* All comparators are *point* measuring devices and, although capable of very precise measurement, are not suited for continual line plotting. For this reason, the analytical plotter was developed to combine the high precision and flexibility of a stereocomparator with the continuous measuring facility of a stereoplotting instrument.

Most analytical plotters have been developed for aerial photogrammetric work and are very efficient in carrying out this task. Because of their theoretical ability to handle virtually any type of photography, they are also ideally suited to close-range photogrammetric work. This is especially the case if the analytical plotter can be fully integrated with a database system so that the collected data are made available for further handling and processing. However, some analytical plotters do not as yet have the necessary software to handle close-range photogrammetric data.

As discussed previously in Chapter 7, the two main classes of non-automated analytical plotter are the Helava type in which the inputs to the computer are the object coordinates, and the alternative type in which the input values are the comparator coordinates measured on the photographs. The differences between these two approaches are of major significance in close-range photogrammetry; the former type is far more flexible, since the machine can be driven under computer control to individual three-dimensional points which have been specified by the operator. If, for instance, a regular grid model is to be generated, then the grid spacing can be entered into the computer and the analytical plotter will be driven to the X, Y coordinate position of each required point in turn, leaving the operator to measure the Z-coordinate in each

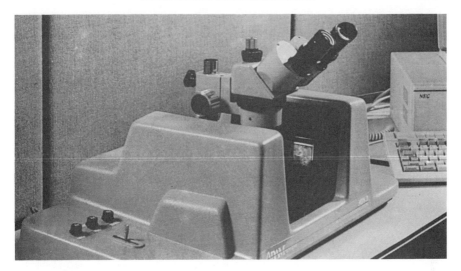

Figure 8.9 Adam Technology MPS-2 analytical stereoplotting instrument.

case. Similarly, the instrument can be programmed to follow specific profiles or cross-section lines across the object being measured.

While the standard type of analytical plotter able to handle large 230×230 mm photographs can obviously be used for close-range photogrammetric work, an interesting development has been the design and construction of low-cost analytical plotters based on the use of small format (up to 70×70 mm) photographs which have been designed with the solution of close-range photogrammetric problems as a primary objective. Examples of such analytical plotters are the Adam Technology MPS-2 (Australia) and the HDF Maco 35/70 (USA).

The MPS-2 (Fig. 8.9) follows the object coordinates primary approach of Helava with the X, Y positions and the height Z of a point being input to the computer via a pair of joysticks. The real-time programs—which solve the collinearity equations, ensure that a continuously oriented stereomodel is formed, and drive the machine to predetermined points—are executed by two internal microprocessors and a dedicated floating point numeric data processor. The remaining software tasks, such as absolute orientation, the control of an output device such as a drum or flatbed plotter, and the management of the measured and recorded data, all fall well within the capabilities of a small computer. The host software is written in FORTRAN and can be installed under a variety of operating systems, so ensuring compatibility with a wide range of machines. Thus the MPS-2 can be attached to a microcomputer such as the IBM-PC or NEC APC, or one of the smaller models within the DEC PDP-11 series of minicomputers.

The HDF Maco 35/70 is, as the name suggests, optimized for use with 35 mm or 70 mm photographs. It adopts the alternative algorithmic approach by which image coordinates are input by the operator, in this case under the control of a tracker ball. The Tektronix 4052 desktop computer used for the computational tasks outputs the small px and py corrections needed to keep the stereomodel in orientation. The system

Figure 8.10 Components of the Rolleimetric-MR system. Reproduced by kind permission of AV Distributors (London) Ltd.

is available complete with a non-topographic control set which provides an accurate control framework for individual stereomodels in the form of metal-framed cubes and crosses with precisely-known dimensions. Like the MPS-2, the HDF-Maco 35/70 is reasonably easy to transport to the site where the object is being measured.

The Rolleimetric-MR system is an even cheaper, more portable solution (Fig. 8.10). The system is designed for use with photography taken with the Rolleiflex range of cameras. The system consists of a digitizing tablet, microcomputer and x-y plotter. Enlarged paper prints are placed on the tablet and measured with a small hand cursor. Rolleiflex cameras are fitted with réseaux, so that by first measuring the images of the réseau crosses, the software will be able to correct for any image distortion introduced by the film and the enlarging and printing processes. The same feature is measured separately on two or more photographs, and a bundle solution is used to compute the object coordinates of the point measured. As no stereoviewing is possible, only well-defined discrete points can be measured. The system is very simple to use, and has the advantage for the non-photogrammetrist that good stereovision is not required.

8.6 Applications of close-range photogrammetry in engineering

It is impossible to include in one small section of a single chapter all the applications of close-range photogrammetry in engineering. Therefore only a few examples will be discussed to illustrate the scope of photogrammetry. It is hoped that the reader will be able to gain an appreciation of the flexibility of this measurement technique and will note a connection between requirements in other fields and some of the examples given below.

8.6.1 *Applications in civil engineering*

Aerial photogrammetry has long been used by civil engineers for providing base maps and digital terrain models for large construction projects, such as dams and road alignments. Conventional aerial photography is not economical, and often not capable of providing the required precision, for detailed analysis of small sites. Helicopters (Stanbridge, 1987) can be used to obtain photography from much lower altitudes than with conventional fixed-wing aircraft, and can be classed as close-range, although the techniques employed may be those of traditional mapping.

The classification of the examples which follow into the various subheadings is somewhat arbitrary, since many of the examples overlap more than one subject.

8.6.1.1 *Geotechnical engineering.* Many of the applications of close-range photogrammetry in geotechnical engineering are based directly on the early use of terrestrial photogrammetry in topographic mapping. Cheffins and Rushton (1970), for example, used an analogue stereoplotting instrument to produce contour plots of a rockface which were subsequently used in the design of stabilization measures. However, analogue methods of measurement are not always possible when photographing hillsides, because it is often very difficult to select camera stations for good coverage of the slope from which the resultant photography would be capable of insertion in an analogue stereoplotting instrument. Analytical techniques where highly tilted and convergent photography can be used are better suited to this application, especially if high accuracy is required and the taking distance, because of terrain limitations, is large. Chandler *et al.* (1987) describe such an approach to the recording of unstable hillslopes in Nepal in which directly measured contours and profiles and digital terrain models were produced.

Figure 8.11 shows a section of the interior lining of a reservoir which has suffered from slumping over a length of 50 m owing to water penetration below the clay lining. In this area three rows of monitoring points were located at approximately 10 m intervals. These had been coordinated by conventional surveying techniques at regular intervals of time to monitor the slumping of the slope. The main drawback of this procedure was that any local slumping occurring between these points would not be detected. Photogrammetry, on the other hand, records the entire area so that any localized movement can be measured. A series of control targets was placed throughout the area and coordinated by electronic tacheometer. The area was photographed with a Zeiss Jena UMK camera from a small dinghy (Fig. 8.12). The instability of the boat platform was not important in this instance, since the photogrammetric evaluation was to be performed analytically. By combining the survey and photogrammetry measurements in a bundle adjustment (8.3.5.3), a precision of ± 5 mm was obtained for the control points. Profiles were measured directly on an analytical plotter and a DTM was produced. These measurements were combined with the results of a borehole study to build up a complete picture of the slope failure and its causes.

In Fraser (1983) and Fraser and Gruendig (1984), the use of photogrammetry for detecting the deformation of Turtle Mountain in Canada is described. In 1903, over 90 million tons of rock collapsed, partially burying the town of Frank. Subsequently much work has been carried out in monitoring the remainder of the mountain to detect any further major instability before it becomes dangerous. This was originally restricted to the use of crack motion detectors, extensometers and levelling to monitor movement at a few points along one specific crack. Any movement occurring elsewhere was

Figure 8.11 Slumping of the interior lining of a reservoir. Part of a photograph taken with a UMK 10/1318 camera from a small dinghy.

Figure 8.12 Zeiss Jena UMK 10/1318 camera mounted in a small dinghy, used to take photograph in Fig. 8.11.

undetectable, so it was decided to use photogrammetry to monitor movement over a much larger area between a series of targets fixed to the mountain. Since the area of interest was around the south peak, aerial photographs were taken at a flying height of about 300 m above the peak. A bundle adjustment procedure was used and, given that the whole of the peak area was suspected of movement, the technique of inner constraints, the so-called 'free' adjustment (8.3.5.3), was adopted, resulting in the detection of deformations of less than 10 mm.

Kennie and McKay (1986) describe two examples where analytical close-range photogrammetry was used to monitor the erosion of chalk cliffs which were proving hazardous to the public, and discuss the results. This paper also includes a review of other literature on geotechnical applications.

8.6.1.2 *Mining engineering.* Many of the techniques discussed in the previous section can also be applied to mining engineering for determining the stability of, and volume of material removed from, open-cast mines and quarries. Whickens and Barton (1971) used photogrammetry to calculate the degree and direction of dip of various rock planes on a quarry face. Photogrammetry is also ideal for measuring tunnel profiles, and this application is one of the special cases of single-photograph photogrammetry mentioned in 8.3.5.2. A laser and rotating prism can be used to 'draw' the required profile line on the tunnel wall. Figure 8.13 shows an example of a disused railway tunnel recorded in this fashion. As can be seen, the laser line picks out many small irregularities in the brick lining of the tunnel wall, thus indicating areas which may need attention. The four triangular marks, two on each tunnel wall, are control point targets in the plane of the profile which are used for resecting the camera station. They are coordinated from traverse stations in the tunnel and are also used for relating the profiles to one other. The marking of the tunnel walls in this way allows the laser to be set up at the same section at a later date so that the deformation of the tunnel can also be measured. The plates were measured monocularly on a stereocomparator, resulting in dimensional information about the tunnel to a precision of ± 5 mm. This technique can also be employed for surveying mineshafts, where the laser is suspended below the camera to define a horizontal section through the shaft.

8.6.1.3 *Structural monitoring.* Conventional geodetic monitoring techniques provide information on movement at only a small number of points on the object because of the work involved in coordinating these points. Photography records the entire structure, and so many more points can be measured efficiently. If results of high precision are required, it is still necessary to fix artificial targets to the object. Figure 8.14 shows a railway viaduct with point targets being photographed for monitoring purposes (Cooper *et al.*, 1984) and Fig. 8.15 illustrates a series of results showing vectors of movement of the targets over approximately yearly intervals.

The measurement of very large objects can also be carried out efficiently. A ground-mounted air-survey camera was used to record the shapes of cooling towers (Chisholm, 1977). By superimposing the resulting contours over the design contours, deviations from the ideal shape were discovered, as were cracks in the concrete surfaces of the towers.

Cooper and Clark (1984) describe research into the use of photogrammetry for detecting deformations in large bridges. It was found possible to detect deformations of 1 mm over a 70 m span using just six photographs, taken in under two hours, and a

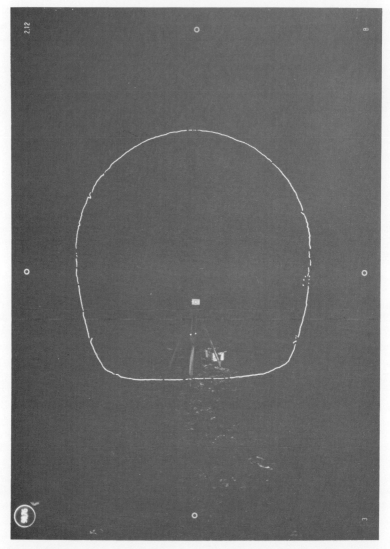

Figure 8.13 Zeiss Jena UMK 10/1318 photograph of laser line defining required tunnel profile.

'free' bundle adjustment. The interesting conclusion was reached that lack of flatness of the photographic plates (8.4.5) was the major factor limiting the accuracy of the method.

The ability of photography to 'freeze' motion is very useful for monitoring the deformations of objects under variable dynamic loading. If three-dimensional movements are required, it is essential that at least two cameras be used which are configured to fire synchronously. Thus, for example, structures which are being tested to destruction, can be photographed at regular intervals, and reliable information on the bending and twisting of the object can be extracted. Welsh (1986) describes a system

CLOSE-RANGE PHOTOGRAMMETRY

Figure 8.14 Railway viaduct being photographed with a Zeiss Jena UMK 10/1318 camera. Note targeted points on face of stonework.

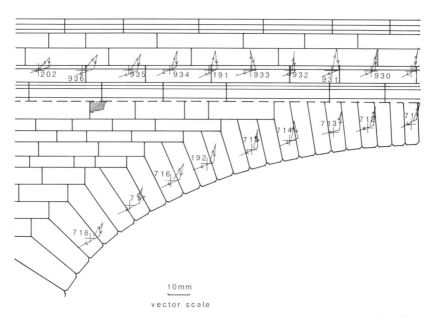

Figure 8.15 Series of displacement vectors showing movement of targeted points on the railway viaduct shown in Fig. 8.14.

used to monitor a test in which a series of steel pipes were bent until they cracked. As well as recording the deformation of the pipe sections, the system also recorded the positions of the slings and pulleys applying the loads and hence the directions of the loading being applied. This paper also describes a single-camera solution where a section of concrete-coated steel pipeline was lifted at one end by a crane, and the photographs used to measure the subsequent bending of the pipe. Because only one camera was used, the assumption was made that the pipe would deform vertically only and that no lateral deflection would take place.

A recent growth in the use of close-range photogrammetry has been for underwater inspection of supports and pipelines for oil rigs and platforms. This has proved to be a very useful technique because, as a result of the rapid data capture possible with photography, the time spent by an expensive diver operating in a dangerous environment has been greatly reduced. Camera systems can also be mounted on remotely operated vehicles (ROVs) which are operated from the surface using a TV link. This application requires an analytical solution, because the distortion caused by refraction of the light rays passing from the water into the camera has to be compensated for. It has been used for the production of design information for a repair patch to a damaged platform leg (Welsh *et al.*, 1980), and for monitoring the growth of corrosion pits on steel components (Leatherdale and Turner, 1983).

8.6.1.4 *Hydraulic engineering*. Photogrammetry, with its ability to freeze movement, can be a very useful tool in hydraulic engineering. Aerial photography can be used to monitor the deposition or erosion of sandbanks, the direction and speed of currents, and dispersion of pollution in rivers, lakes and seas. Close-range techniques have been used to measure wave heights (Mitchell, 1983). In order to freeze the wave movement, two cameras were fired simultaneously. Madejski and Lewandowski (1984) used phototheodolites positioned high on the sides of a Spitsbergen fjord to determine surface water currents by measuring the positions of ice on the surface.

8.6.1.5 *Architectural and archaeological recording*. One of the most common uses of close-range photogrammetry is in recording and measuring buildings, particularly historic buildings. One of the primary reasons for its popularity in this application is that, in many cases, architectural photogrammetry is simply aerial mapping 'tipped on its side'. Because the facades of many buildings are relatively flat compared with the taking distance, conventional mapping techniques using analogue plotting machines can be employed. In this case, the result is often a line 'map' of the building elevation. Figure 8.16 shows a section of wall with the corresponding plot from an analogue stereoplotting instrument in Fig. 8.17. This illustrates the problem involved in deciding how much detail should be plotted. The stonework at the bottom and the brickwork at the top of the wall have been plotted almost in their entirety, but in the rubble section in the middle of the wall only certain large stones have been plotted, because many of the edges of stones in this region are not clearly defined. In this case, the plot and photograph should be used in conjunction with each other, the plot providing dimensional information and the photograph providing the fine detail. Dallas (1983) gives a summary of architectural photogrammetry in the UK.

Although most architectural photogrammetry is carried out using analogue stereoplotting instruments, there is an increasing use of analytical plotters in this field. If an analogue instrument is used, then the photograph axes have to be positioned

Figure 8.16 Zeiss Jena UMK 10/1318 photograph of a section of London Wall.

normal to the facade of the building, and this often requires a hoist to raise the camera up the face of the building to obtain suitable coverage. With the analytical plotter's ability to use tilted photography, it is now possible to obtain coverage of many buildings by simply tilting the camera axes up from their ground-based positions (McKay et al., 1985).

As well as recording buildings, photogrammetry can be used to provide dimensional information for manufacturing components or replacements. Examples of the latter include determining the sizes of openings for windows and doors. As mentioned previously (8.4.4); Thompson (1962) used old photographs to provide dimensional information for the reconstruction of the dome at Castle Howard which had been destroyed by fire.

Photogrammetry is widely used in archaeology for recording sites and artefacts. The rapid data capture provided by photography is especially important in 'rescue' archaeology where speed is essential. Often a site is uncovered during excavation for a new development, and great pressure is placed on the archaeologist to record the information as quickly as possible so that construction, halted upon discovery, can recommence quickly. Instead of the slow manual measurement, recording and drawing of the site, a series of photographs, together with suitable control points can be taken in a short time, and drawings of the site produced later, often after the site has been buried again (Muessig, 1984). Fragile artefacts can also be recorded before they are moved and exposed to the associated risk of breakage.

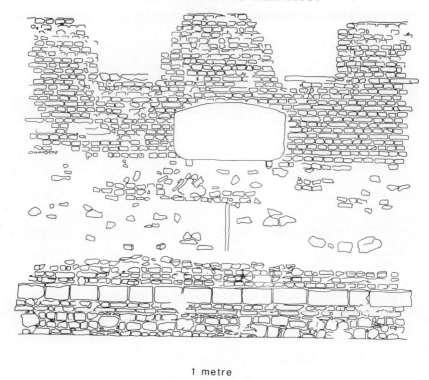

Figure 8.17 Section of plot of part of London Wall. Plotted on an analogue plotter from a stereopair, one photograph of which is shown in Fig. 8.16.

8.6.1.6 *Laboratory and model testing.* Close-range photogrammetry has been widely used in civil engineering for the recording of laboratory tests. Working under laboratory conditions, it is generally easier to select the optimum camera positions than in general site work.

In geotechnics, photogrammetry has been used for recording sand tank tests, often using sequential exposures with a single camera, and analysing pairs of photographs by time-parallax techniques. In this way, the two-dimensional movements of the sample behind the plane glass front of the tank result in an apparent three-dimensional image, thus enabling isolated areas of movement to be easily visible (Wong and Vonderhoe, 1981; Davidson, 1985). El-Beik (1973) actually built two UMK cameras into a soil centrifuge, analysis of the resultant stereopairs producing settlement and heave curves.

In structural engineering, photogrammetry has been used to monitor testing of sample materials (Toz, 1986) and scale models. Single-camera and time-parallax techniques can be used to measure deflections of beams, pipes and cables (Fig. 8.18) under test, as long as any movement towards the camera out of the plane parallel to the photograph is expected to be negligible. Multi-station bundle techniques can be used to provide precise three-dimensional movements of samples.

The ability of stereophotogrammetry to record complex three-dimensional shapes has been employed to translate physical models into digital form. Figure 8.19 shows an

Figure 8.18 Zeiss Jena UMK 10/1318 photograph of section of steel bridge cable under test. This is an example of single-camera photogrammetry where movement normal to the plane of the photograph is expected to be negligible.

architect's model of a proposed cable net structure, with complex and constantly changing curved surfaces. In order to carry out a structural analysis of the net, it was necessary that a digital model of the structure be produced. The model was photographed with a UMK camera mounted on scaffolding above the model. Control points and distances were provided by a grid drawn on the model baseboard and machined steel bars of known length. The photographs were measured in a stereocomparator. Figure 8.20 shows a computer-generated image of the model produced in a structural analysis package.

In hydraulic engineering, photogrammetry has been used to measure scour patterns in channel tests, where the non-contact aspect of the technique is of prime importance since mechanical probes inserted into the tank can alter the flow and thus affect the experiment. It has also been used to measure wave surfaces in the testing of model breakwaters and harbour complexes. In this instance, some method is required to make the water surface visible on the photographs. Matchsticks or other small floating objects can be used to provide a fusable stereoimage. Adams and Pos (1981) added oil to the water, which turned it milky white, and then projected a random pattern on to the white surface to create the necessary texture.

8.6.2 *Applications in mechanical engineering*

Close-range photogrammetry has a number of applications in mechanical engineering which are similar to some of those described previously. Fraser and Brown (1986) used multi-station bundle techniques to measure the surface form of radio antennae to a precision of ± 0.01 mm. Even this extremely high precision was not as good as expected,

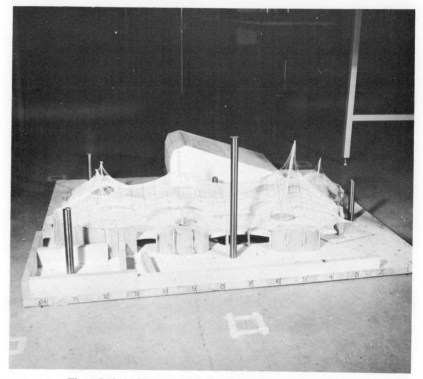

Figure 8.19 Architect's model of a proposed cable net structure.

and this was attributed to slight thermal variations in the antennae over the time taken to obtain the photographs. The same paper describes how photogrammetry has been used to build up the XYZ database of a component for the controlling of a robot milling machine to a precision of $\pm\,0.1$ mm. Photogrammetry can also be used to monitor the wear of tool jigs (8.6.3), and in producing design drawings from master models (8.6.4).

8.6.3 *Applications in aeronautical engineering*

Simple photogrammetric techniques have been employed in the aerospace industry for many years. For example, phototheodolites and cine cameras have been used to measure the take-off and landing runs of prototype aircraft and in recording airflows during wind tunnel testing. Kilburg and Rathburn (1984) describe how these techniques were improved to give high-precision results for the deformation of a model aircraft during aerodynamic loading in a wind tunnel. The authors also describe how photogrammetry was used to compare as-built aircraft sections with design sections. This detected a difference in wing incidence on a full-size aircraft compared with the designed incidence angle. A similar technique was used to measure the deformation of the wing under simulated loading applied by hydraulic jacks. The same paper discusses how photogrammetry can be used to check for wear of production tooling without the need to remove the jig from the assembly line. Fraser and Brown (1986) employed

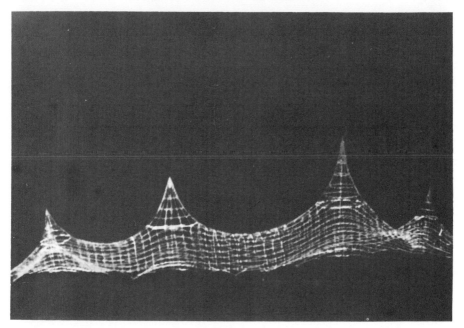

Figure 8.20 Digital model of the physical model shown in Fig. 8.19 produced photogrammetrically.

photogrammetry to provide input into a CADD* system to produce design information on the existing engine nacelle of a DC-8 aircraft. This was subsequently used to design a conversion kit to reduce engine noise emission.

Rolls-Royce have developed a photogrammetric system using X-ray photography to study the dimensional changes which take place inside aero engines as they are running (Stewart, 1979).

8.6.4 *Applications in automobile engineering*

Photogrammetry has been employed for crash testing in the car industry for many years, normally using high-speed cine-cameras and fairly crude analysis techniques. High-precision applications have been more recent and include the production of tooling drawings or digital data from models and components (Cooper, 1979; Wahl, 1983). The car manufacturer Renault uses photogrammetry to produce data on various driving positions by photographing a wide variety of people seated in a car, and then uses this information to improve seat design. Soft components such as padded dashboards and seats can be measured without deformation of the object, which can occur when conventional 3-D probes are used. Wahl (1984) also mentions the problem of various parts which do not fit when brought together for assembly. By photographing the individual components at each stage of assembly, the source of the problem can be detected. A major German manufacturer has recently installed an analytical plotter with digital image correlation (8.7.1) to produce design and production drawings from models.

*Computer Aided Design and Draughting.

8.6.5 *Applications in marine engineering*

Close-range photogrammetry is now a standard production tool in shipyards around the world. In his brief survey of photogrammetry in US shipyards, Chirillo (1984) notes that photogrammetry is used for the dimensional checking of major subassemblies of aircraft carriers and submarines before they are brought together. Another US shipyard has used photogrammetry for controlling the mating of the two halves of a number of supertankers, with a significant saving in time over more conventional methods (Newton, 1974; Peel, 1984). Photogrammetry has been used to record ship launches dynamically, to measure the location of the ship on the slipway, the heel of the ship, its velocity of movement and the horizontal and vertical deformations of the ship during launch (Szczechowski, 1982). A feasibility study carried out in the UK (Cooper, 1979) showed that photogrammetry could be used to measure the size and shape of the initial casting of ships' propellers to give information for machining to the final designed shape. As mentioned previously (8.6.1.3), photogrammetry has been used for the inspection and measurement of underwater structures in the North Sea and the production of information for prefabricating repair patches (Welsh *et al.*, 1980; Leatherdale and Turner, 1983).

8.6.6 *Applications in process engineering*

Oil refineries and chemical plants are large complex structures which have proved ideal subjects for the use of close-range photogrammetry. A major problem in modifying existing plants is to obtain accurate 'as-built' information, because this is often significantly different from the original design information. A photogrammetric system developed at ICI has been used to produce a three-dimensional database of various sections of plant in a form compatible with a standard CADD pipe design package (Bracewell and Klement, 1983). Precision multi-station bundle techniques have been used to measure the deformations of vessels and pipes due to thermal variations (Fraser, 1985) and other causes. Photogrammetry has also been used to study the flow of liquids and the effects of air bubbles in flows (Stewart, 1986; Welsh, 1986). The advantage of using these systems over conventional instrumentation is that there is no invasion, and hence alteration, of the flow being studied.

8.7 Automation of the photogrammetric process

Although photogrammetry is a very rapid form of data capture, some organizations have been reluctant to adopt close-range photogrammetry as a production tool because of the time interval between taking the photographs and obtaining the results. The film or plates have to be processed, dried and then measured. If these facilities are not available on site, then the delay can become unacceptable. Even with facilities on site, a delay of even one hour between photographing an object for, say, quality-control purposes and obtaining data for correcting the component can be too long. Systems are now under development which will reduce this delay or remove it altogether, and this is often described as *real-time photogrammetry* (see 8.7.2).

8.7.1 *Digital image correlation*

Some systems simply speed up the measurement process by using scanning systems to digitize the photographic image. The digital images from two or more photographs are then compared by a computer and correlated to produce three-dimensional coordinates of the object photographed. *Digital image correlation* is based on the numerous image processing techniques which have been developed for analysing remotely-sensed data from Earth satellites and aircraft-based non-photographic systems, such as thermal infrared sensors and radar. These have been described in more detail in Chapter 7.

The simplest form of instrument for this process is a single video camera, normally of the charged-coupled device (CCD) type, which scans a photograph in a raster pattern to build up a digital model of the photograph. A CCD camera has an array of discrete light sensitive elements in its focal plane. The light falling on each element creates a charge proportional to its intensity, which is first converted to a voltage and then, via an analogue-to-digital converter, into a numerical value for computer processing. The use of CCD arrays in remote sensing is discussed under sections 6.3.1.1, 1.5.1.3 and 7.5.6.1. A photograph such as Fig. 8.16 contains a vast amount of data which has to be correlated with similar amounts of data from other photographs, so increasing the computing requirements and time taken (Ackerman, 1984). One solution to this problem is to take photographs which contain greatly reduced amounts of data. The system developed by Geodetic Services Inc. (Fraser and Brown, 1986), for instance, utilizes photographs taken in a darkened environment using retro-reflecting targets on the object and a ring flash around the camera lens. In this way, the resulting photographs contain relatively few images, each with a high contrast, which are ideal for digital image correlation. With the GSI Autoset, it is claimed that five times the accuracy can be obtained in a tenth of the time than with manual measurement. A system such as this will produce very precise coordinates of discrete targeted points Although not suitable for mapping or architectural drawing applications, it is ideal for structural monitoring or the surface modelling of components.

For the digital analysis of more conventionally illuminated photographs, the analytical plotter (8.5.2.2) has proved to be an ideal instrument with its computer-controlled driving of the photograph stage plates. A number of analytical plotters have been modified to include two CCD cameras. Each camera digitizes the small portion of each photograph visible to the operator through the optics. The operator has to prepare the instrument manually so that the two cameras are viewing the same areas of the object. In this way, only relatively small amounts of data have to be correlated at a time. The system will then drive the stage plates automatically, and the correlation is carried out in 'real time'. The operator can monitor the procedure through the eyepieces and intervene if the system becomes confused. In this way, problems in correlation because of perspective or illumination (Torlegård, 1986) can be corrected by the operator.

8.7.2 *Real-time' photogrammetry*
The conventional photographic film camera can be dispensed with altogether and CCD cameras used to image objects directly, and this, when linked with correlation techniques, can produce coordinates immediately, in 'real time'. When CCD cameras or arrays are used to digitize a photographic image the relatively low resolution is not of major significance since only small areas are digitized at a time under magnification, resulting in a sufficiently fine model of the object. However, when they are used to digitize objects directly, their low resolution becomes significant. In all major tests

reported to date (El-Hakim, 1986; Haggrén, 1986) they have been used on objects with relatively large, high-contrast targets. Haggrén notes that conventional photogrammetric techniques can achieve accuracies of one part in 250 000 of the size of the object, whereas at present 'real-time' systems only yield accuracies of 1:5000. Haggrén suggests that a laser can be used to produce a target spot on the object which can then be moved across the surface of the object, its coordinates constantly being computed from the CCD-generated images. 'Real-time' techniques are at present confined to laboratories, although applications are envisaged using CCD images to control robots on production lines (Pinkney and Perratt, 1986; Real, 1986). Continued developments in CCD technology and image processing techniques may make 'real-time' photogrammetry more accurate and flexible.

8.8 Conclusions

Close-range photogrammetry is a very powerful and flexible measuring technique for the surveyor and engineer, capable of producing very precise results in a wide variety of applications. With the advent of cheaper and faster computers and simpler measuring systems, more and varied uses of close-range photogrammetry will no doubt appear. It has to be admitted, however, that electronic tacheometers and electronic coordinate determination systems (Chapter 1) have displaced close-range photogrammetry in a number of organizations because of the time delay between taking the photographs and obtaining the results. 'Real-time' photogrammetric techniques may, however, prove more effective in the long term, when the resolution of these systems improves considerably. However, if a large amount of dimensional data is required, then close-range photogrammetry can compete effectively against other measuring techniques in terms of efficiency and accuracy. Its competitiveness is even greater if photogrammetric and electronic tacheometric measurements are combined using bundle techniques (8.3.5.3), resulting in a very rapid and flexible data acquisition system. Only photogrammetry can 'freeze' motion, and record an image in an instant in time. Furthermore, photographs provide a *permanent record*. They can always be re-examined in the future if more information is required or more measurements need to be taken. Photogrammetry is unique in this respect compared to other measuring processes.

Bibliography

Atkinson, K.B. (ed.) (1980) *Developments in Close Range Photogrammetry—1*. Applied Science Publishers, Barking, 222 pp.
Burnside, C.D. (1985) *Mapping from Aerial Photographs*. 2nd edn., Collins, London, 348 pp.
Karara, H.M. (ed.) (1979) *Handbook of Non-topographic Photogrammetry*. American Society of Photogrammetry, Falls Church, Virginia, 206 pp.
Slama, C.C. (ed.) (1980) *Manual of Photogrammetry*. 4th edn., American Society of Photogrammetry, Falls Church, Virginia, 1056 pp.

References

Abdel-Aziz, Y.I. and Karara, H.M. (1971) Direct linear transformation from comparator coordinates into object space coordinates in close range photogrammetry. *Proc. American Society of Photogrammetry Symp. on Close Range Photogrammetry*, Urbana, 1–18.

Ackerman, F. (1984) Digital image correlation, performance and potential applications in photogrammetry. *Photogramm. Record* **11** (64) 429–439.
Adams, L.P. and Pos, J.D. (1981) Model harbour wave form studies by short range photogrammetry. *Photogramm. Record* **10** (58) 457–470.
Bracewell, P.A. and Klement, U.R. (1983) The use of photogrammetry in piping design. *Proc. Inst. Mech. Engrs.* **179**a (30), 14 pp.
Burch, J.M. and Forno, C. (1982) Preliminary assessment of a new high precision camera. *Int. Arch. Photogramm.* **24** (V/1) 90–99.
Chandler, J.H., Clark, J.S., Cooper, M.A.R. and Stirling, D.M. (1987) Analytical photogrammetry applied to Nepalese slope morphology. *Photogramm. Record* **12** (70) 443–458.
Cheffins, O.W. and Rushton, J.E.M. (1970) Edinburgh Castle Rock, a survey of the north face by terrestrial photogrammetry. *Photogramm. Record* **8** (46) 417–433.
Chirillo, L.D. (1984) The history of photogrammetry in U.S. shipbuilding. *Close Range Photogrammetry and Surveying: State of the Art. Proc. of part of the American Society of Photogrammetry American Congr. on Surveying and Mapping 1984 Fall Convention*, 754–757.
Chisholm, N.W.T. (1977) Photogrammetry for cooling tower shape surveys. *Photogramm. Record* **9** (50) 173–191.
Cooper, M.A.R. (1979) Analytical photogrammetry in engineering: three feasibility studies. *Photogramm. Record* **9** (53) 601–619.
Cooper, M.A.R. (1985) Photogrammetric measurement of deformation. *Proc. 2nd UK Nat. Conf. on Land Surveying and Mapping*, Paper F3.
Cooper, M.A.R. and Clark, J.S. (1984) Final report on a feasibility study of the use of photogrammetry for health monitoring of bridges. *TRRL Contract Report* TRR/842/392, UK Transport and Road Research Laboratory, Crowthorne, Berks., 75 pp.
Cooper, M.A.R., Lindsey, N.E. and Stirling, D.M. (1984) Measuring the three-dimensional movement of a large stone structure. *Int. Arch. Photogramm.* **25** (A5) 214–222.
Dallas, R.W.A (1983) Plumb-bob to plotter: developments in architectural photogrammetry in the United Kingdom. *Photogramm. Record* **11** (61) 5–27.
Davidson, J.L. (1985) Stereophotogrammetry in geotechnical engineering research. *Photogramm. Eng. and Remote Sensing* **51** (10) 1589–1596.
El-Beik, A.H.A. (1973) Photogrammetry in centrifugal testing of soil models. *Photogramm. Record* **7** (41) 538–554.
El-Hakim, S.F. (1986) A real time system for object measurement with CCD cameras. *Int. Arch. Photogramm.* **26** (5) 363–373.
Fraser, C.S. (1982) Film unflatness effects in analytical non-metric photogrammetry. *Int. Arch. Photogramm.* **24** (V/I) 156–166.
Fraser, C.S. (1983) Photogrammetric monitoring of Turtle Mountain: a feasibility study. *Photogramm. Eng. and Remote Sensing* **49** (11) 1551–1559.
Fraser, C.S. and Gruendig, L. (1984) The analysis of photogrammetric deformation measurements on Turtle Mountain. *Proc. Engineering Survey Conference, FIG Commission 6, Washington*, 90–103.
Fraser, C.S. (1985) Photogrammetric measurement of thermal deformation of a large process compressor. *Photogramm. Eng. and Remote Sensing* **51** (10) 1569–1575.
Fraser, C.S. and Brown, D.C. (1986) Industrial photogrammetry: new developments and recent applications. *Photogramm. Record* **12** (68) 197–217.
Gates, J.W.C., Oldfield, S., Forno, C., Scott, P.J. and Kyle, S.A. (1982) Factors defining precision in close-range photogrammetry. *Int. Arch. Photogramm.* **24** (V/I) 185–195.
Granshaw, S.I. (1980) Bundle adjustment methods in engineering photogrammetry. *Photogramm. Rec.* **10**(56) 181–207.
Haggrén, H. (1986) Real-time photogrammetry as used for machine vision applications. *Int. Arch. Photogramm.* **26** (5) 374–382.
Kennie, T.J.M. and McKay, W.M. (1986) Monitoring of geotechnical processes by close range photogrammetry. *Proc. 22nd Ann. Conf. of the Engineering Group of the Geological Society*, Plymouth Polytechnic, September 1986.
Kilburg, R.F. and Rathburn, S.J. (1984) Application of close-range photogrammetry at General Dynamics Fort Worth Division. *Close Range Photogrammetry and Surveying: State of the Art. Proc. of part of the American Society of Photogrammetry American Congress on Surveying and Mapping 1984 Fall Convention*, 705–714.

Leatherdale, J.D. and Turner, D.J. (1983) Underwater photogrammetry in the North Sea. *Photogramm. Record* **11** (62) 151–168.

Madejski, P. and Lewandowski, M. (1986) The measurements of the surface water currents of Hornsund Fjord by photogrammetric methods. *Int. Arch. Photogramm.* **26** (5) 217–224.

McKay, W.M., Dowman, I.J. and Farrow, J.E. (1985) Experiences with analytical plotters. *Photogramm. Record* **11** (66) 729–743.

Mitchell, H.L. (1983) Wave heights in the surf zone. *Photogramm. Record* **11** (62) 183–193.

Muessig, H. (1984) Mapping archaeological sites using ground-based photogrammetry. *Close Range Photogrammetry and Surveying: State of the Art. Proc. of part of the American Society of Photogrammetry American Congress on Surveying and Mapping 1984 Fall Convention*, 394–404.

Newton, I. (1974) Dimensional quality control of large ship structures by photogrammetry. *Photogramm. Record* **8** (44) 139–153.

Oldfield, S. (1986) Photogrammetric plate measuring facilities at NPL. *Int. Arch. Photogramm.* **26**(5) 541–545.

Peel, D.D. (1984) Joining ships built in halves using close range photogrammetry. *Close Range Photogrammetry and Surveying: State of the Art. Proc. of part of the American Society of Photogrammetry American Congress on Surveying and Mapping 1984 Fall Convention*, 722–740.

Pinkney, H.F.L. and Perratt, C.I. (1986) A flexible machine vision guidance system for 3-dimensional control tasks. *Int. Arch. Photogramm.* **26** (5) 414–423.

Real, R.R. (1986) Components for video-based photogrammetry of dynamic processes. *Int. Arch. Photogramm.* **26** (5) 432–444.

Shortis, M.R., 1981. Computer aided orientation of terrestrial models on the Zeiss (Jena) Topocart B. *Photogramm. Record* **10** (58) 481–491.

Stanbridge, R. (1987) Close-range photogrammetric techniques. *Land and Mineral Surveyor* **5** (4) 178–184.

Stewart, P.A.E. (1979) X-ray photogrammetry of gas turbine engines at Rolls-Royce. *Photogramm. Record* **9** (54) 813–821.

Stewart, P.A.E. (1986) The non-invasive measurement of void fraction and velocity in two-phase flow using high-speed photography and videophotogrammetry. *Photogramm. Record* **12**(67) 5–24.

Szczechowski, B. (1982) Photogrammetric study of the course of a ship launching. *Int. Arch. Photogramm.* **24** (V/2) 497–506.

Thompson, E.H. (1962) Photogrammetry in the restoration of Castle Howard. *Photogramm. Record* **4** (20) 94–119.

Torlegård, A.K.I. (1986) Some photogrammetric experiments with digital image correlation. *Photogramm. Record* **12** (68) 175–196.

Toz, G. (1986) The applicability of close-range photogrammetry in structural model testing. *Int. Arch. Photogramm.* **26** (5) 250–257.

Wahl, M. (1983) Photogrammetry at Regie Renault. *Photogramm. Record* **11** (62) 195–201.

Wahl, M. (1984) Industrial photogrammetry at Renault. *Close Range Photogrammetry and Surveying: State of the Art. Proc. of part of the American Society of Photogrammetry American Congress on Surveying and Mapping 1984 Fall Convention*, 741–749.

Welsh, N., Leadbetter, I.K., Cheffins, O.W. and Hall, H.M. (1980) Photogrammetric procedures for a North Sea oil rig leg repair. *Int. Arch. Photogramm.* **23** (B5) 474–483.

Welsh, N. (1986) Photogrammetry in engineering. *Photogramm. Record* **12**(67) 25–44.

Wickens, E.H. and Barton, N.R. (1971) The application of photogrammetry to the stability of excavated rock slopes. *Photogramm. Record* **7** (37) 46–54.

Wong, K.W. and Vonderhoe, A.P. (1981) Planar displacement by motion parallax. *Photogramm. Eng. and Remote Sensing* **47** (6) 769–777.

PART C

DEVELOPMENTS IN COMPUTER BASED MAPPING, TERRAIN MODELLING AND LAND INFORMATION SYSTEMS

The previous two parts of the book have dealt primarily with the instrumentation used for the acquisition of survey data together with the processing techniques required to provide 3D positional information, mainly of objects in the terrain. This final part also deals with the instrumentation used for the acquisition of survey data, in this case from existing map sources, together with the technology and techniques used for the display, management and output of this information.

The use of computer technology in mapping leads to a number of obvious advantages. Computer systems enable maps and other products to be produced more quickly, with less reliance upon manual draughting skills. Modifications are easily made and can be used to satisfy specific user needs, perhaps varying scale or content, or to allow experimentation with graphical presentation of data. The revision of existing maps can also be carried out successfully using digital methods and, in many cases, lead to significant cost savings in comparison with manual methods. Using computer technology for these operations is not a recent phenomenon and the production of maps by digital methods has been carried out in the UK, for example, for almost twenty years. What is relatively new is the ability to perform these operations using lower cost computer systems, with consequent benefits in increased production rate and reduced costs. The other major area of development has been in methods of analysing geographically referenced data and the provision of tools to enable users to store, manipulate and retrieve such information from a variety of sources and to answer questions based on this data.

In common with the introduction of computers into many other activities there is often a degree of mystique associated with their use. Blind acceptance of the results may also be a problem, often caused by a confused and superficial understanding of the capabilities of the systems. In many cases, this confusion is exacerbated by the technical literature on the subject which frequently suffers from excessive dependence upon technical jargon and ill-defined terminology. Thus this part of the book is an attempt to provide a more ordered and systematic approach to the basic principles of computer technology as it relates to digital mapping, terrain modelling and land information systems. It is perhaps useful at this stage to differentiate between the topics to be discussed in this part of the book.

Digital mapping is concerned with automated methods of producing topographic maps, generally at medium to large scales. The terms computer cartography, automated cartography, computer assisted cartography or simply automated mapping (AM), are also used to describe this process. The term actually employed tends to vary according to the country and to the discipline within which it is used. For example, computer cartography is more common in geographical circles and is often used to describe automated thematic mapping generally at small scales; digital mapping tends to be the preferred term in surveying and engineering practice in the UK.

Chapter 9 (Digital Mapping Technologies, Procedures and Applications) provides a comprehensive review of the hardware used for digital mapping. The principles involved in the operation of the digitisers used to acquire coordinate data from existing maps, the display devices, and the cartographic plotters used to produce hard copy maps are all considered. A description of a number of typical digital mapping systems is then presented. This section illustrates the varied approaches to the acquisition of the original survey data, the manner in which they can be processed and the range of applications of the data. The emphasis throughout the chapter is on the technologies and procedures used for the production of large and medium scale topographic mapping, primarily for engineering purposes.

Digital terrain model (DTM) A DTM consists of a set of discrete height points, referenced to a national grid or some other geographic reference system which defines the Earth's surface. The process of modelling also involves the use of mathematical techniques to generate surfaces representing the irregular surface of the Earth or of proposed engineering designs. A DTM with the associated software for manipulating such data is now a well established element of the Computer Aided Design (CAD) packages and procedures used for the planning and detailed design of most major civil engineering projects.

Chapter 10 (Digital Terrain Modelling) concentrates on the fundamental principles of terrain modelling. The first part of the chapter reviews the alternative approaches which can be adopted for the creation of terrain models and the interpolation of contours from either gridded or randomly located data sets. The second half of the chapter deals with specific systems and with applications of terrain modelling in the area of civil engineering.

Land/geographic information systems (LIS/GIS) These have much in common with digital mapping and terrain modelling and to those unfamiliar with the subject, may appear to duplicate the previous two topics. The most significant distinguishing features of LIS/GIS are: the inclusion of a database to store both graphic and attribute data about individual features or areas present in the landscape; a database management system (DBMS); and the requisite software or tools to analyse the data held in the system. Systems which are designed for or applied primarily to small scale, mainly area-based, thematic data (e.g. soil types, agricultural or population census data, socio-economic classifications) are usually termed *geographic information systems (GIS)*. The term *Land information system (LIS)* tends to be associated with those systems which are based on the more detailed and more accurately surveyed point and line information required for cadastral and administrative purposes, or for the support and management of the facilities provided by public utilities. It must be said, however, that this distinction between GIS and LIS is by no means universally applied.

The final chapter in the book, Chapter 11 (Land Information Databases) deals with this rapidly developing field of land and geographic information systems. In particular, it discusses the data structures, file structures and databases used for handling topographic and attribute data acquired by many of the instruments described in the first two parts of the book. The latter part of the chapter highlights some typical systems and a number of applications related particularly to land registration and administration and the information used in the management of public utilities.

9 Digital mapping technology: procedures and applications

G. PETRIE

9.1 Introduction

The introduction of digital mapping techniques to engineering surveys began in the early 1970s. At that time, the use of these techniques was confined to a few large organizations which had the financial resources available to meet the cost of acquiring and operating the then very expensive computer hardware (mainframe computers and minicomputers) and associated peripherals (digitizers, graphics terminals, automated coordinatographs) and also had the volume of work to justify and realize the benefits of the very large capital investment required. Since then, both the technology and the methodology have developed apace. The result of this development has been a huge increase in the number of organizations employing digital mapping methods, largely as a result of the dramatic reduction in cost and the increased computational power of the new technology. In 1988, with the availability of inexpensive microcomputers and powerful graphics workstations and the widespread knowledge as to how to use these devices in a fairly sophisticated manner, there are few organizations engaged in survey work for engineering purposes which do not employ computer-based methods of processing the survey data and delivering it to the client in graphic or digital form.

In parallel with this development, digital mapping techniques have also been used widely, in for example thematic mapping carried out by geographers; for production of multicoloured topographic, soil and geological maps by government mapping agencies; and for production of multiple-page atlases by cartographic publishing houses. However, the cartographic design and thus the techniques used in such mapping operations are often very different to those used in engineering survey work. For example, the monochrome polygon-based choropleth maps, based on the use of dot or cross-hatched patterns, which are produced by geographers, or the multiple-colour area-filled maps of a school atlas based on the use of variable-density tint screens, use quite different techniques to those employed in engineering survey work. Thus they will not be covered in this chapter. Instead, the primary area of concern will be the high-accuracy surveys carried out specifically for engineering projects and the large-scale topographic maps produced by national survey organizations and private survey companies on which the planning of civil engineering projects is usually based.

9.1.1 The justification for digital mapping
The reasons for adopting digital mapping techniques vary widely from one organization to another, but there are certain objectives which are shared. The first is the desire

to speed up the process of map production so as to shorten the period between the initial data collection in the field or in the photogrammetric machine, and the availability of the resulting map in digital or hard-copy form for use by engineers, architects or planners. In this respect, it must be noted that the development of automated computer-based techniques for mapping is also partly a response to the vastly increased rates of survey data collection. For example, the development and adoption of electronic theodolites and distance-measuring equipment for field survey work, the widespread use of stereoplotting instruments in aerial survey work, the development of continuous-logging echo-sounders and electronic position-finding systems for hydrographic survey operations have all resulted in a several-fold increase in the speed of measurement or acquisition of survey data. In order to cope with the increased flow of data, automated techniques of processing and outputting this data were a necessity.

Closely associated with these considerations is the desire to reduce or even eliminate much of the tedious yet demanding cartographic work, such as compilation, draughting, scribing, mask-cutting, lettering and symbol generation and placement, which requires highly skilled personnel who are often difficult to find. The objective of reducing the requirement for large numbers of skilled draughtsmen is of course one that is shared by most other areas in civil engineering which require large numbers of drawings, for example those associated with construction of buildings, roads, dams, sewers, and water, gas and oil pipelines. In many cases, the automation of these draughting activities, often conducted on a massive scale, has been one of the first areas where engineering organizations have implemented computer-based techniques. Since many, though not all, of the techniques used in computer-aided design and draughting (CADD) are similar to those used for digital mapping work, the experience gained in the one area could be deployed in the other. In this respect, the fact that the terrain data generated by the survey and mapping operation is the input to the engineering design stage has often resulted in a close integration of the respective computer-based systems.

An area of great activity which is very closely related to engineering and has a very large requirement for maps is the industry concerned with the provision of water, gas, electricity, telephone and sewerage services—the so-called public utilities. Their networks of pipelines and cables are numerous, widespread and very complicated in structure, and their exact location and function need to be known for planning, operational and maintenance purposes. Much of this data needs to be recorded and displayed on maps and kept up to date for management purposes. Thus the public utilities have always been large customers for plans and maps. However, they also need their maps to be revised continually, both in terms of the basic topographic information which they contain and in respect of the specialist information drawn, overlaid or annotated on them. The promise of being able to achieve this desired currency of information by adopting digital mapping techniques has therefore been a considerable factor in causing these large, technically-aware and capital-intensive industries to become interested in digital mapping. This has extended both to putting pressure on national mapping organizations to provide their basic topographic information in digital form and in playing a prominent role in the development and implementation of digital mapping systems.

Another declared objective for adopting digital techniques is the reduction in the *cost* of map and plan production. In practice, this has often been very much harder to achieve than the increased speed of map production. The capital costs of purchasing, installing, operating and maintaining computer-based mapping equipment have fallen, but are still quite high, especially if high accuracy is to be maintained throughout the production process. Furthermore, highly trained (and expensive) specialist personnel capable of operating, programming and maintaining the equipment need to be acquired and retained within the surveying or engineering organization. However, the situation regarding costs will vary greatly from country to country. In less developed countries, the availability of large numbers of comparatively low-paid staff who can be trained to carry out traditional draughting techniques using simple, inexpensive equipment creates a very different situation to that in highly developed countries where high labour costs and shortages of skilled staff are often encountered. Given the fact that the less developed countries also experience an acute shortage of capital and foreign exchange and often lack the essential infrastructure (reliable power supplies, ready availability of spares and service, widespread knowledge and understanding of computer-based technology and techniques) required to implement digital mapping techniques successfully, obviously there is less justification or pressure to adopt these techniques.

9.1.2 *Relationship of digital mapping with geographic and land information systems and digital terrain modelling*

A matter of considerable importance to those carrying out digital mapping operations, especially those concerned with data acquisition, is the fact that the data acquired during field survey and photogrammetric operations or derived from existing maps may also be used as the geographically referenced data which forms the basis of a geographical information system (GIS) or land information system (LIS). This is a structured collection of digital data (the database) which is controlled by a database management system (DBMS) and is usually envisaged or implemented as a query answering system. Burrough (1986) defines such a system as one which has a powerful set of tools for collecting, storing, retrieving at will, transforming and displaying spatial data from the real world for a particular set of purposes. In the case of a GIS or LIS, the emphasis will be on geographically referenced data which can be displayed graphically, i.e. in map form. Such systems are being implemented by government planning agencies, public utilities, and cadastral or land registration services. If the data collected by surveyors, photogrammetrists and cartographers is to be used in such an information system, then it will have to be arranged or structured in a manner which suits the purpose of the particular GIS or LIS system with which it is associated.

Alternatively, that part of the measured topographic data which describes the terrain morphology—principally the measured spot heights, contours and breaklines—may be used to carry out digital terrain modelling. This can be used directly in aircraft simulators or, in the context of this present volume, input as basic data to engineering design packages used for the design of roads, dams and reservoirs, where earthwork quantities need to be generated, and the impact of the finally designed structures on the surrounding landscape may need to be visualized before their actual construction.

The specific areas of digital terrain modelling and of geographical and land information systems will be covered in Chapters 10 and 11. This chapter will concentrate on the technology used in all these closely related areas and will deal specifically with its application to high-accuracy digital mapping as defined above.

9.2 Data acquisition for digital mapping

A matter of fundamental importance to any type of digital mapping is the acquisition of coordinate data in computer-compatible form which has been measured with sufficient accuracy and which has been encoded, structured and labelled with appropriate descriptive codes and attributes to enable it to be used in the available system. Essentially there are three main methods by which this data may be acquired for input to a digital mapping system. These are:

(i) Ground survey methods, preferably based on the use of electronic tacheometers
(ii) Aerial survey methods based on the use of stereo-plotting instruments
(iii) Graphics digitizing methods by which the data contained in existing plans and maps may be digitized, i.e. converted to digital form.

Table 9.1 sets out the main characteristics of each of these methods of data acquisition for a digital mapping system in the specific context of its application to engineering survey work.

The instrumentation and methods used for digital data acquisition by field survey and photogrammetric instruments have already been covered, in Chapters 1 and 7 respectively. Thus only the technology and procedures used to digitize existing cartographic material, i.e. plans and topographic maps, will be covered in this chapter. This will be followed by a description of those graphics output devices—including both display and hard-copy devices—which are commonly used in all digital mapping operations and in the closely associated terrain modelling and geographical information systems described in Chapters 10 and 11. Finally, a few representative examples of digital mapping systems will be described.

9.3 Cartographic digitizing technology and procedures

From the digitizing point of view, the features contained in existing plans and maps can, in the first instance, be divided into two types of data:

(i) Point data—for example, the individual objects depicted by cartographic symbols, spot heights, or the locations of names
(ii) Line data—for example, continuous linear features such as roads and railways, rivers and canals, vegetation and forest boundaries, walls and fences, building outlines, administrative boundaries and contours.

Normally the amount of point data present in a large-scale plan or map comprises only a tiny proportion of the total data contained in the document. Furthermore, this type of data does not lend itself to automated methods of data capture due to its intrinsic nature and its representation in cartographic form. Symbols, names, and spot height values almost always require human interpretation and the entry of descriptors or codes to enable the computer programs to handle the data. Point data is therefore almost invariably entered by a human operator who will measure the coordinate position using a manual digitizing device and enter the associated text, codes or attribute information via a keyboard or menu.

By contrast, line data forms by far the largest proportion of the data which can be derived from plans and maps. It may be digitized either manually by a human operator (as for point data) or by an automated or semi-automated device. Contrary to many people's expectations, it is still, at the present time, economic to carry out the digitizing

Table 9.1

Source of digital mapping data	Instrumentation and methods used	Accuracy of digitized data	Areal coverage	Typical applications
Field survey	Electronic tacheometry using total or semi-total stations equipped with digital data collectors	Very high—measurements made in static mode only	Limited to specific sites of small areal extent	Small area site planning and design—housing estates; individual large buildings; road improvement schemes
Photogrammetric measurements	Analogue or analytical stereo-plotting machines equipped with encoders and recorders. 3-D coordinates (X, Y, Z or E, N, H) are recorded digitally	(i) High—if discrete point measurements made in a static mode. (ii) Lower—if continuous plotting of line detail and contours carried out in a dynamic mode	Larger area projects, especially in rough terrain	Large engineering projects of considerable areal extent—dams, reservoirs, roads, railways, open cast mining
Existing survey plans and topographic maps	(i) Manual point and/or line following digitizing (ii) Semi-automatic line-following digitizing (iii) Fully automatic raster scan digitizing	Lower—quality dependent on the scale and accuracy of the existing plan or map. Field survey may be necessary for map revision (updating). Supplementation of height and contour information may be required—carried out by field or photogrammetric methods	From small areas covered by large scale maps to extensive areas usually covered by maps at medium scales	Preliminary planning and design based on existing plans and maps. Land acquisition. Also landscape representation and visualization e.g. for environmental impact studies

of line features manually. This is especially the case with the planimetric detail present in highly developed urban areas which is of a high density and great complexity in terms of land use and building type. Furthermore, many of these linear features split up and re-combine and many others cross or intersect. In addition, discontinuous lines, pecked and dotted lines, or lines of different colour may have to be digitized. All of these

Table 9.2

Line following ⟶ Manual operation
Raster scanning ⟶ Automatic/semi-automatic operation

(with cross-links between the two pairs)

Table 9.3

Mode	Measurement	Features	Characteristics of	
			(i) Manual operation	(ii) Automatic operation
(a) Line following	Points measured along the lines only	1. Selective- only the required lines are measured 2. Less data to be recorded and stored 3. The length of time required for measurement is related to the total length of line to be digitized	1. Low speed of measurement 2. Relatively inexpensive hardware 3. Relatively easy feature coding 4. Low speed data recording	1. Very high speed measurement and data recording 2. Expensive hardware 3. Operator intervention required for coding, etc.
(b) Raster scanning	The whole area of the map is scanned	1. Not selective- whole sheet is scanned and measured irrespective of whether data is present or not 2. Very large amounts of data need to be recorded and stored, much of it a zero record 3. The length of time required for measurement is related to the size of map sheet and the resolution of each scan line	1. Low speed 2. Enormous time required NB: not practicable to implement	1. Very high speed measurement and data recording 2. Very expensive hardware 3. Need for a separate feature coding operation 4. Very extensive post-measurement processing including editing of artefacts and vectorization of data

are better handled by a human operator equipped with a manually-controlled digitizer. By contrast, other types of line feature, such as contours which do not exhibit these complexities of line crossing, interpretation and classification are more susceptible to automated or semi-automated methods of digitizing. This can be important in hilly terrain and at medium scales where contours may comprise 70% or more of the topographic map detail requiring conversion to digital form.

Basically there are two main methods by which the line data on a map may be measured and digitized. These are (i) line following, and (ii) raster scanning.

Combining these with the two possibilities of manual and automatic or semi-automatic operation gives four possible techniques for the digitizing of line data on plans and maps, as shown in Table 9.2.

A summary of the main characteristics of each of these four possibilities is presented in Table 9.3. It will be seen from these characteristics that only three of the four possible methods of digitizing the detail on existing topographic maps can be implemented in practice. These are:

(i) Manual point and line-following methods of digitizing
(ii) Automatic or semi-automatic line-following digitizing methods
(iii) Automatic raster-scan digitizing of complete map sheets.

9.3.1 *Manual point and line-following digitizers*
A very large number of manually-operated digitizers are available on the market, but these may conveniently be regarded as falling into one of two classes: mechanically-based digitizers or tablet digitizers.

9.3.1.1 *Mechanically-based manual digitizers.* Originally, most manual digitizing was carried out using digitizers utilizing mechanical slides equipped with a measuring cursor and linear or rotary encoders to generate the rectangular (x, y) coordinate positions of the planimetric and contour data (Figs. 9.1, 9.2). The operator then carried out the measurement of the positions of individual point features using a cursor equipped with a measuring mark. In the case of linear features, each line was also measured individually. Either the operator decided the locations of the successive points required to best represent the specific line-feature—the so-called 'point mode' of digitizing—or the selection of the positions along the line was executed automatically by the digitizer electronics operating in the so-called 'stream mode'. In this latter mode, the interval between the recording of successive points would be pre-set by the operator to work either on a time base (e.g. 0.1 to 10 seconds) or a distance base (e.g. 0.1 to 2.0 mm). The coordinate data would usually be recorded on paper or magnetic tape.

A special type of mechanically-based digitizer employing cross-slides which was widely used for digitizing data on existing maps in the late 1970s was the *Pencil Follower*, originally manufactured by d-Mac (later part of Ferranti-Cetec). In this type of device, the cross-slide and its supporting rails were positioned below the surface on which the map was placed (Fig. 9.3). The measurement of the line detail was carried out by the operator using a cursor equipped with cross-hairs, around which a field coil was placed. This generated an electric field which was picked up by sensors (pick-up coils) mounted on a trolley, which in turn was mounted and could move along the cross-slide. When the cursor was located exactly above the trolley, the signals received by the opposite coils were equal, since they were in balance. As the cursor was moved by the

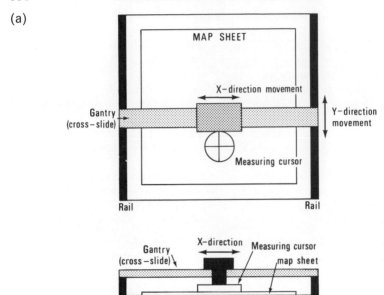

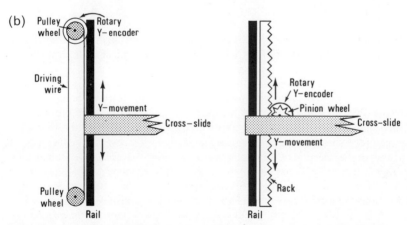

Figure 9.1 (*a*) Cross-slide digitizer; plan view (upper); section (lower) (*b*) Linear-to-rotary converters—using pulleys and wire (left); using rack and pinion (right).

operator to follow the line features, the signals detected by the pick-up coils went out of balance. This signal then activated a motor which moved the trolley via a pulley and wire system, with virtually no delay, so that it was again stationed below the new position of the cursor. A rotary encoder mounted on the other pulley wheel generated the coordinate position which was passed to the output electronics. The great advantage of this arrangement was that the surface remained completely clear of all

DIGITAL MAPPING TECHNOLOGY

Figure 9.2 Mechanically-based cross-slide digitizer.

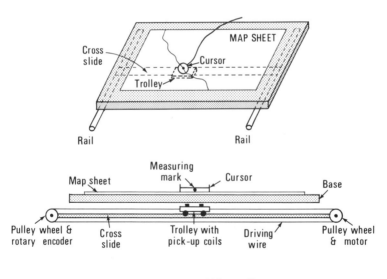

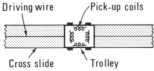

Figure 9.3 Diagram of Pencil Follower: plan view (top); section (middle); plan view (bottom) of trolley.

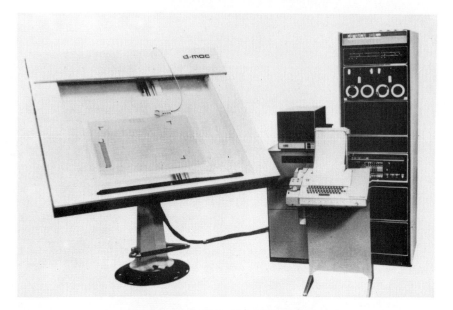

Figure 9.4 D-Mac Pencil Follower.

devices (such as the cross-slide system) which might obstruct the operator's view of the map sheet (Fig. 9.4).

9.3.1.2 *Tablet digitizers.* Relatively few such mechanically-based devices are still in use, and nowadays the process of manually digitizing point or line features for the purpose of acquiring digital cartographic data is almost always carried out using a solid-state *digitizing tablet* in which two sets of position-locating wires (one set in the x-direction, the other in the y-direction), forming a measuring grid, are embedded in a matrix of fibreglass, epoxy resin or plastic material (Fig. 9.5). In some recent examples, the grid is manufactured as a large printed circuit board. The tracing/measuring of the map data is carried out using a cursor around which a field coil is located. This produces a signal which can be picked up by the sensing wires of the grid below. The exact position of the cursor between the wires is interpolated electronically. Originally most of these tablets were incrementally based, and a loss of contact between the cursor and the map surface resulted in a loss of count in the coordinates. However, current designs are now absolutely encoded and virtually impervious to any loss of count.

The cursor may have an attached *keypad* (Fig. 9.5) by means of which text and numerical values or feature codes may be entered. Alternatively, this may be done via a separate *alphanumeric keyboard* or by allocating part of the tablet surface to act as a *'menu'* in which a series of individual boxes are set aside to provide feature coding or headers for specific classes of feature.

A feature of the new digitizing tablets is that most of them are 'smart' in that they have built-in *microprocessors* with PROMs (programmable read-only memories) which allow specific functions, such as the time and distance modes of stream digitizing, line length and area measurement, and a shift of origin, to be

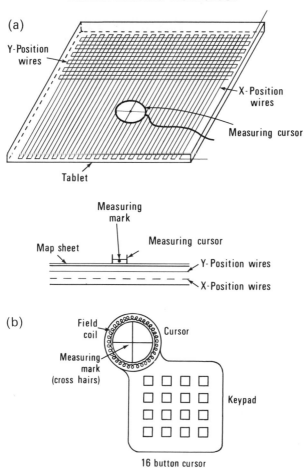

Figure 9.5 (*a*) Tablet digitizer: in plan view (upper) and section (lower); (*b*) cursor and keypad.

implemented as integral features of the digitizer. Also, most have inexpensive LED or LCD displays of the measured coordinate values. Thus these tablets can be used off-line as separate units, with the coordinate data being recorded on a digital cassette or cartridge drive (Fig. 9.6). This procedure allows data to be collected without tying up a microcomputer or a substantial part of the capacity of a minicomputer. It is usually termed '*blind*' *digitizing*, since the measured data is not displayed or plotted out. This procedure does not permit editing or correction of errors during measurement; this must be carried out later on a minicomputer or mainframe machine equipped with interactive graphics editing facilities.

9.3.1.3 *Graphics digitizing/editing stations.* The alternative procedure of providing an on-line display of the digitized features and powerful interactive editing facilities is becoming increasingly popular with the advent of powerful *graphics workstations*

Figure 9.6 Blind digitizing using GTCO digitizer and tape cassette recorder.

equipped with a tablet digitizer as an integral part of the system. An example is the LaserScan LITES 2 system (Fig. 9.7) which is based on the use of a DEC VAX station graphics workstation equipped with a high-resolution (960 × 840 pixel) monochrome VDU display and a large-format GTCO tablet digitizer. This has full windowing, manipulation and editing facilities. The digitized map data is drawn on the screen and can be manipulated in terms of the different data types such as straight lines, arcs, curves, building, symbols, text, etc. The editing software allows specific features to be added, modified, copied or deleted. Large numbers of these LITES digitizing/editing stations are in use in national mapping agencies such as the Ordnance Survey (OS) and the Mapping and Charting Establishment (MCE) and Hydrographic Department of the Ministry of Defence in the UK.

It is a point for some debate whether it is economical to use such a powerful and still relatively expensive on-line system on a task such as the manual digitizing of point and line data. Within smaller surveying, mapping and engineering companies, there is therefore considerable interest in the development of similar but lower-cost alternatives based on the use of small, powerful *microcomputers* linked to tablet digitizers. An example of such a system is that developed by Map Data Management (Fig. 9.8) based on the use of an Apricot Xi or Xen or an IBM PC-AT microcomputer with a medium-resolution (800 × 400 pixel) screen linked to a tablet digitizer and costing between one-third and one-quarter as much as the LITES system, albeit with lower functionality and capability. It is, of course, easy to network these microcomputer-based digitizing stations to a local file server and to transfer the edited, corrected terrain information to a mainframe or minicomputer on which the actual digital mapping, terrain modelling or GIS/LIS operations will be carried out.

DIGITAL MAPPING TECHNOLOGY 341

Figure 9.7 LaserScan LITES 2 graphics digitizing/editing station.

Figure 9.8 Map Data microcomputer-based digitizing station.

9.3.1.4 *Positional accuracy of manual digitizing.* The accuracy with which the point and line data required for digital mapping, terrain modelling and GIS work may be measured manually is a matter for debate. The resolution (least count) of the commonly used tablet digitizers lies in the range 50–100 μm; the actual accuracy of the x/y coordinates produced by the device itself will be a little lower. To these figures must be added the errors made by the operator while measuring the point data or tracing the lines, and any distortion or change in shape present in the map document itself. The latter can easily amount to several millimetres across a large-format sheet; much will depend on whether the map containing the required information is available on a stable plastic base or on paper. To overcome some of these difficulties, measurement of the grid corners is usually undertaken, with a subsequent affine transformation of the measured data to remove any regular change in scale between the x and y coordinate directions. Even with such precautions and processing, it is obvious that the accuracy (root mean square error) of the final positional coodinate values generated by manual digitizing of the point and line data contained on map will be of the order of $\pm\,100$–$250\,\mu$m (0.1–0.25 mm), equivalent to $\pm\,0.1$ to 0.3 m at the standard 1:1250 scale of Ordnance Survey (OS) plans covering urban areas and to $\pm\,1-2.5$ m at the scale 1:10 000 at which contours first become available on an OS series.

9.3.2 *Automatic and semi-automatic line-following digitizers*
As with manually controlled digitizers, it is convenient to classify automatic or semi-automatic line-following digitizers into two categories; those which are mechanically based and those which are not.

9.3.2.1 *Mechanically-based automatic line-following digitizers.* Orginally, these systems were based on the use of some type of sensing device which illuminated and scanned a small area to establish the presence and direction of a line and so allowed its continuous tracking and measurements. The sensing head was mounted on a mechanical cross-slide arrangement; high-accuracy drafting machines (flatbed plotters), such as those from Calcomp, Gerber or AEG, were often adapted to form the basis for such automatic line followers (Fig. 9.9). The tracking movement was implemented using stepping motors actuating either a lead screw or a rack and pinion arrangement attached to the slide defining an individual coordinate axis. Any movement of either of these slides would then be measured by linear or rotary encoders in much the same manner as in the equivalent manually-controlled digitizers. Because of the frequency of branching or crossing lines, some type of video display, for example using a closed-circuit television (CCTV) camera, to view the map sheet, was a complete necessity to allow operator intervention. This allowed him either to guide the system along the correct path in a difficult or confusing area, or to accurately re-position it at a branch or junction. Typical of these systems were the AEG/Aristo device used by one or two agencies in the UK and the similar Gerber 32 device used by the Royal Australian Survey Corps (RASVY) digital mapping unit in the 1970s. The expense and complication of these systems, and the need for an operator to be present throughout the digitizing process, if only for monitoring purposes, meant that such mechanically-based automatic line-following systems enjoyed little popularity among users. Well-executed manual digitizing was only a little slower and very much cheaper because of the lower capital cost of the equipment used.

Figure 9.9 AEG/Aristomat Geameter mechanically-based automatic line-following digitizer with operator and video monitor in use at the Experimental Cartography Unit (ECU).

9.3.2.2 *The Fastrak and Lasertrak line-following digitizers.* This situation has, however, been transformed with the development of the Fastrak and Lasertrak digitizing systems by LaserScan of Cambridge. The origin of these devices is the Sweepnik device originally developed at the Cavendish Laboratory, Cambridge, for the measurement of nuclear bubble-chamber photography (Davies *et al.*, 1970). This used a laser beam steered by a computer-controlled mirror to follow the tracks appearing on the bubble chamber film. The basic computer-controlled laser deflection technology was then used to produce the HRD-1 display/plotter (Woodsford, 1976; Bell and Woodsford, 1977). In the Fastrak or Lasertrak device (Howman and Woodsford, 1978), the laser beam is steered to scan the line needing digitizing in a local raster scan (Fig. 9.10). The actual pattern varies in its vertical and horizontal components as the line changes direction (Fig. 9.11). The crossing of the line by the laser beam is detected on a reduced scale (A6 size) negative of the document being digitized at 10 to 15 μm intervals (Fig. 9.12). This produces x and y coordinates of points on the line at intervals of between 50 and 70 μm at the original map scale at speeds up to 500 points per second.

The system is highly interactive with operator intervention in the form of backtracking, redirection of the scan or fully manual digitizing if the automatic line following should falter or fail. Operator monitoring is achieved through the display of the lines being digitized on a large viewing screen, 1 m × 700 mm in size. The operator also selects features, adds coding such as contourline values or descriptors of specific

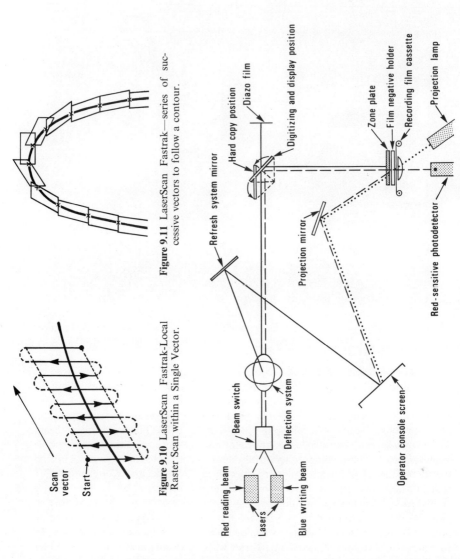

Figure 9.10 LaserScan Fastrak-Local Raster Scan within a Single Vector.

Figure 9.11 LaserScan Fastrak—series of successive vectors to follow a contour.

Figure 9.12 LaserScan Fastrak—diagram of overall system.

DIGITAL MAPPING TECHNOLOGY

Figure 9.13 LaserScan Lasertrak—showing display screen, control keyboard and tracker ball, and graphics display terminal.

types of line features, and carries out any required editing with the aid of a keyboard and a tracker ball positioning device (Fig. 9.13).

The control computer is normally one of the more powerful models in the Digital (DEC) VAX-11/700 or MicroVAX 2 series, incorporating a high-speed floating point arithmetic unit. Average rates of 500 line inches (12.5 m) per hour of fully coded and edited digitizing are achieved on contour and culture-type maps. Plans and maps up to A1 (594 × 841 mm) in size can be handled by the Fastrak or Lasertrak in a single pass. Graphical output for verification purposes, archival use and final plotting can be carried out using diazo-based microfilm. Digital output is normally in the form of magnetic tape.

The digitizing of contour lines from existing topographic maps is particularly suited to the Fastrak and Lasertrak systems, especially if the original contour-only sheet is available for the purpose, since the crossing and branching of planimetric line detail which gives so much difficulty to any form of automatic line-following device is largely eliminated. Difficulties may of course still arise in areas where the contours lie very close together or are replaced by cliff symbols, as may happen in steep terrain, in which case the operator will have to take over the digitizing operation and carry it out manually for this particular area of the sheet.

Large national mapping agencies such as the Ordnance Survey (OS), the Mapping and Charting Establishment (MCE) of the UK Ministry of Defence, the United States Geological Survey (USGS) and the Japanese Geographical Survey Institute (GSI) have the volume of graphic digitizing to have been able to justify the purchase of Fastrak or

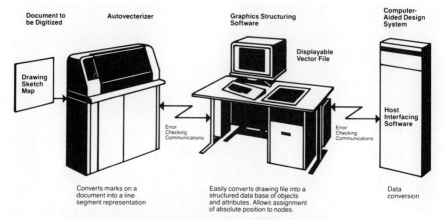

Figure 9.14 Tektronix 4991S raster-scan digitizing system.

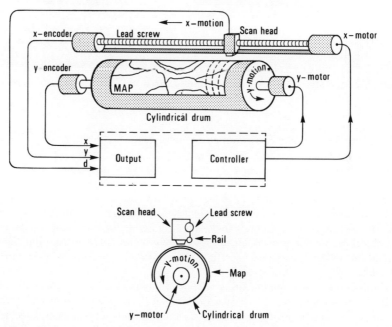

Figure 9.15 Diagram of raster-scan digitizer: perspective view (upper) and cross-section (lower).

Lasertrak devices with the certainty that they would be used on a full-time production basis. For those engineers, landscape architects and planners who may need to digitize a large amount of graphic data, occasionally or even just once, for example to form the topographic and locational base for their design activities, resort can be made to LaserScan's service bureau in Cambridge or to private surveying and mapping agencies equipped with Lasertrak equipment which also offer this service.

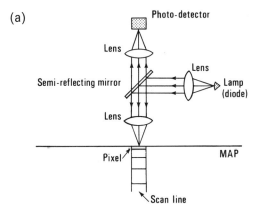

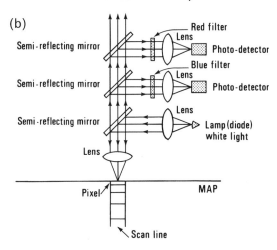

Figure 9.16 Diagram of photodetector head of raster-scan digitizer: (*a*) scan head for monochrome image; (*b*) scan head for colour image.

9.3.3 *Automatic raster-scan digitizers*

The raster-scan digitizers available for the digitizing of large graphic documents, such as maps and engineering drawings, are fully automatic devices capable of producing and storing a file of raw coordinate data for all the lines and symbols contained in the sheet. Till recently, most of these devices were *drum scanners*, for example those manufactured by Optronics (used by Intergraph), Scitex (the Response series) and Tektronix (Model 4991) (Fig. 9.14). In each of these, the map or graphic document is wrapped round a drum which is rotated at a constant speed below a photodetector head which is moved continuously forward in steps along the axis of the drum (Fig. 9.15), the step size defining the width of the scan lines. With these devices, it is possible to digitize either black-and-white (monochrome) sheets, as in the case of colour-separated contour sheets, or full-colour maps. In the former case, a single photodetector will be used; in the latter, different filters are used to perform colour separation of the detail contained in the map, each colour-separated channel being sensed for the presence or absence of line data by its own individual photo detector (Fig. 9.16).

The alternative type of raster-scan digitizer is the so-called *flatbed scanner*. Two examples of such digitizers are the SysScan KartoScan (West Germany), and the Broomall Scan Graphics System (USA). In both of these devices, the map sheet is laid flat on a base board and is then scanned by a cross-slide or gantry which rapidly traverses the sheet from top to bottom (Fig. 9.17). Alternatively, the gantry may be held fixed and the sheet passed below it, as in the newer types of KartoScan devices (Fig. 9.18). The elements which sense the presence or absence of lines on the sheet are an array of solid-state photodiodes in the case of the KartoScan device, and a laser source travelling bi-directionally across the gantry in the case of the Scan Graphics equipment. In each case, the coordinate data is read out continuously on a line-by-line basis and stored on disc or tape. In the case of the Scan Graphics device, the maximum area which can be scanned is 24 × 50 inches (1.12 × 1.65 m); with the largest KartoScan (Model FE), the maximum imaging area is 47 × 64 inches (1.2 × 1.6 m).

Recently, various alternative types of low-cost, small-format raster-scan digitizers have appeared on the market which utilize linear or areal arrays of charge coupled devices (CCDs) to scan graphics documents. The arrangement of such a device utilizing an areal array, together with a lens to form a CCD camera, is shown in Fig. 9.19. Since the current state of this technology allows only relatively small areal arrays to be manufactured, it is necessary to scan a large-format document such as a plan or map in sections, i.e. patch by patch. This will result in an increased time in the subsequent data processing, so that the individual scanned patches may be joined together to reconstitute the original plan or map. Software must also be available to reconcile the inevitable mismatches between the same lines occurring on adjacent patches.

These scanners utilizing linear arrays include the Topaz Picture Scanner (UK) and the Eikonix E-Z Scan Model 4434 (USA). In both cases, the map is scanned by the linear CCD array which is stepped across the image field of view behind a lens (Fig. 9.20), instead of the map sheet being passed under the linear CCD array of detectors, as in the smaller KartoScan CC and CE scanners.

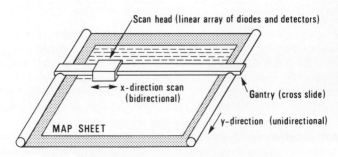

Figure 9.17 Diagram of flatbed raster-scan digitizer.

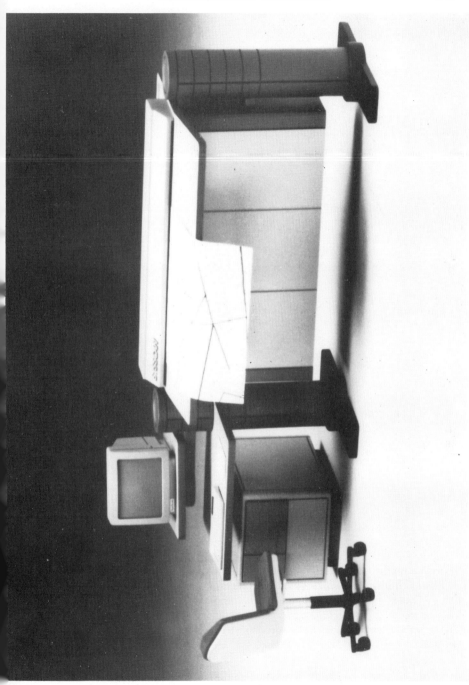

Figure 9.18 SysScan KartoScan digitizer.

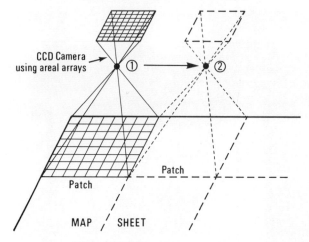

Figure 9.19 CCD areal array scanner.

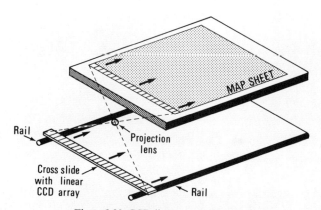

Figure 9.20 CCD linear array scanner.

9.3.3.1 *Characteristics of raster-scan digitizing.* The advantage of this type of raster-scan digitizing is the sheer *speed* of the digitizing process, the time taken being a function of the size of the graphic document and quite independent of the length of the line to be digitized, which is the main factor in both manual and semi-automatic line-following methods. Using a fairly fine resolution (e.g. 25 μm), a large-format map may typically be scanned in 0.5–1 h. If the map data digitized by the raster scanner does not need alteration, if the specific features present in the data do not need to be identified or labelled, and if there is no need to re-order or structure the data, then raster scanning is an ideal method of map digitizing. In this case, essentially it is providing the map detail as a backcloth on which engineering design data can be placed.

However, in other situations, line-following has the advantages of economy (in that only those lines and areas which exist are measured) and of selectivity (since only those features required for a specific purpose need to be measured). Furthermore, where something more than a backcloth map image is required, the time and expense needed to edit and process the digitized raster data must be considered as well. With complex maps containing numerous graphic symbols, it will certainly be necessary to supplement the raster-scanning with manual digitizing to pick up all the individual point symbols and text. Furthermore, display and editing of the digitized raster data to remove unwanted artefacts which have been digitized, labelling and addition of descriptive header codes and attributes, and structuring of the data to conform to the topological relationships required by the specific mapping system will also need to be carried out later on an interactive graphics workstation as an off-line process. All these processes add considerably to the time and expense of the raster scanning/digitizing operation, and reduce its basic simplicity and speed.

9.3.3.2 *Raster-to-vector conversion.* Consideration must also be given to the conversion of the digitized data from the scanned raster format to a line or vector format which must be implemented both to ensure some degree of data compaction and to give a more convenient format which is required by many display or plotting devices. Software must be written or acquired to carry out the following two basic steps in this conversion operation.

(i) *Line thinning:* the process of reducing the lines to unit thickness (i.e. to single pixel width) at a given resolution
(ii) *Line extraction,* which involves the identification of a continuous series of coordinated points which constitute an individual line or line segment as portrayed in the original map.

The detailed procedures (including various alternative solutions) are described in papers by Peuquet, 1981 and Baker *et al.*, 1982.

One solution to the problem of *line thinning* is the so-called medial axis approach, which is based on the calculation of all those points (pixels) lying at the maximum distance from all edges of the original contour line which may be several pixels thick (Fig. 9.21). In a first pass through the digitized raster data set, each pixel is inspected in

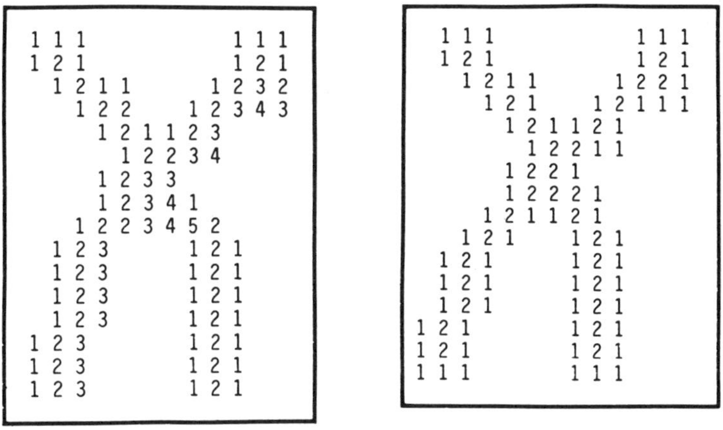

Figure 9.21 Line thinning operation.

turn in scan-line order (i.e. from left to right) starting from upper left corner of the rasterized data.

Each occupied pixel is assigned a value equal to the minimum value of its neighbours directly above and immediately to the left, plus one. In a second pass, the resulting matrix of pixel values is processed in reverse order, assigning to each pixel the minimum value among itself, its neighbour to the right plus one, and its neighbour below plus one (Peuquet, 1981). The resulting skeleton of maximum value pixels represents the centre line of the line at unit thickness (Fig. 9.21).

The second step, that of *vectorization* or *line extraction*, involves the scanning of the whole line data set initially around the edges of the digitized area to pick up the starting point of each line. Once such a line is found, it is then followed, typically by using a 3 × 3 pixel template to follow the line to its terminal position. Once this has been done, the data comprising the reconstructed line is eliminated from the data set. The starting point of the next line is then picked up and again followed to its terminus. This process is continued until all the digitized raster data has been picked up and assigned to a specific line feature.

As discussed above, a vast amount of *data processing* is required both for this raster-to-vector conversion process and the supplementary manual digitizing required to edit, label and structure the scanned line data. This tends to offset the very rapid initial raster-scanning of the basic map document, which preferably has to be available in the form of the separation transparencies used for colour printing if high-quality results are to be obtained. Enthusiasts for raster-scanning are mainly to be found among the academic community (e.g. Boyle, 1980) and these tend to minimize the difficulties and expense encountered with this method. However, at the present time, given the enormous investment needed for the purchase and implementation of these fully automatic raster-scan systems, only the largest mapping organizations with a dedicated long-term commitment to map digitizing on a huge scale are likely to contemplate their acquisition.

9.4 Digital map data structures

The discussion conducted above highlights some of the basic differences in data structure between the raster and vector data required for digital mapping purposes. This subject is dealt with in detail by Burrough (1986); only a very brief outline will be included here.

A *raster data structure* consists of an array of grid cells often described as picture elements (or pixels). Each cell or element within the array is defined by its row and column number and will contain the specific value of the information being mapped. In the case of the raster scan digitizer being described above, this will simply be the occurrence or otherwise of a point or line within the pixel which is easily represented by the simple codes 0 or 1 (Fig. 9.22). In this case, a point is defined by an individual cell or pixel; a line by a connected series of adjacent pixels; and an area by an areal agglomeration of adjacent pixels. The resolution of the raster data is defined by the size of the pixel element used in the scanner, typically 25–50 μm. This must be multiplied by the scale factor of the scanned map to give the equivalent ground resolution.

By contrast, in the *vector data structures* used in digital mapping, a point is defined by a single x, y coordinate pair. Where no specific point data exists, no attempt is made to define any point—a great contrast to the situation in raster digitizing where all points

```
0 0 0 0 0 0 0 0 0
0 0 0 0 0 0 0 0 0
0 0 0 0 0 0 1 1 1
0 0 0 0 0 0 1 0 0
0 0 0 0 0 0 1 0 0
0 0 0 0 0 0 1 0 0
0 0 0 0 0 1 1 0 0
0 0 1 1 1 0 0 0 0
0 1 0 0 0 0 0 0 0
1 0 0 0 0 0 0 0 0
0 0 0 0 0 0 0 0 0
```

(a) Raster Data Structure-Line Represented by a Connected Series of Adjacent Pixels (shown by Code 1).

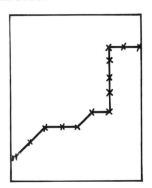

(b) Equivalent Vector Data Structure for the same line

Figure 9.22 Raster data structure.

occurring within the map are included in the data set. Also in vector data, a line in its simplest form is made up of a straight line defined by the x, y coordinates of its start and end points. An individual sinuous line may be subdivided, defined by a series of such straight-line segments or vectors, again defined by the coordinates of their end points. An area is defined by a continuous line or a set of connected line segments which closes back on itself.

A description of the specific type of point, line or area which has been digitized must be given by assigning a specific code or attribute with the coordinate pair or set of coordinates. Also the topological relationships between all these individual data elements (point, lines and areas) measured during digitizing must be defined. A simple example of these relationships defined for a vector data set, based on that provided by Leberl and Olsen (1982), is given in Fig. 9.23. This shows a polygon data structure. A series of curved line segments is defined by a set of coordinated points; the ends of these are termed nodes and are numbered 1–10. It will be seen that an individual node can be the starting or end point for a number of line segments: node 5 is the starting or end point for line segments (5, 2), (5, 4), (5, 6) and (5, 8). In turn, each of the areas defined by a closed polygon is defined by the connected series of line segments enclosing it. The overall topological structure of the polygons and line segments may be represented as a planar graph. Such polygon-based data structures are very commonly used in digital mapping systems. To this basic topological structure must be added the descriptors and plot codes which constitute the attributes and allow the digitized map data to be processed, displayed and output (plotted) in graphical form (Fig. 9.24).

9.5 Digital map data processing

After the various data sets required for a specific map have been acquired, coded and structured, it is unlikely that they will be free from error. Thus the detection and the elimination or correction of such errors is a vital part of the process that is carried out using an interactive graphics editing station. This may be based on the use of a graphics terminal linked to a minicomputer, a stand-alone graphics workstation, or a microcomputer equipped with a graphics card and screen. Further processing will then

354 ENGINEERING SURVEYING TECHNOLOGY

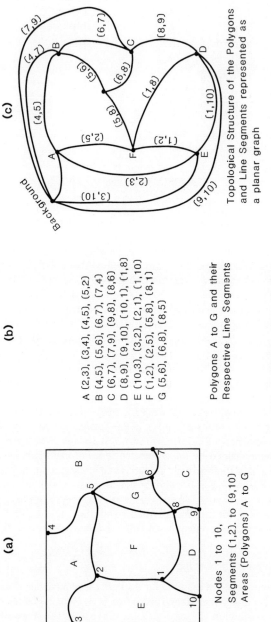

Figure 9.23 Vector data structure.

DIGITAL MAPPING TECHNOLOGY

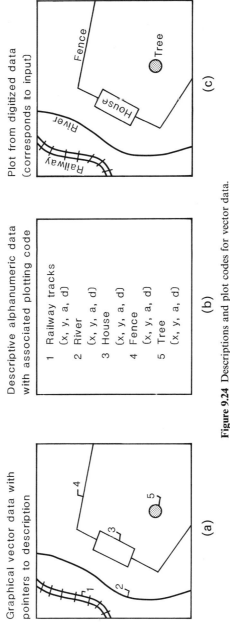

Figure 9.24 Descriptions and plot codes for vector data.

be carried out. Typical of such operations are one or several or all of the following.

(i) Merging of individual data sets acquired by different methods, different organizations and at different times
(ii) Selection of the data so that only a limited part of the available data set or only certain types or classes of map data will be displayed or plotted
(iii) Generalization of the map data so that only the essential elements can be displayed or plotted at a much smaller scale
(iv) Transformation of the data into a different map projection or grid system or into a different format so that it is compatible with a specific engineering design or modelling package
(v) Revision or updating of the map data base as changes in the terrain situation are surveyed in the field and the new data incorporated in the digital mapping system.

Obviously software which will allow all of these operations (and many others) to be executed by the computer will form a vital part of every digital mapping system.

9.6 Cartographic display and plotter technology and procedures

The final phase of the digital mapping process will be the display and output of the digitized and processed data in cartographic form, either visually on the screen of a graphics display or in hard-copy form as a map plotted on film or paper. The devices giving such outputs are both numerous and very varied in terms of their capabilities. If a correct choice of output device has to be made for use in a digital mapping system, it is therefore essential that the various possibilities offered by each device and also its limitations should be understood.

9.6.1 *Graphics display devices*

These devices permit the visual display of digital map data on a screen. Although alternative display technologies, such as liquid crystal displays (LCDs) and plasma panel displays, do exist, the vast majority of graphics display devices used in the digital mapping field are based on the use of a cathode ray tube (CRT). This may take any one of several different forms; for example:

(i) A *monitor*, a simple display device on which the graphical image is written on the screen of the CRT
(ii) A *graphics terminal*, which has an additional keyboard with which the operator may interact with the controlling computer, and may also have a joystick, tracker ball or digitizing tablet to control the position of a measuring cursor or of the image itself
(iii) A *graphics workstation* in which a very high-resolution graphics terminal is closely coupled to a powerful, dedicated processor and display memory, and the whole integrated unit is optimized for the production of high-quality graphics.

It is usual to classify the various graphics display devices into two main categories on the basis of the manner by which the graphics image is written on the screen. These comprise those which carry out the direct drawing of *vectors* or lines on the display screen, and those which use a *raster*-based scan pattern to generate the graphical image on the display screen. In addition to this basic classification, each of these two types of display can be categorized further into two subtypes: the *refresh* type of display in which

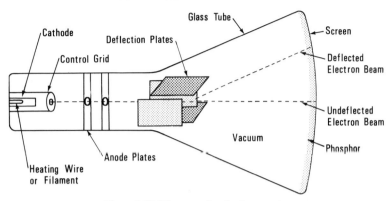

Figure 9.25 Diagram of cathode ray tube.

the image has to be continually refreshed for it to remain visible on the screen of the display device; and the *storage* type of device in which the graphical image, once written on the screen, remains there as long as the device remains continuously powered up. Thus four main types of display device can be recognized:

(i) vector refresh
(ii) vector storage
(iii) raster refresh
(iv) raster storage.

9.6.1.1 *Cathode ray tube.* The basic principle and operation of the cathode ray tube (CRT) is shown in Fig. 9.25. It comprises a conical sealed glass tube at one end of which is an electron gun comprising a cathode, a metal control grid and an anode which are mounted within the CRT tube. The heated cathode emits a continuous stream of negatively charged electrons which pass out through a hole in the control grid, are attracted by the positive anode placed in front of the cathode; and are accelerated and focused so that they arrive at a central position on the glass screen of the CRT. This screen is coated with a thin layer of phosphor. At the position where the beam of high-speed electrons strikes this layer, a series of electrons is displaced so that this point on the layer becomes positively charged and glows brightly. As long as the stream of electrons continues to strike the screen, the spot will continue to glow. Any required variation in the brightness or intensity level of the spot can be achieved by modifying the rate of flow of the electrons in the ray beam by altering the potential, i.e. voltage level, of the control grid. Indeed, the flow of electrons can be cut off completely, as for example at positions where no image has to be written on the screen.

The position of the bright spot may be altered by a deflection system consisting of two pairs of coils or plates. If current is applied to a pair of these coils, a magnetic field is produced which deflects the stream of electrons so that it hits a new position on the phosphor screen. The amount of this deflection is proportional to the input voltage. Thus all CRT-based displays contain a controller which converts the output signals from the computer into the corresponding deflection voltages to be applied to the coils. One pair of coils or plates controls the deflection of the electron beam in the horizontal (x) direction; the other controls the vertical (y) direction. When used in combination, the two pairs of coils allow the positioning of the beam (and hence that of the bright spot) at any desired position on the screen. The whole surface of the screen can in fact be scanned by the beam in a fraction of a second. Through a series of successive commands translated into the corresponding voltages applied to the deflection coils, a bright image of a digitized map or survey plan can therefore be written on the screen.

However, the bright spot will glow for only a tiny fraction of a second after the electron beam has swept past, since its charge is soon lost, in which case, the spot will disappear. Thus if a graphical image such as a map is to be observed over a period of time, it must be regenerated continuously. The length of time over which the spot will glow is dependent on the type of phosphor used. There are short- and long-persistence phosphors, but whichever is used, the image still needs to be regenerated.

9.6.1.2 *Vector refresh displays.* In the case of the vector refresh type of display, a series of vectors is written on the screen to form the graphical image. The list of coordinates, called the *display list*, which defines the map image, is usually stored in a special display memory (or frame buffer), or else a portion of the main memory is reserved for the purpose. On the highest-quality vector refresh displays, typically 30 000 to 80 000 vectors are written within a period of 1/30th to 1/60th second. Since these are being continually refreshed, any change in the coordinate list will result in a corresponding change in the map image. Thus any change resulting from the editing of the data will immediately be apparent on the screen.

Graphics terminals of this quality and performance, such as those manufactured by Evans & Sutherland, Megatek, or Adage, are very expensive, but such is the quality of the image that they have been extensively developed and used for CADD work in the aerospace industry and in mechanical engineering. Several types of vector refresh

Figure 9.26 Graphics terminal utilizing vector refresh display.

graphics terminals such as the Sanders Graphic 7 and 9 (used in SysScan mapping systems) and the Imlac (used in the Kern Maps 300 work station) have also been used extensively in the surveying and mapping industry (Fig. 9.26).

A limitation of vector refresh graphics terminals is that they are not well suited to the generation of coloured areas, a requirement for certain types of map; this is because these areas must be filled by a very large number of vectors. If too many vectors need to be written within the very short refresh period, this manifests itself as 'shaking' or flickering of the image. Thus the quantity of map data that can be written on the screen is restricted both by the size of the display memory and by the rate at which this data can be transmitted and written on the screen.

9.6.1.3 *Vector storage displays.* On the vector storage type of display, it is not necessary to refresh the graphical image. Instead, the map image, once written on the screen, is continuously stored on the face of the CRT itself without any requirement to be refreshed. As can be seen from Fig. 9.27, the vector storage display has two additional electron guns, called flood guns, constantly emitting low-velocity electrons over the face of the tube which is coated with an extremely long-persistence phosphor. These electrons have too low a velocity to dislodge electrons on unwritten areas, and these remain undisturbed. However, the written lines or areas, being positively charged, attract these electrons at a sufficiently high velocity that secondary electrons are dislodged and the phosphor target remains positive and the spot bright. In fact, most vector storage tubes have a fine wire mesh grid coated with dilectric mounted just behind the phosphor-coated screen. The pattern of charges is written on this grid and then transferred to the phosphor by the electrons from the flood guns (Newman and Sproull, 1979). In this way, the lines and symbols of the map image written on the screen remain bright.

Since the storage type display does not require a display memory and does not need to be continually refreshed at high speed, these characteristic requirements of the vector refresh type are avoided. Instead, the map image may be written over a period of several seconds; thus a slow and less expensive line can be used to transmit data between the computer and the display.

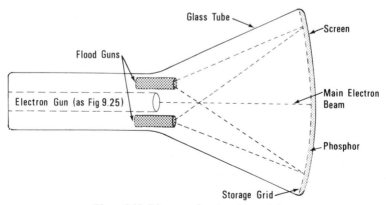

Figure 9.27 Diagram of vector storage tube.

It should also be noted that, on a vector storage display, the stored image is always displayed free from shake or flicker. However, a disadvantage of this type of display is that it is impossible to erase a specific part of the image as is required, for example, in map editing operations. Instead, if a change has to be made other than a simple addition to an existing displayed image, the whole image has first to be erased and then redrawn from the altered display list held in the computer. A single manufacturer, Tektronix, dominates the supply of this type of display, offering its 4000 and 4100 series of terminals in a series of screen formats and resolutions from small (7×10 inches) in the Model 4006 to very large (23×23 inches) in the Model 4016, the last in combination with an image resolution of 4000×4000 pixels. However, although this type of display is now being replaced by newer types of high-resolution, raster-based displays, it has been used extensively in the surveying and mapping industry in the past, and many instruments remain in service.

A vector storage display which utilizes a totally different technology is the *Laser-Scan HRD-1* (Woodsford, 1976). This provides the same display and plotting capabilities as those of the Fastrak and Lasertrak digitizing devices described in section 9.3.2.2. Thus the map image is written by the same computer-controlled mirror deflection mechanism and laser used in the Fastrak and Lasertrak. The image is written and stored on orange photochromic film, the laser beam causing an immediate blackening of the lines, symbols and text as they are being written on the film. The map image is then projected on to the same type and size ($1 \text{ m} \times 70 \text{ cm}$) of translucent screen as is used on the Fastrak to give a $10 \times$ enlarged image. Again the image can be recorded permanently as a negative on diazo-based microfilm. Full screen erase is carried out by advancing the photochromic film, either manually or under computer control. The very fine detail ($20 \, \mu\text{m}$ spot size), high accuracy ($20 \, \mu\text{m}$), high resolution (equivalent to 7000×5000 pixels) and the fast writing speed ($4.5–5 \text{ m}$ per second) provide the HRD-1 with the highest resolution in a commercially available display at the present time. Digital mapping users of the HRD-1 include Rijkswaterstaat, the Dutch state public works and engineering organization.

9.6.1.4 *Raster refresh displays.* This type of display, like its vector equivalent, operates by continuously writing and refreshing the image displayed on the CRT screen. However, as the name suggests, the raster-based device writes its image by systematically scanning the whole of the screen in a parallel series of horizontal scan lines in the manner made familiar by domestic television sets (Fig. 9.28). The map image is built up by the generation of a series of bright spots (pixels), located at specific positions along each scan line. A series of adjacent pixels located on different scan lines can be illuminated to form a vector or line. The complete raster image may be refreshed at rates of 25 to 60 times per second, i.e. at 25 to 60 Hz.

Usually if the refresh rate of the device is low, then only every second line, e.g. the even-numbered lines, of the displayed image may be written every 1/50th second. During the next period of 1/50th second, the intervening lines, i.e. the odd-numbered lines, will be written. Each of the two sets of interlaced lines persists long enough on the screen to give the impression of a single continuously displayed image. The individual alternating images are termed *fields*; a single complete image constitutes a *frame*. In the example given, 50 fields are written per second, but only 25 complete frames or map images are displayed per second.

The *resolution* of the displayed raster image is often quite low—until recently 512

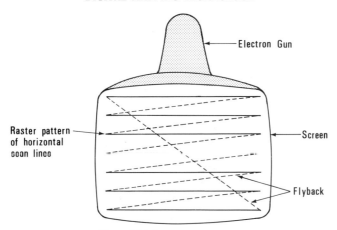

Figure 9.28 Raster scan pattern.

× 512 pixels has been a common standard. Thus for map images, either only a small part of the complete map image can be displayed at its inherent resolution, or the whole image can be displayed, but at a much reduced resolution. As with the vector refresh type of screen, a frame buffer stores the list of coordinate values of the points to be displayed. As the cost of memory has fallen in the last few years, larger memories can be provided so that the resolution of the newest types of graphics terminals and workstations has been increased greatly—a display resolution of 1024 × 1024 pixels is now quite common. Obviously the rates at which this increased amount of data has to be transmitted to and then displayed on the screen must rise proportionally if the benefits of the higher resolution are to be achieved. In fact, the required increase in performance is being achieved, and non-interlaced, high resolution, raster-based displays refreshing the screen at 50 or 60 Hz are becoming more common in the graphics terminals and work stations used for mapping.

The advent of inexpensive *memory* has also been utilized in other ways, in particular to supply several planes of memory at the required resolution. These can be utilized in any one of several different ways.

(i) They allow the use of *multiple image planes*, each storing the complete image of a specific class of map feature. For example, a single plane can store all the contour data; a second, all the water features; a third, all the communications features; etc. Each of these individual images may be displayed in monochrome separately, or they may be combined to display any combination of these features as a single composite image.

(ii) Each memory plane can be used to control the writing of a coloured map image, with the addition of three electron guns arranged in a triangular configuration, a shadow mask comprising a metal plate pierced with small holes on which all three electron beams are focused, and the appropriate arrays of tri-colour dots on the face of the screen used on a *colour CRT* (Fig. 9.29). Using the three standard additive colours (red, green and blue) of such a CRT, $2^3 = 8$ individual colours may be generated on such a display.

(iii) An alternative use of multiple image planes (Fig. 9.30) is to control the intensity, i.e.

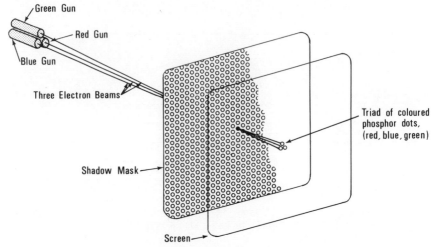

Figure 9.29 Shadow mask-based colour CRT display.

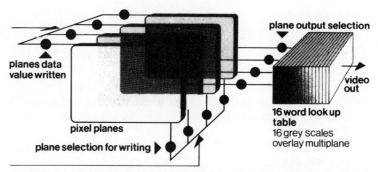

Figure 9.30 Multiple-plane configuration of Sigma T5670 raster graphics display.

the brightness, of each pixel. Thus the use of four memory planes could be used to give any one of $2^4 = 16$ brightness levels in a monochrome image display.

(iv) By using a still larger number of image planes, it is possible to control both the colour and the intensity variations at each pixel position over the whole area of the display screen. This gives the possibility to display coloured map images with variable intensities of colour within specific areas, or the remotely-sensed colour and false-colour images discussed in Chapter 6. For example, using eight memory planes, $2^8 = 256$ different colours may be displayed on the colour CRT display for this particular application.

Within the present context of digital mapping, where vector images are commonly required to be stored and used both for display and as hard-copy images, it should also be mentioned that it is necessary to rasterize the vector-based image before it can be displayed on a raster-based display. This is carried out by purpose-built electronic hardware which forms an integral part of the display device.

Since much of the raster-based CRT display technology is akin to that of domestic television receivers, the availability of many mass-produced components from the latter type of device has also helped to drive down the cost of raster refresh CRT-based graphics displays. However, there are also some notable differences in that the domestic television set utilizes analogue TV technology which can handle the broadcast luminance/chrominance signals and must have the capability of receiving these signals at different transmitted frequencies and wavelengths. The graphics display monitor or terminal does not receive these broadcast UHF/VHF signals, and instead handles the stream of digital data sent by the computer which is translated into the red, green and blue (RGB) signals used by most raster-based colour displays or terminals.

9.6.1.5 *Raster storage displays.* The technology used in raster storage displays is not based on the use of a CRT, but utilizes either *plasma panels* or *LCD displays*. The former are flat transparent displays comprising two parallel sheets of glass with a parallel pattern of thin electrodes attached to their inner faces and covered by a dilectric material. On one of the glass sheets, the electrodes are set in a horizontal pattern; in the other they are set vertically. The narrow space between these two sheets is filled with neon gas and sealed. This gas can be made to behave as if it was divided into tiny cells, each of which can be made to glow individually if required. By applying voltages between the electrodes, the appropriate cells can be addressed to create a bright image of the map on the screen of the panel. As long as a sustaining signal continues to be sent across the electrodes, the map image will continue to be displayed. At present, this display technology is rather limited in its application to maps due to its limited resolution. Much the same remarks can be made regarding the alternative LCD displays familiar from their use in clocks, watches, pocket electronic games and calculators.

9.6.2 *Hard-copy output devices (plotters)*
Essential though graphics displays are for editing and displaying the map data produced by a digital mapping system, inevitably they need to be supplemented by devices which can produce maps in hard-copy form on paper or stable polyester film, so that the map can be included in reports, drawn upon or annotated, or used in the field. A bewildering variety of plotters have been devised for the purpose based on the use of different technologies, media and formats, thus allowing the user to select a specific plotter suitable for his own particular requirements from a wide range of alternatives.

9.6.2.1 *Basic plotter configuration.* All plotters consist of three basic elements: an input device, a controller, and the actual plotting or draughting device (Fig. 9.31).

(i) The *input device* may be a minicomputer or microcomputer to which the plotter is coupled direct for on-line operation. Alternatively, the input device may be a

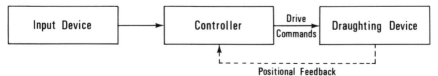

Figure 9.31 Basic elements of a plotter.

magnetic tape unit which allows the map data stored on cassette, cartridge or tape reel to be plotted in an off-line operation.

(ii) The *controller* comprises all the electronic circuitry required to control the operation of the plotter. First it reads and stores all the coordinate data and plotter control commands which are sent to it from the input device. It then issues the signals which actuate and control the acutal operation of the plotter. These include the commands to the motors which drive the actual plotting device to the desired coordinate position, those which actuate the mechanism which select the appropriate drafting tool, and those controlling the action of raising and lowering the pen or of switching on or off the lamp necessary to expose a photographic film. The controller may be make use of hard-wired electronic circuitry or of a microprocessor in conjunction with firmware or some combination of these elements to interpret the input data and to control these various plotter operations.

(iii) The third element is the *plotting device* itself which will comprise a drawing surface, a mobile drive and positioning system and some type of draughting tool. The drawing surface may be of either the flatbed or drum type on which the plotting media (paper, polyester draughting film, scribing or masking material or photographic film) is mounted. The *drive and positioning system* will typically comprise digital stepping motors or analogue servo motors as the driving elements, to which lead screws, a rack and pinion mechanism or a wire and pulley system are attached as transmission devices. Finally, the *draughting tool* may be mechanically based, as in the case of pens, scribing points, and mask-cutting tools, or photo-optical, when a lamp, laser or light-emitting diode (LED) is used to expose an image on light-sensitive material.

Just as with digitizing devices and the graphics display devices, the most basic division of hard-copy plotting devices is into those which are vector-based (i.e. they plot lines) and those which are raster-based and cover the whole map image in a parallel series of scans during their plotting action. A further subdivision is into *flatbed* plotters, in which the paper or film is laid flat on a plane surface during the plotting action, and *drum* plotters in which, as the name suggests, the plot material is mounted on a cylindrical drum on which plotting takes place. This results in four main classes of plotting device:

 (i) Vector flatbed plotters
 (ii) Vector drum plotters
(iii) Raster flatbed plotters
(iv) Raster drum plotters.

Each has its own advantages and disadvantages for different types of digital mapping application.

9.6.2.2 *Vector flatbed plotters.* From the point of view of its basic mechanical design, this type of plotter is very similar to the mechanically-based, line-following digitizing devices discussed above in section 9.3.1.1. Thus typically it will comprise the flatbed drawing surface over which is set a cross-slide or gantry which carries the draughting tool (Fig. 9.32). The controller will send out the drive commands to the x and y coordinate motors, the latter to drive the cross-slide or gantry as a whole, while the

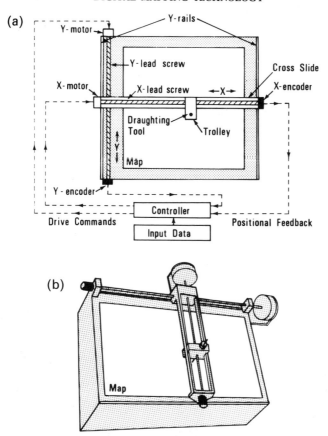

Figure 9.32 (a) Diagram and (b) perspective view of vector flatbed plotter.

former drives the draughting tool which is mounted on a small trolley that can be moved along the cross-slide itself. In the more expensive, high-accuracy type of flatbed plotter often used for large-scale engineering survey plots, lead screws will be used as the driving/transmission elements, often equipped with rotary encoders to establish the exact position of the draughting tool.

In this case, the controller also acts as a type of *comparator*. It compares the x and y coordinate values from the input device with the actual x and y coordinates of the current position of the draughting tool as given by the encoders. On the basis of this continuous comparison of coordinate values, signals will continue to be sent to the driving motors until the comparison gives rise to a null difference between the desired and the actual coordinate position. At this point, the motion stops and a command will be sent to actuate the operation of the draughting tool.

Typical of this type of large-format, high-accuracy type of vector flatbed plotter (Fig. 9.33) are those manufactured by the well-established suppliers of surveying and photogrammetric equipment, such as Wild (Aviotab series), Zeiss Oberkochen (DZ series) and Kern (GP-1 plotter). Many of these are of course directly attached to the

366 ENGINEERING SURVEYING TECHNOLOGY

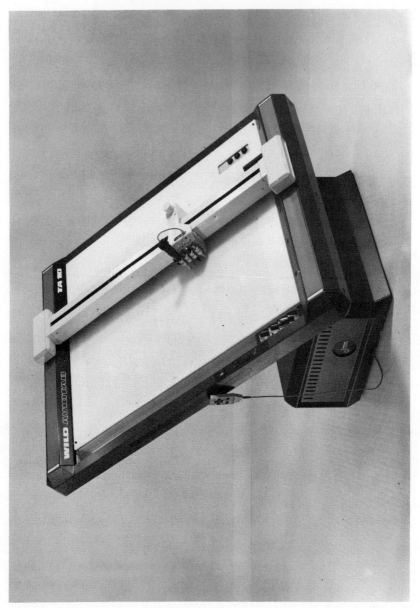

Figure 9.33 Wild Aviotab TA-10 plotter.

analogue and analytical stereo-plotting instruments described in Chapter 7, but they can also be used as stand-alone devices for use with data derived from survey measurements or cartographic digitizing. However, specialist plotter manufacturers (Calcomp, Gerber, Kongsberg, Glaser) also supply high-quality flatbed plotters of this type which also find extensive use in engineering CADD work and in the printed circuit-board (PCB) industry.

At the other end of the scale are numerous inexpensive, lightly built, small-format flatbed plotters, which, like their larger brethren, are of the line-drawing type. Usually these will have simpler driving elements using digital stepping motors without the use of rotary encoders and the coordinate comparison carried out in the more expensive type, and a less accurate transmission system such as a wire and pulley arrangement instead of lead screws. Overall accuracy will be much lower: ± 0.2–0.5 mm instead of the ± 25–50μm of the large high-accuracy machines; plotting speeds are often higher; and the range of drawing tools is much more limited. Thus the application of these lower-specification flatbed plotters will be limited to the production of edit plots and maps of limited coverage where accuracy is not required. Numerous manufacturers (Houston Instrument, Bryan Instruments, Watanabe, Rikadenki, Mutoh) supply these devices to a large market, which extends to business graphics, where graphs and diagrams need to be drawn, besides the relatively much smaller digital mapping market.

There do exist other mechanical arrangements for vector flatbed plotters. One such design is that of the Ferranti Master Plotter which has been used extensively for the production of high-quality maps and plans in the UK and elsewhere in Europe, besides being used in CADD work. The mechanical arrangement (Fig. 9.34) features a fixed gantry instead of the moving cross-slide described above. A trolley carrying the draughting tool moves across the gantry in the y-direction as before. However, in the x-direction, the flatbed drawing surface is moved instead of the cross-slide or gantry. This arrangement is intended to eliminate the cross-coupling effects of the moving gantry

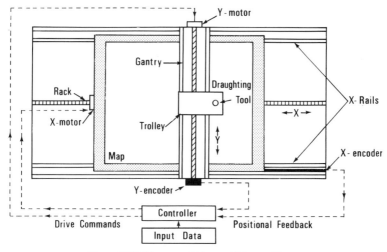

Figure 9.34 Diagram of Ferranti Master Plotter.

Figure 9.35 Ferranti Master Plotter.

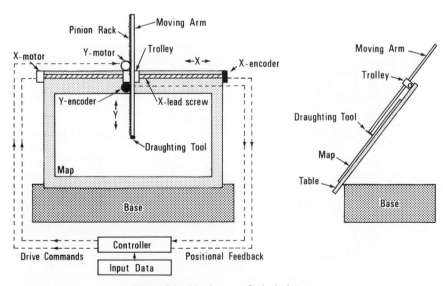

Figure 9.36 Moving-arm flatbed plotter.

design. Also, the fixed gantry can better support the heavy and elaborate photo heads which are used on many Master Plotters. However, the effect of this design is also to increase the physical size of the plotter for a given format as compared with the more conventional arrangement of the moving cross-slide or gantry (Fig. 9.35). This results from the need to provide an x-axis which must be twice as long as the plotter format, so that all points which have to be plotted on the map or plan can be reached by moving the flatbed drawing surface.

Another alternative design of flatbed vector plotter which has a widespread use in digital mapping is the *moving arm type*. This design is associated particularly with the firm of Datatek which has been the principal manufacturer of this design, although often it has been re-badged and supplied by Kern, Galileo, Ferranti and other surveying and photogrammetric manufacturers. The design has also been implemented by Zeiss Jena. As can be seen from Fig. 9.36, a plotting arm replaces the normal cross-slide. This arm can be driven in the x-direction, as in the conventional cross-slide design, but the arm can also be driven up and down in the y-direction to give the new position along this axis of the end of the arm which carries the draughting tool. This type of medium accuracy device has had quite extensive application as an intelligent (microprocessor-controlled) plotting table attached to stereoplotting instruments equipped with encoders. The sloping position of this lightly-built table is especially convenient to allow direct observation of the plotted map detail by the photogrammetric operator.

A very different vector plotting technology which eliminates the high-precision mechanical elements of the flatbed plotters described above is the *Laserplot* photo plotter (Fig. 9.37a). As the name suggests, this device is manufactured by LaserScan and uses some of the laser writing elements already discussed in respect of the Fastrak and Lasertrak digitizing devices (section 9.3.2.2) and HRD-1 Display Plotter (section 9.6.1.3). An HeCd laser provides the fine writing beam which is steered using deflection mirrors (Fig. 9.37b) under computer control with a positioning accuracy of $3\,\mu m$ to write the map image direct on photographic film. The Laserplot is enclosed in a light-tight box and accepts a film size up to $400 \times 300\,mm$ with a writable area of $300 \times 230\,mm$, slightly greater than standard A4 size. The resulting negative film can be enlarged quite readily in a suitable enlarger or process camera to give A0-size prints of the map while still retaining good quality and accuracy. A well-known user of the Laserplot device in the UK is the Mapping and Charting Establishment of the Ministry of Defence.

9.6.2.3 *Vector drum plotters.* This type of plotter has a widespread application in digital mapping, particularly for the production of edit plots. Various types exist, some with single, others with twin drums. Some can utilize flat sheets of paper or film, others require media with rows of perforated holes located along the edges. Since the media is invariably in motion across the drum while plotting is in progress, accuracies are generally lower than in the flatbed type. Hence operations such as scribing, mask-cutting and the plotting of colour-separated originals on photographic materials for the eventual production of multicoloured maps by offset lithographic printing are not carried out on vector drum plotters, though, as will be seen later, the last of these operations can be executed on raster drum plotters.

The most common design of vector drum plotter is shown in Fig. 9.38. The cylindrical drum has a series of pins located at exact intervals around the ends of the

(a)

(b)

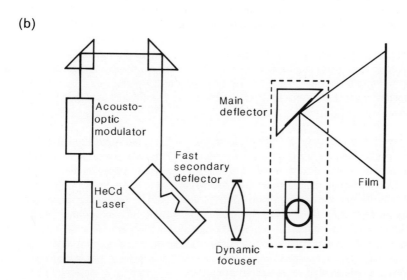

Figure 9.37 (*a*) LaserScan Laserplot and (*b*) Laserplot writing mechanism.

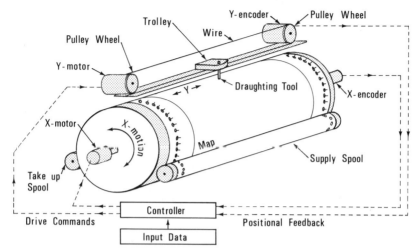

Figure 9.38 Diagram of vector drum plotter.

drum. These project above the drum surface to engage in the perforated holes which are pre-punched along the edges of the roll of paper or film used as the draughting material. The drum can be rotated bi-directionally forward or backward under the action of a stepping motor to change the x-coordinate position of the material under the fixed gantry carrying the pen used for plotting. A second motor drives a wire and pulley mechanism to which the trolley carrying the plotting pen is attached. The dual action of the moving plot material in the longitudinal x-direction and the moving pen in the y-direction along the gantry ensures that all the map detail can be plotted.

This particular type of vector drum plotter is made by a large number of manufacturers, especially Calcomp, Benson and Houston Instrument, and in a large number of different widths—11, 19, 24, 36, 42 and 48-inch wide drum plotters can all be encountered in digital mapping operations. This width defines the format of the map in the y-direction. The x-direction can of course be as long as the roll of material used as the draughting medium.

In recent years, a slightly modified version of this design has become popular in digital mapping work, in which the pins along the edges of the cylinder have been eliminated and the roll of perforated draughting material is usually substituted by a flat sheet of paper or film. In this case, the drum is driven bi-directionally in the x-direction by a motor as before, but the draughting material is caused to move by a grit-impregnated wheel bearing down along each edge of the material. The pen is still driven along the fixed gantry in the y-direction by the motor and its wire and pulley mechanism. This type of drum plotter was originally introduced by Hewlett–Packard and Houston Instrument (Fig. 9.39), but has been manufactured by a wide range of suppliers since then. The ability to use sheets of standard paper or film for the plotting of maps instead of specially manufactured rolls of accurately perforated material is a major advantage to many users in the digital mapping field. On the other hand, the continuous roll type allows a whole series of maps to be plotted as a single batch job without an operator having to be in attendance to replace individual sheets as they are completed.

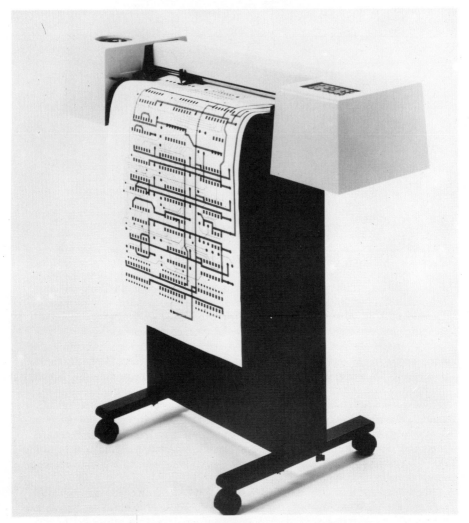

Figure 9.39 Hewlett Packard Draftpro sheet feed drum plotter.

Yet another variation of drum plotter design is the *twin-drum* model manufactured by Houston Instrument (Fig. 9.40). The plotting media is stretched out between the two drums or rollers so that the plotted map can be more readily seen by the operator, the bi-directional rotational motion of the two drums being exactly synchronized. This design has been developed still further in the popular Calcomp 1070 Beltbed machine in which the two drums are connected by a wide and very fast-moving belt. This can be operated in either of two modes: cut-sheet mode, in which an individual sheet of film or paper can be mounted on the belt to produce the plot of a single sheet, and drum plotter mode, by which a whole series of maps can be plotted as a single-batch job on a continuous roll of paper or film. Changing modes is an easy operation, and the plot

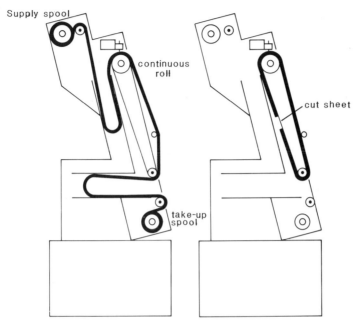

Figure 9.40 Diagram of twin drum plotter.

surface is normally mounted in vertical position both for economy in space and for easy observation by the map user or operator.

9.6.3.3 *Draughting tools associated with vector plotters.* A matter of great importance in all forms of digital mapping is the availability of a range of *draughting tools* of the specific types required to output the final map in the form and on the material desired by the user. A great variety of tools exist which are used on different types of vector plotter. These may be classified into five groups: pencils; pens; scribing tools; mask-cutting tools; and light-spot projectors.

Obviously each of these is associated with a specific type of draughting material: pencils and pens with paper and draughting film; scribing tools with specially coated film; mask-cutting knives with masking material; and light-spot projectors with photographic film. Only a very few plotters can accommodate all five types of tool; many plotters can only utilize one or two possible types.

Pencils are seldom used nowadays since the very rapid up/down action of the plotting mechanism quickly causes blunting of the point or a broken lead.

Pens are the most commonly used draughting tool, especially in engineering survey applications. A great variety of pens exist, each with its own specific merit. *Ballpoint pens* are widely used on both flatbed and drum plotters. The quality of line may not be high either in terms of consistent width or colour, but they are used especially for edit plots or final maps of a lower cartographic quality. *Fibre-tip pens* also have a wide currency, especially on smaller-format pen plotters. They give a high-quality line when new, but the continuous hammering of the pen through the actions of the rapid up/down mechanism causes the tip to flatten, in which case the line width increases and the quality of the plotted line decreases. Also, these pens have a tendency to dry out if not capped. Finally, the *liquid ink type* is used especially if a high-quality, final plot is desired. Maintenance of the line quality may be difficult; in particular, problems may arise from the drying out of the ink at the tip of the pen, so that the ink does not flow evenly or consistently. Care taken with the capping of the pen helps to overcome these difficulties. However, another problem with very high-speed plotters is that the ink release from a liquid pen may be too slow for the production of consistent line quality. This has been overcome by the development of pens with pressurized ink flow.

One of the standard requirements for complex maps is to improve the clarity of their representation through the use of several colours. Thus many vector plotters have special *multiple pen holders* and complex mechanisms to allow automatic changing of pens during plotting so that the plotting process does not have to stop, nor is it necessary for an operator to be in attendance to physically change pens, as is the case if only a single pen-holder has been provided. These multiple pen holders and changing mechanisms may take any one of several different forms (Fig. 9.41). Two, three or four pens may be mounted in line on the moving trolley, each set at a precise distance from its neighbour so that allowance may be made for the offset during the plotting of the map. Alternatively, if a larger number of pens of different widths and colours are required, a turret or carousel may be provided on the trolley carrying the draughting tools, the rotational movement needed to select the appropriate pen being provided by a stepping motor. Yet another alternative used in small-format flatbed plotters is to mount the pens in a line of stalls located along one edge of the plotter surface parallel to the cross-slide. With vector drum plotters, the stalls or carousel will be located at the ends of the fixed gantry. Users of plotters must have regard to the specific number of pens, the location of each pen and the type of pen-changing mechanism when carrying out the programming and plotting of the map.

Scribing tools and *mask-cutting knives* share many features in common and may be treated together. Essentially, the use of these tools is limited to those large, heavy-duty vector flatbed plotters designed specifically for use in mapping or in the printed circuit-board (PCB) industry where negative scribes and masks are also widely used. Again multiple tool holders need to be provided, for example to allow a change in line width through the use of scribe points of different diameter. It may also be necessary to provide a rotational control of the cutting knife or chisel-pointed scribing tool so that the blade is always tangential to the line being cut. Provision has also to be made to clear away all the swarf (debris) generated by the cutting process. Typically this is

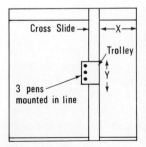 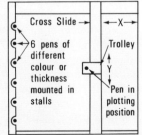 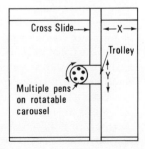

Figure 9.41 Multiple pen holders.

DIGITAL MAPPING TECHNOLOGY

Figure 9.42 Ferranti light-spot projector.

achieved using a small suction pump to which is attached a tube whose end is placed just behind the actual cutting tool.

Light-splot projectors are again limited to use on large heavy-duty flatbed plotters such as the Kongsberg, Gerber and Ferranti Master Plotters described above (Fig. 9.42). However, these have been extensively used for the production of the final, high-quality cartographic output on photographic film. They are often very complex in design and costly to purchase. Nevertheless, they have found a wide application within national mapping agencies and in the larger private surveying companies. Most designs (Fig. 9.43) use interchangeable glass optical discs containing either the apertures of various sizes needed to produce different line widths, or the letters, numbers and symbols of differing size, font and style required to produce text and symbols on the final map. Provision will also be made for the optical rotation of each individual character or symbol using a dove prism. Illumination of the disk aperture or the character or symbol is provided by a powerful lamp, diode or laser. Final plotting on light-sensitive photographic material must of course be carried out in a darkroom, and appropriate photographic developing facilities need to be available.

It will be obvious to the reader that the complexity of programming, of data provision, labelling and structuring, and of the actual procedures associated with the operation of these high-quality vector plotters requires skilled and highly trained personnel.

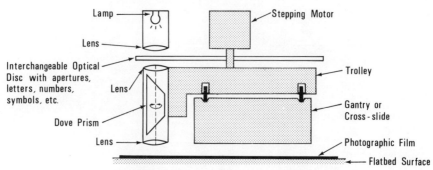

Figure 9.43 Diagram of light-spot projector.

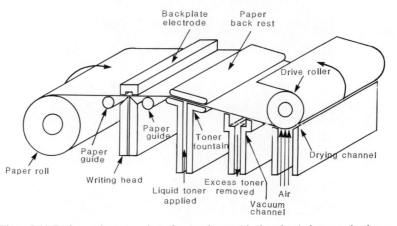

Figure 9.44 Design and construction of monochrome (single colour) electrostatic plotter.

9.6.3.4 *Raster flatbed plotters.* A feature of all raster plotters is that plotting often proceeds simultaneously along a series of parallel lines. A consequence of this type of plotter action is that it does not permit the use of pencils, pens, scribe points and mask-cutting blades, as is possible with the different types of vector plotter described above in which the basic action is one of plotting individual lines sequentially. A result of this fundamental characteristic is that raster plotters use totally different technologies, operational modes and media, quite apparent from the obvious fact that the coordinate data must be supplied to the plotter in rasterized form, i.e. in scan line order, instead of as a series of successive vectors.

A most important type of raster-based flatbed plotter is the *electrostatic plotter*, manufactured by a number of companies—Calcomp, Versatec (USA), Benson (France) and D-Scan/Seiko (Japan)—and in widespread use in digital mapping, especially where high speed of output is of prime importance. Both monochrome and colour electrostatic plotters are available from each of these manufacturers.

The basic design and construction of a monochrome (single colour) electrostatic plotter is given in Fig. 9.44. The paper or film is held in a roll and passes over a series of guide rollers and then between a writing head and its backplate electrode. This imparts

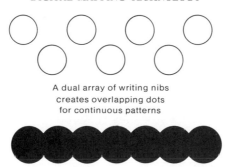

Figure 9.45 Writing nib patterns on an electrostatic plotter.

a series of very fine electrostatic charges to the surface of the plot material wherever a dot (equivalent to a pixel) needs to be plotted. The whole of an individual line lying at right angles across the parallel series of lines constituting the raster plot pattern is written simultaneously in parallel by the nibs of the writing head (Fig. 9.45). Once the charges have been written in the appropriate positions, the paper or film, being in continuous motion, passes over a liquid toner (ink) solution which adheres to the material wherever it has been given a charge. Next the paper or film passes over a vacuum channel where any excess toner is removed. This is followed by the drying channel where hot air is blown continuously on to the material to dry the ink. Finally, the plotted material is stored on a take-up spool off which the plotted maps can be cut into sheet form.

The plot resolution, in this case the number of dots that can be written by the nibs simultaneously across the width of the writing head, is typically 200 or 400 dots per inch (8 to 16 dots per mm). The range of models available for example from Versatec have plot widths of 11, 22, 36, 42 and 72 inches respectively. The rates of passing the material past the writing head is high: 1 inch per second for the 11- and 22-inch wide models and 0.75, 0.5 and 0.25 inches per second respectively for the other three wider models. From these resolution values, plotting widths and speeds, it is obvious that raster coordinate data has to be supplied in parallel to electrostatic plotters at an enormous rate if satisfactory operation is to be achieved. Thus it is necessary first to transform the vector data into an appropriate raster format in a pre-processing operation. This can be done either by software in a computer (which is usually a slower operation) or to use a purpose-built hardwired rasterizing device (which will give higher speeds). It will be obvious too that the rasterization process to produce plot data cannot begin until the whole of the vector data is available for processing.

The use of these large-width electrostatic plotters (Fig. 9.46) is common in the digital mapping and modelling operations carried out by large oil companies and by geophysical survey companies, which have available the very powerful computers necessary to reduce and process geophysical data, need high-speed output, and possess the large amounts of capital necessary to purchase and operate large-format electrostatic plotters. But the smaller-format types are used also in the mainstream surveying and mapping industry as edit plotters to provide check plots before the final map is output on accurate, high-quality plotters, as for example in the production of colour-separated film transparencies.

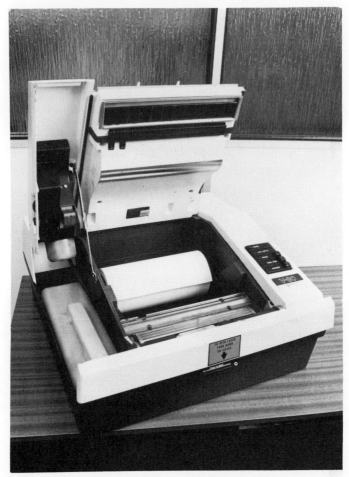

Figure 9.46 Versatec electrostatic plotter.

A recent development has been that of *colour electrostatic plotters*. Both multipass and single-pass models have appeared. The design of the multipass model is shown in Fig. 9.47. From this it will be seen that, instead of a single toner channel, there are four, which allow the application of cyan, magenta, yellow (the subtractive colour primaries) and black toners to the material to allow the production of a multicolour map or the perspective plot of a digital terrain model. In this multipass type of colour electrostatic plotter, there is still only a single writing head, and so only one colour can be plotted in an individual pass. The paper or film must then be rewound back to the starting position defined accurately by registration marks. Then the next colour is written and applied to the material, and so on, until the required four passes have been completed. Obviously very accurate positioning of the material must be achieved to give good results. In the single-pass type, there are four writing heads instead of one, these being located in appropriate positions between the respective toner channels. It should be

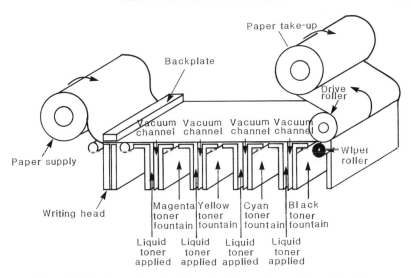

Figure 9.47 Design and construction of multipass colour electrostatic plotter.

recognized that the coloured map produced by all of these colour electrostatic plotters is a single multicoloured original: in effect it is a colour proof copy of the map. If, however, the printing of a considerable number of copies of this map is required, then colour-separated film transparencies must be produced using some type of photoplotter. From these transparencies, the necessary printing plates can be made for the lithographic printing of the required number of copies of the map.

9.6.3.5 *Raster drum plotters.* From the constructional point of view, this type of plotter is virtually identical to that of the automatic raster drum scanner already described in section 9.3.3 (Fig. 9.14). The only major change is the substitution of the scan-head with its photodetector by a light source, usually a laser or light-emitting diode, which exposes the map image on to the photographic film which is almost invariably used as the recording material. Thus for example the Scitex Response machine (Fig. 9.48) can act either as a raster drum scanner or as a plotter. In such devices, the photographic film is wrapped around the cylindrical drum which is spun up to a constant high speed. The film is then exposed in a series of parallel scans, the illuminating source being switched on and off at high speed to give exposed and unexposed pixels along each track as required.

As will have become apparent from the discussion above, the vector line data of the map will again have to be converted to raster format before photo plotting can commence. Once all the individual pixels of a map have been exposed on this very high-resolution type of plotter, the adjacent pixels will combine together to form a continuous line. A special feature of this type of raster plotter is that the aperture of the illumination source can be controlled to give precisely formed dots of a specific size. This lends itself to the direct plotting of tint screens of a given dot size, interval and

Figure 9.48 Scitex Response raster drum plotter.

angular direction, as required for filling those areas of a map, such as forests, or water areas, which need to be depicted or represented by a specified colour or pattern. As a result of this capability, a number of larger mapping organizations with a requirement to produce high-accuracy colour separations for multi-colour map production have adopted this type of plotter in spite of its very high cost. For smaller organizations with a similar requirement, access to Scitex and other comparable raster drum photoplotters can be obtained via the services of a bureau, as is done with automatic digitizing equipment.

9.7 Applications of digital mapping

Having reviewed the main types of digitizing, display and hard-copy devices which are available for digital mapping operations, it is instructive to see how they can actually be deployed and integrated into a mapping sysem and how they have been applied in actual practice. Since a very large number and variety of digital mapping systems exist which are used in an almost bewildering range of applications, only a very small sample of these can be described and discussed in the remaining part of this chapter. This will be done in terms of systems in which the input data has been acquired primarily from (i) field survey, (ii) photogrammetric, and (iii) cartographic sources respectively.

9.7.1 *Digital mapping based on field survey data*

Over the last few years, a large number of software packages have been devised, written and implemented specifically for use with digital field survey data with the production of a large-scale map one of the principal final products. One of the most widely used systems in the UK, with 70 users, is the *Eclipse system* (Gould, 1986), originally devised and introduced to the market by a civil engineer, Malcolm Grant, in 1978. This may explain its specific orientation towards engineering survey work and the fact that numerous associated modules exist for terrain surface modelling, volumetric analysis, road design, building layout, drainage design and quantities take-off, besides those for survey observations and processing and the production of digital maps which are the main concern of this chapter. Thus the Eclipse package aims to achieve a high degree of integration between land survey operations and civil engineering design, although many surveying companies utilize it purely as a surveying and mapping package which supplies data to a wide range of CADD systems. Another notable feature of the Eclipse system is that it has always been oriented towards the use of small desktop computers, especially the Wang 2200 series, and their successors, microcomputers such as the ubiquitous IBM–PC machine and its clones.

The field survey control and detail observations can be keyed in manually, as in the case of dimensions or offsets measured by tape, or they can be input directly from a wide range of electronic theodolites and tacheometers equipped with solid-state data loggers or hand-held field computers such as the Husky Hunter. The field survey data can first be displayed in alphanumeric form on a terminal, and specific observations can be deleted, altered and inserted. Other editing functions include automatic relabelling of features. The actual editing operations are similar to those carried out during word-processing and spreadsheet operations. Once the data has been edited, the computation and the adjustment of the control survey can be implemented including traversing, triangulation, intersection, resection, trigonometric heighting and levelling. The resulting control information including coordinates and station descriptions can be stored and recalled later by station name or number.

Once the control information has been established, the detail survey observations can be processed. Either polar or radial observations from optical or electronic tacheometric observations or the offset and dimensional measurements made by tapes or any combination of these observations can be handled by the program. Data from existing plans and maps can be digitized and merged with the data produced from the field survey measurements. On completion of the detail survey processing, the results can be displayed as a map on a graphics monitor or terminal, either in monochrome or in colour. A wide choice of line styles, text fonts and symbols is available. Pan and zoom operations can be executed to allow the detailed inspection of the map or plan, and interactive editing may then be carried out with the aid of a cursor, including selective erasure of parts of the graphics image. When editing has been completed, the data is stored as an Eclipse Drawing File. This can then be used to drive any one of a wide range of plotters to give the final hard-copy output (Fig. 9.49).

It is also possible to use the Eclipse package for the measurement, editing, manipulation and storage of digital map data derived from Ordnance Survey (OS) large-scale maps. Indeed, a few private survey companies use Eclipse systems to carry out contract digitizing for the OS as part of the massive task of converting all the existing 220 000 large-scale plans and maps of Great Britain which is currently in progress and will be discussed in section 9.7.3.

Although Eclipse has been discussed here in some detail as an example of a field-survey-oriented package or system used for digital mapping, it must be noted that many other comparable packages exist including those supplied by the principal survey instrument manufacturers, such as Wild (Geomap), Sokkisha (Survis) and Kern

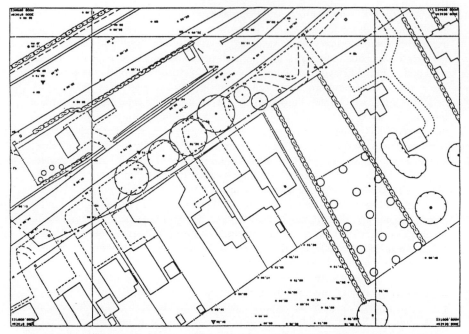

Figure 9.49 Large-scale plan generated from field survey observations using the Eclipse package.

(Infomap), and by software houses or systems suppliers, such as Optimal Software, Hasp Associates and many others.

9.7.2 *Digital mapping based on photogrammetric data*

A digital mapping system in which the main sources of data are photogrammetric and cartographic in origin is that installed by Mark Hurd Aerial Surveys of Minneapolis, and described by Leberl and Olsen (1982). The overall configuration is shown in Fig. 9.50. A number of stereoplotting instruments are attached to microcomputers which act as local workstations. The digital coordinate data which is logged and stored on these machines is then downloaded to a central Digital VAX 11/780 minicomputer which acts as the central processing engine of the overall mapping system. Data from existing maps can also be input to the system through tablet digitizers to which graphics terminals are attached in order to display what has been digitized. Yet another source of cartographic data is an MBB KartoScan raster flatbed digitizer which is used principally for the digitizing of contour line sheets to give the basic height information required for the generation of digital terrain models.

On the output side of the system, a number of Sanders Graphic 7 vector refresh displays equipped with tracker balls and lightpens are also connected to the central VAX computer, and act as interactive editing stations to correct erroneous data, to insert any omissions or revisions and to carry out the labelling and structuring of the data. Once these editing operations have been completed, an edit plot can be produced in four colours using either a Zeta vector drum plotter or a Calcomp beltbed plotter. Further editing can then follow on the interactive display terminals. Finally, the digital

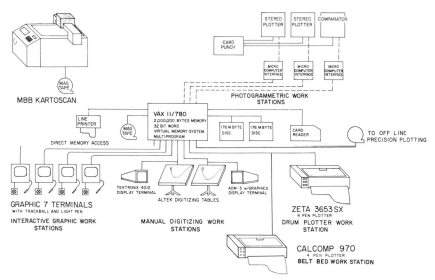

Figure 9.50 Photogrammetrically-based digital mapping system of Mark Hurd Aerial Surveys.

map data can also be recorded on magnetic tape for the plotting of the map later on a high-quality vector flatbed plotter or raster drum plotter as an off-line operation. Alternatively, the tape can be supplied direct to the client. Although the specific items of hardware all come from different manufacturers, these have been integrated into a single digital mapping system including the provision of software, by the system supplier, in this case SysScan. This is the situation with most digital mapping systems, although a few survey companies have integrated the hardware and written the software themselves.

Another example of a mainly photogrammetrically-oriented digital mapping system is that installed at BKS Surveys Ltd of Coleraine, Northern Ireland, which has been described in papers by Byrne and Neil (1983) and by Byrne (1986). The system has been installed by Wild under the name Wildmap. Basically it comprises the Synercom Informap system which is used for its interactive editing and database functions, with Wild supplying the encoders and digitizing hardware, and Synercom the CAP/IN (Cartographic and Photogrammetric Input) software to allow the Wild A8 and A10 analogue stereoplotting instruments to be used for photogrammetric data capture. These have been supplemented recently by a Wild BC-2 Aviolyt analytical plotter. Each instrument also has its own Wild Aviotab vector flatbed plotter for the production of hard-copy plots while digitizing takes place. The data goes directly to a Digital VAX 11/780 minicomputer which acts as the main digital map data processor and the database management machine. The display devices are Tektronix vector storage terminals and the hard-copy output device yet another Wild Aviotab vector flatbed plotter. While the specific items of hardware used are different and there are some differences in configuration, many of the overall concepts are the same as those employed in the SysScan system used by Mark Hurd Aerial Surveys.

Wildmap uses a hierarchical structure to organise and manage the data held in the system. As can be seen in Fig. 9.51, there are three levels of organization, the first being

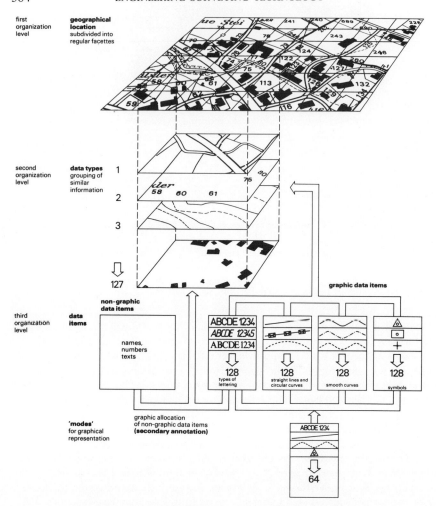

Figure 9.51 Digital database structure of the Wildmap system.

the geographical location of the data. The area being mapped is sub-divided into a grid of regular shapes (squares or rectangles)—the so-called *facettes*. All the data within an individual facette is placed in a separate file, thereby allowing direct access of this particular data set without the need to search all the data of the area in question. At the second organizational level, within each facette, the data is divided into 127 different *data types*. This gives the user rapid direct access to all the data falling into an individual category within each facette. At the third level of organization, each individual data type can contain up to 128 different *data items* which are divided into non-graphic items (e.g. names of streets, buildings, etc., numbers, text) and graphic items (e.g. straight lines and circular curves, smooth curves, symbols). The non-graphic data items are allocated to specific graphic items and are automatically marked on the final maps or plans. All data of a specific data type or a combination of data types can be displayed on a graphics terminal or plotted on the Aviotab plotting table.

A special package, MAP/IN, monitors and supports the input of all data, (e.g. from the CAP/IN software) and structures and organizes the data to form the database. A further package, INFORM, carries out the actual database management and prepares the required data for display or plotting in map form, based on any desired combination of data types for the defined area. A final package, COGO/1, allows the calculation of standard geometric problems required by surveyors and engineers. In summary, Wildmap can best be described as a mapping information system which is distinct from the geographic and land information systems (GIS/LIS) to be discussed later in Chapter 11.

The main usage of the BKS Wildmap system has been for the digital production of large-scale maps and the creation of digital data base information for various countries in the Middle East, especially in some of the Gulf States. In this case, the photogrammetric data has been captured digitally, plotted and then field checked. The revisions derived from the field completion and verification were then entered into the database via the interactive graphics workstations. The corrected final data set was then used to produce maps at 1:500, 1:1000 and 1:2000 scales. Based on these digitally-produced base maps, a series of specialist maps were produced. These included a series a cadastral maps which are essentially overlays on the topographic base map which are used for land registration purposes. A further series of digital map overlays were produced to cover six required utilities (water, electricity, telephone, irrigation, sewage and storm drainage) to show the location of pipelines, cables, plant installation, manholes and wells, all referenced to the features contained in the topographic database.

9.7.3 *Digital mapping at the Ordnance Survey*

The Ordnance Survey (OS) has been using digital mapping techniques for the production of its standard large-scale 1:1250 and 1:2500 plans of Great Britain since 1972. Accounts of the initial trials and considerations which led to its adoption of digital mapping techniques are given by Irwin (1970) and Gardiner-Hill (1972, 1973). More detailed descriptions of the methods and experience gained in the actual production of these maps are given by Atkey and Gibson (1975) and Thompson (1978).

Initially most of the OS effort went into the actual production of its 1:1250 and 1:2500 scale plans using digital methods. Data was measured in the field using conventional survey techniques such as optical tacheometry and tape measurements, and plotted manually on a stable plastic sheet. Alternatively, a photogrammetric plot executed on a stereoplotting instrument and field-completed using conventional surveying techniques, formed the basic graphic document. A third source of map data was the huge archive of existing OS plans at 1:1250 and 1:2500 scales, kept continually revised by small teams of field surveyors stationed in local offices throughout Great Britain. Irrespective of the actual methods used for the original survey, the OS has concentrated on the conversion (i.e. digitizing) of these various graphics documents as the main source of input to its digital mapping system. Only very recently has any substantial amount of digital data been acquired directly using electronic tacheometers and stereoplotting instruments equipped with encoders (Newby and Walker, 1986).

9.7.3.1 *Ordnance Survey digital mapping procedures.* Digitizing is carried out not on the original document (the Master Survey Drawing) but on a stable film negative copy of the map enlarged by a 3:5 ratio from the original scale. Manual digitizing of the map detail is carried out on this sheet using a tablet digitizer. A menu consisting of a strip of plastic with a series of clearly labelled boxes, one for each possible class of feature contained in OS maps, is also mounted on the digitizer alongside the actual map sheet. The feature code required to describe a specific object (house, fence, road edge, etc.) being digitized is entered by placing the measuring cursor within that box and pressing the recording button, the resulting coordinates being converted to the appropriate numerical code during subsequent computer processing.

During the processing, the digitized map data is also transformed from digitizer to National Grid coordinates, and any scale distortion which may have been present in the original document is removed using a suitable geometric transformation. The digitized data is then checked for accuracy, correctness and completeness by producing a four-colour ballpoint pen plot on paper using a fast vector plotter and comparing this plot on a light table with the original film negative used for digitizing. This comparison identifies omissions, errors in measurements and feature coding, and wrong spelling or locations of text. Editing operations to rectify these faults are carried out using interactive graphics workstations such as the LITES 2 system described in section 9.3.1.3.

Once the editing has been completed, the digital map data is stored in the main OS database, from which either a new edition of the map can be generated or the data supplied in digital form to users on a magnetic tape. For the production of a new edition of the map, the data, including the grid, symbols and text, is plotted out on photograpic film using Ferranti Master Plotters equipped with light spot projectors. Very recently, the OS has begun to replace these machines with a Scitex raster drum plotter which can of course plot tints and patterns directly on the film material. This eliminates the process by which masks had to be cut manually or digitally and the tints or pattern generated by subsequent photographic processing. The final result of these plotting operations is a negative film transparency from which a printing plate can be produced for the offset litho printing of the new edition of the map.

OS digital map data is also supplied to customers on magnetic tapes, using a simple character format known as the *Ordnance Survey Transfer Format* (OSTF). A typical map sheet in OSTF occupies about 0.3 Mbytes, but in dense urban areas this figure can increase to about 0.6 Mbytes. Up to 60 large-scale maps or plans can be accommodated on a single standard high-density 2400 ft magnetic tape. A new National Transfer Format (NTF) has recently been announced and will be used in the future. The OS also makes a program known as DO9 available to users of its digital map data to enable them to plot the data at a specific scale selected by the user; to plot only certain classes of data defined by their feature codes; and to produce plots of specified areas other than those defined by the OS National Grid Sheet lines.

9.7.3.2 *Current applications of Ordnance Survey digital map data.* Many local government bodies, such as district, borough and county councils and public utilities, make use of existing OS digital map data. An example of a local government organization is Dudley Metropolitan Borough Council, in the West Midlands of England, which operates its own digital map production system based on the DO9 program to generate master copies on film of 1:500, 1:1250 and 1:2500 scale plans from the OS data set covering the area of the borough. Also the Council has a separate data set defining land parcels only, again produced by the OS. Dyeline copies made from the master film transparencies can be supplied to all the departments of the Council.

Dudley is also the location of a joint trial (Dudley Digital Records Trial) carried out

DIGITAL MAPPING TECHNOLOGY

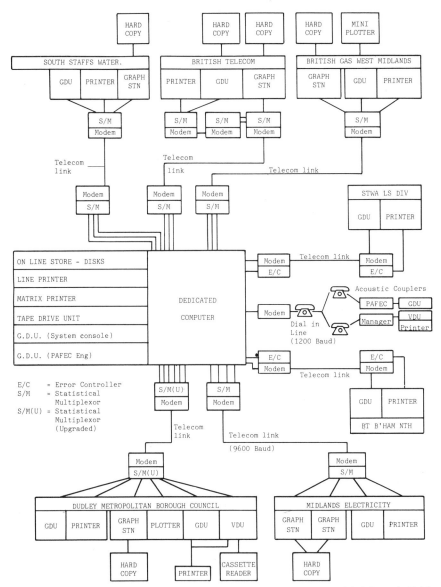

Figure 9.52 Configuration of the computer system used in the Dudley Digital Records Trial showing the links to each utility office.

since 1982 by four major public utilities (water, gas, electricity and Telecom) which form the National Joint Utilities Group (NJUG). The OS digital map data of the area forms the main geographical location database on which all plant, cable and pipeline information has been digitized and combined to form a single combined database (Fig. 9.52). The digitized records also include sewer and highway information supplied

Figure 9.53 Interactive map editing station used in the OS digital field update system.

by the Borough Council. The system is designed around a mapping package, DOGS Mapping, which is based on a computer aided design and drafting (CADD) package called DOGS (Drawing Office Graphics System) produced by Pafec of Nottingham. As Fig. 9.52 shows, this package is installed on a Prime mini-computer located at one utility's offices. As the diagram also shows, each of the other participants have a graphics workstation and peripherals installed at their offices and linked to the main computer. Either high-quality multicolour maps can be produced centrally using pen plotters, or graphics terminals can be used in each utility office to display the information with lower quality plots produced locally using monochrome electrostatic plotters.

Information is organized into different drawing layers depending on its source, content and ownership. The spatial data is held in a spaghetti model (see Chapter 11) both within a layer and between layers. All basic topographic map details are divided into just three layers: roads, property, and other details. The Borough Council is responsible for maintaining this information. Each utility is allocated approximately six layers in which to record their own information. One of these is accessible to be viewed by all other users; the remaining layers are private to the utility concerned. A major aim of the trial is to operate the statutory exchange of map-based information required by the Public Utilities Street Works Act (PUSWA) using digital mapping techniques instead of the traditional approach utilizing annotated OS maps or overlays which has been employed up till now.

9.7.3.3 *Planned developments in Ordnance Survey digital mapping coverage and procedures.* By 1987, 16 000 of the OS 1:1 250 scale map series covering urban areas (amounting to 29% of the total number of maps in the series) had been converted to digital form, together with 13 000 of its 1:2 500 scale map series covering periurban and rural areas (amounting to 8% of the total). Recently, additional funds have been made available by central government to accelerate the programme of digitizing these large-scale map series. As a result, a substantial amount of digitizing is now being undertaken by private sector surveying companies working under contract to the OS. These companies often employ hardware and procedures different to those used by the OS, but the work is carried out to the same specifications for coding, accuracy and completeness as those applied by the OS for its own in-house operations. Priority has been given to digitizing all the maps covering the major urban ares of Great Britain, a task which is expected to be completed in 1995.

The OS has also started to introduce its *digital field update system* (Coote, 1986) into certain of its regional survey offices (Fig. 9.53). This allows digital map data which has been measured during OS continuous revision operations to be captured, stored and made available very soon after it has been surveyed. The data is acquired by field survey either using electronic tacheometers or using tape and graphical field survey methods which are plotted on plastic material and converted locally using tablet digitizers. The digital map data can then be edited and plotted in the regional office using a high-resolution graphics workstation to which is attached a tablet digitizer and a fast vector flatbed plotter. The data is also sent over a communications link to the mainframe computer located in the OS headquarters in Southampton where it can be merged and used to update the main database. Once this has been done, the new centrally-held data can be downloaded to the regional office as required.

9.8 Conclusion

It is obvious that digital mapping technology and techniques are now very well established in a wide range of surveying and mapping organizations. Furthermore there are very few organizations in the civil engineering, building and construction industry which have not been affected by the developments outlined in this chapter, either through the supply of digital map data from specialist surveying firms or a national mapping organization as an input to their planning and design processes, or by the adoption of digital mapping techniques within the engineering organization itself. Closely linked to these developments are those concerned with digital terrain modelling. These will be described in the next chapter.

References

Atkey, R.G. and Gibson, R.J. (1975) Progress in automated cartography. *Conf. of Commonwealth Survey Officers, 1975: Report of Proceedings, Part II*, Paper J3.

Bell, S.M.B. and Woodsford, P.A. (1977) Use of the HRD-1 laser display for automated cartography. *Cartographic J.* **14**(2), 128–134.

Boyle, A.R. (1980) The present status and future of scanning methods of digitization, output drafting and interactive display and edit of cartographic data. *Int. Arch. Photogramm.* **23**(B4), 92–99.

Byrne, S.T. and Neill, L. (1983) Application of the Wildmap system in a production environment. *Photogramm. Record* **11**(61), 47–52.

Byrne, S.T. (1986) Digital mapping: some commercial experiences. *Photogramm. Record* **12**(68), 143–154.

Coote, A.M. (1986) The digital field update system (an application of the graphics workstation at the Ordnance Survey). *Paper Presented at a Joint Meeting of the British Cartographic Society, Photogrammetric Society and Remote Sensing Society*, University of Glasgow, 27th February 1986.

Davies, D.J.M., Frisch, O.R. and Street, G.S.B. (1970) Sweepnik: a fast semi-automatic track-measuring machine. *Nucl. Instr. Meth.* **82**, 54–60.

Gardiner-Hill, R.C. (1972) The development of digital maps. *Ordnance Survey Prof. Pap., New Series*, **23**, HMSO, London.

Gardiner-Hill, R.C. (1973) Automated cartography in the Ordnance Survey. *Conf. of Commonwealth Survey Officers, 1971: Report of Proceedings, Part I*. Paper E3: 235–241.

Howman, C. and Woodsford, P.A. (1978) The Laser-Scan Fastrak automatic digitizing system. *Presented Paper, 9th Int. Conf. on Cartography (ICA)*, Maryland, U.S.A.

Irwin, M. StG. (1970) Automated Cartography in the Ordnance Survey. *Chartered Surveyor* **102**(10), 467–473.

Leberl, F.W. and Olsen, D. (1982) Raster scanning for operational digitizing of graphic data. *Photogramm. Eng. and Remote Sensing* **48**(4), 615–627.

Newby, P.R.T. and Walker, A.S. (1986) The use of photogrammetry for direct digital data capture at Ordnance Survey. *Int. Arch. Photogramm.* **26**(4), 228–238.

Newman, W.M. and Sproull, R.F. (1979)*Principles of Interactive Computer Graphics*. McGraw-Hill, New York.

Peuquet, D.J. (1981) An examination of techniques for reformatting digital cartographic data—Part I: The raster-to-vector process. *Cartographica* **18**(1), 34–48.

Woodsford, P.A. (1976) The HRD-1 laser display system. *Computer Graphics (SIGGRAPH/ACM)* **2**(10), 68–73.

10 Digital terrain modelling

T.J.M. KENNIE and G. PETRIE

10.1 Introduction

Digital terrain modelling is a particular form of computer surface modelling which deals with the specific problems of numerically representing the surface of the Earth. The initial concept of a digital terrain model (DTM) originated in the USA during the late 1950s (Miller and LaFlamme, 1958). Since then, considerable advances have been achieved, particularly in the methods of acquiring and processing terrain information. This chapter will therefore consider the major developments in this area and examine some typical surveying and engineering applications of DTMs.

The term DTM originally referred to the use of cross-sectional height data to describe the terrain. Nowadays, however, the definition is more general and includes both gridded and non-gridded data sets. Several other terms are also used to describe essentially the same process. Among the more common are Digital Elevation Model (DEM), Digital Height Model (DHM), Digital Ground Model (DGM), and Digital Terrain Elevation Model (DTED).

10.2 Data acquisition

Since data acquisition is so important to all practitioners of terrain modelling, this immediately poses the question as to which techniques should be considered for use in the collection of elevation data. As discussed previously in the context of digital mapping (section 9.2), the three main methods which can be used to acquire elevation data are:

(i) Ground survey methods normally using electronic tacheometers, or total stations with data collectors
(ii) Photogrammetric methods based on the use of stereoplotting instruments
(iii) Graphics digitizing methods by which the contours shown on existing topographic maps are converted to strings of digital coordinate data and the required elevations derived from them.

Table 9.1 in the previous chapter sets out the main characteristics of these three alternative techniques as applied to digital mapping. These characteristics also apply to terrain modelling.

10.2.1 Ground survey methods

In the case of ground survey, the data will nowadays normally be acquired using the electronic tacheometers described in Chapter 2. However, it is also possible to acquire the data for DTM work using the traditional optical instrumentation—theodolites and tacheometers, or even surveyor's levels to carry out grid levelling—though hardly with the same speed or efficiency.

The accuracy of all these field survey methods is very high, but they are only really practical and economic to implement over relatively small areas of terrain, for example over specific sites of limited areal extent on which buildings are being planned for construction, or new roads, sports grounds or quarries are being developed. If, however, DTM data has to be acquired over large areas, then it will be obtained either using photogrammetric methods or by digitizing the contours on existing topographic maps.

10.2.2 Photogrammetric methods

The acquisition of DTM data by photogrammetric methods is carried out using the stereoplotting instruments described in Chapter 7. The height measurements are carried out either manually by an operator or automatically by an electronic correlator, in the latter case usually under computer control. The latest development in this area is of analytical stereoplotting instruments in which the optical or mechanical models are replaced by equivalent purely mathematical or numerical solutions, executed in real time by a suitably programmed high-speed minicomputer (see section 7.5.3).

One of the major advantages of using an analytical plotter for the acquisition of terrain elevation data is that it can be programmed to drive to any desired position or series of positions corresponding to the required pattern and density of points in the stereomodel. As soon as the operator has measured the height of an individual terrain point, the computer automatically moves the measuring mark to the next point in the stereomodel which has to be measured. Thus, with the use of analytical plotters, the rate of acquisition of the terrain elevation data is greatly increased as compared with that of analogue stereoplotting instruments. However, the latter continue to be used because of the large capital sums required to purchase analytical plotters at the present time. As time goes on and the cost of computers continues to fall, the use of analytical stereoplotters must become more widespread.

10.2.3 Graphics digitizing methods

The acquisition of terrain model data over large areas, as required for aircraft simulators, landscape modelling and vizulisation, is normally carried out by digitizing the height information contained in existing topographic maps using the technologies outlined in section 9.3. Since these maps contain very few spot heights or elevations, essentially one is dealing with the measurement of contour lines so that they are represented by suitably structured strings of digital coordinate data. Subsequently the actual DTM spot height or elevation data is derived by interpolation from the digitized contour lines.

It must be recognized from the outset that such a procedure will never produce the same metric accuracy as the direct measurement of spot heights carried out by field survey or photogrammetric means. Typically the accuracy of such contours is only one-third of that of spot heights measured photogrammetrically. It follows that the

DIGITAL TERRAIN MODELLING

Table 10.1 Accuracies of DTM elevation values derived from the digitizing of contours on topographic maps. (Factor 10 = high accuracy; 1 = low accuracy)

Maps	Spot heights or contours	Accuracy factor
	1. Photogrammetric spot heights—measured in a stationary mode.	10
Medium scale topographic maps (e.g. 1:10 000 scale)	2. Map contours by photogrammetry—measured in dynamic mode.	3
	3. Spot heights at grid nodes—derived by interpolation from digitized contours.	2
Small-scale topographic maps (e.g. 1:50 000 scale)	4. Contours generalized from medium-scale maps.	2
	5. Contours measured by field survey in 19th century.	1
	6. Spot heights at grid nodes derived by interpolation from digitized contours.	0.5–1

individual spot heights derived by interpolation from such contours are still less accurate. Table 10.1 summarizes the relative accuracies of the heights obtained from a range of sources.

However, as long as this limitation in accuracy is realized by and is acceptable to users, the utilization of this lower-quality data is quite acceptable. Certainly, many users, such as regional planners, landscape architects, geologists and geophysicists, aircraft simulator operators and civil engineers engaged in preliminary project or reconnaissance studies, find it quite adequate for their purposes.

The measurement of contour lines is normally carried out manually using a solid-state tablet digitizer (Fig. 9.5) on which each contour is traced by an operator using a cursor around which a field coil is placed (see section 9.3.1.2 and Fig. 9.5). This produces a signal which can be picked up by the measuring wires embedded in the tablet, which give the x and y coordinates of the successive positions measured along an individual contour. Highly automated line-following devices such as the LaserScan Fastrak or Lasertrak (section 9.3.2.2 and Fig. 9.13) are used in large national mapping agencies to speed up the task of contour digitizing, but the enormous cost of such systems precludes their use by surveyors or civil engineers, except where their services can be obtained through bureau digitizing facilities.

10.3 Measurement patterns

The required terrain elevation information may be obtained in any one of several sampling patterns.

10.3.1 *Systematic sampling*

A systematic pattern of spot heights may be measured in a regular geometric (square, rectangular, triangular) pattern (Fig. 10.1) as specified by the client. Such an approach is that favoured in any type of photogrammetric operation which is either fully or partially automated, where the locations of the required grid node points can be programmed and pre-set to a specific interval and then driven to under computer control.

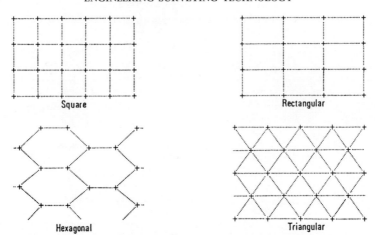

Figure 10.1 Regular grid patterns.

However, the obvious shortcoming of the regular grid-based approach to data acquisition is that the distribution of data points is not related to the characteristics of the terrain itself. This has the consequence that the finer but perhaps significant terrain features will not be measured specifically. On the other hand, if the data-point sampling density is high enough to portray accurately the smallest terrain features present in the area being modelled, then the density of measured data will be too high in many other areas of the model, in which case, there will be an embarrassing and unnecessary data redundancy in these areas. In this situation, filtering of the measured data may need to be carried out as a pre-processing activity before the terrain model can be defined.

10.3.2 *Progressive sampling*
Since this grid-based approach is most easily implemented in computer-controlled photogrammetric instruments such as analytical plotters, a solution to the above-mentioned shortcomings has come from photogrammetrists in the form of progressive sampling, originally proposed by Makarovic of the I.T.C., The Netherlands (Makarovic, 1973, 1975). Instead of all the points in a dense grid being measured, the density of the sampling is varied in different parts of the grid, being matched to the local roughness of the terrain surface (Fig. 10.2).

The basis of the method is that one starts with a widely spread (low-resolution) grid which will give a good general coverage of height points over the whole area of the model. Then a progressive increase in the density of sampling (or measurement) takes place on the basis of an analysis of terrain relief and slope using the on-line computer attached to the photogrammetric instrument. Thus the basic grid is first densified by halving the size of the grid-cell in certain limited areas based on the results of the preceding terrain analysis. Measurements of the height points at the increased density are carried out under computer control only in these pre-defined areas. A further analysis is then carried out for each of these areas for which the measured data has been increased or densified. Based on this second analysis, an increased density of points may be prescribed for still smaller areas. Normally, three such runs or iterations are sufficient to acquire the terrain data necessary to define a satisfactory model. In this

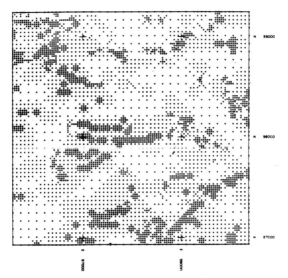

Figure 10.2 Progressive sampling.

way, the progressive sampling technique attempts to optimize automatically or semi-automatically the relationship between specified accuracy, sampling density and terrain characteristics.

10.3.3 *Random sampling*

An alternative approach, which is widely used by field surveyors and sometimes employed by photogrammetrists, is to measure heights selectively at significant points only—at the tops of hills, in hollows and along breaks of slope, ridge lines and streams. Thus all the points to be measured are identified by the surveyor or photogrammetrist on the basis of his inspection and interpretation of the terrain features. Therefore all the measured points will be randomly located, that is, an irregular network of points will result from the measurements. The consequence of this is that more thought must be given to the structuring and arrangement of the measured data than if a purely grid-based approach to data acquisition had been adopted.

10.3.4 *Composite sampling*

Quite often, the approach to data acquisition taken by photogrammetrists will combine the elements of both of the above approaches, especially if a non-automated, operator-controlled type of stereo-plotting instrument is being used for the height measurements. This approach is often referred to as composite sampling (Makarovic, 1977). The basic grid-measuring pattern will be supplemented by the measurements made at significant points in the terrain, e.g. on hilltops, along break lines and streams, as mentioned in 10.3.3. Furthermore, if a road line has already been selected, then additional points will be measured both along the centre line to give a longitudinal profile and at right angles to give a series of lateral cross-sections to provide accurate estimates of earthwork quantities (cut and fill) for alternative designs.

10.3.5 *Measured contours*

The final (alternative) approach is to systematically measure contours in a stereo-model or from an existing topographic map over the whole area to be modelled, outputting the measurements in the form of strings of digital coordinate data. Again, this may be supplemented by spot heights measured along terrain break lines or along streams.

10.4 Modelling techniques

As will be shown in the later part of this chapter, a very large number of program packages have been devised and written for terrain modelling applications in surveying and civil engineering. However, in spite of this diversity, when their characteristics are analysed it can be seen that basically they follow one or other of two main approaches:

(i) They make use of height data which has been collected or arranged in the form of a regular (rectangular or square) grid
(ii) They are based on a triangular network of irregular size, shape and orientation, based on randomly-located height data.

As Fig. 10.3 shows, these two approaches can be conducted either wholly independent of one another, or they can be combined to give a composite or hybrid approach to terrain modelling and contouring.

10.4.1 *Grid-based terrain modelling*

The first approach is in many ways the simplest in that the data comprising the terrain model may have been measured or collected in the form of a regular grid. In this situation, direct modelling of the grid can take place. However, as noted above, digital terrain data will often have been collected at specific locations, either in the field using optical or electronic instrumentation, or by photogrammetric methods. If a grid-based terrain modelling package is to be used with such specific but randomly located points, then a preliminary random-to-grid interpolation must be carried out which converts this measured data to a suitably dimensioned regular grid.

Usually the following interpolation methods are distinguished:

(i) Pointwise methods

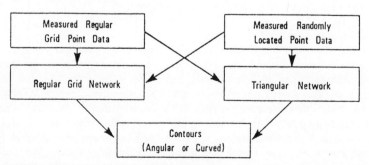

Figure 10.3 Overall relationship between measured point data, networks and contours in terrain modelling.

(ii) Global methods
(iii) Patchwise methods.

10.4.1.1 *Pointwise methods.* These involve the interpolation of the values of the terrain elevation at each specific grid node from its neighbouring randomly-located measured height points. Since the height of each point or node on the final grid is determined independently of that for any other node, it has no effect or impact on the adjacent points in the terrain model, and a continuous surface may be generated through all the derived grid nodes without any discontinuities or boundary problems.

Almost all the algorithms used for the determination of the height of each individual grid point are based on a search for the set of nearest neighbours, followed by the averaging of their heights weighted inversely by some function of their respective distances d from the position of the grid node. This weight $w = 1/d^m$ where m is the power used, typically in the range 0.5 to 4.

The search for the nearest neighbours may involve a simple area search (Fig. 10.4a) with a circle of predefined radius or a box of predefined size used to select the data points from which the grid point value is determined. A variant of this technique, in which the n nearest neighbours (Fig. 10.4b) are searched for, may also be used, n being user-defined but usually in the range 6–10. Since these two methods offer no control over the distribution of the points used for the interpolation of the grid node height, a further variant is the sectored nearest-neighbour technique in which the area around the grid point or node is divided into equal sectors—usually four (quadrants) or eight (octants), with the nearest two or four neighbours being searched for in each (Fig. 10.4c).

If the measured terrain model data takes the form of contours, another form of search may be implemented via the so-called sequential steepest slope algorithm described by Leberl and Olsen (1982) (Fig. 10.5). In this procedure, a search is made along each of the four lines passing through the required grid node and oriented along the grid directions (VV and HV) and their bisectors (UU and GG). The intersection of each of the eight directions with the nearest contours is established and the slope of each of the four lines calculated. The line with the steepest slope is then selected, and the value of the elevation of the grid node established by linear interpolation along this line—for instance, in the example shown in Fig. 10.5, search line GG is the steepest, and the height

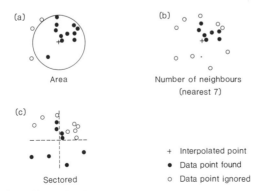

Figure 10.4 Pointwise interpolation: area search techniques.

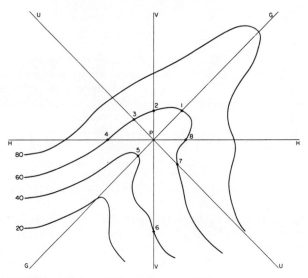

Figure 10.5 Sequential steepest slope algorithm showing cross-sections HH, VV, UU, and GG, and the intersection points on the contours (Leberl and Olsen, 1982).

of the grid node P is derived from

$$H_P = [H_1 - H_5]/(1,5)[(P,5) + H_5] \qquad (10.1)$$

10.4.1.2 *Global methods.* These involve the fitting of a single three-dimensional surface defined by a high-order polynomial through all of the measured randomly-located terrain height points existing within the model (Fig. 10.6). Once this global surface has been defined and the specific values of the polynomial coefficients have been determined, the values of the heights for each grid node point can then be interpolated. Difficulties may arise if many terms are used (as will be the case with higher-order polynomials), and if a large data set is required to form the model, resulting in large amounts of computation time. Also, the unpredictable nature of the oscillations produced by such high-order polynomials may produce poorly interpolated values for the grid node points.

10.4.1.3 *Patchwise methods.* These lie in an intermediate position between the pointwise method and the global method. The whole area to be modelled is divided into a series of equal-sized patches of identical shape. The shapes of each patch are regular in form, typically square or rectangular. Quite separate mathematical functions are then generated to form the surface for each patch. Thus separate sets of fitting parameters are computed for each individual patch. Once each of the patches has been defined, the elevation of all the grid points falling within each individual patch can be interpolated using these parameters.

(i) Exact-fit patches (Fig. 10.7a) may be defined in which each patch abuts exactly on to its neighbours. The difficulty that may result from the use of such patches is that they may result in sharp discontinuities along their junctions, which show up

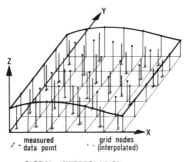

GLOBAL INTERPOLATION

Figure 10.6 Global interpolation.

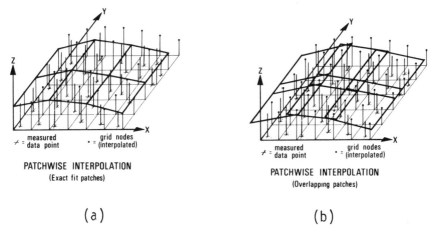

Figure 10.7 Patchwise interpolation: (*a*) exact fit patches; (*b*) overlapping patches.

markedly when the isolines or contours are finally produced. The patches are larger than the cells defined by the grid nodes, so normally several grid nodes will fall within a single patch.

(ii) The alternative is to use an arrangement of overlapping patches (Fig. 10.7b), in which case there will be common points lying within the overlap which will be used in the computation of the parameters for each patch and, indeed, can be used to ensure a smooth continuity or transition between adjacent patches.

The advantages of using patchwise methods over global methods are that quite low-order terms (parameters) can be used to satisfactorily describe each patch. So only a few unknowns need to be solved, via simultaneous equations using least-squares methods for each patch. Also, once the unknown parameters have been solved for, it is easy to calculate the derived points, that is, the grid nodes, by back-substitution in the functions or equations describing the patch.

However, there are also some disadvantages of the patchwise method. In the first place, it needs much more organization of its data and of its processing than pointwise or global methods. Also, the subdivision of the model surface into patches needs to be carried out with care. If the data is poorly distributed towards the patch corners, then this affects the computed parameters and, in turn, the accuracy of the subsequent heights determined for the grid node points.

10.4.1.4 *Polynomials used for surface representation.* As noted above, polynomial equations are used to represent the terrain surfaces in the global and patchwise methods of interpolation. The basic general polynomial equation used is

$$Z_i = a_0 + a_1 X + a_2 Y + \cdots$$

as shown in Table 10.3,

where Z_i is the height value of an individual point i
X_i, Y_i are the rectangular coordinates of the point i
a_0, a_1, a_2, etc., are the coefficients or parameters of the polynomial.

One such equation will be generated for each individual point i with coordinates X_i, Y_i, Z_i, occurring in the terrain model. In the first step, the values of X, Y and Z are known for each measured point present in the overall data set or patch. Thus the values of the coefficients $a_1, a_2, a_3 \cdots$ can be determined from the set of simultaneous equations which have been set up, one for each data point. Once the values of the coefficients $a_1, a_2, a_3 \cdots$ have been determined, then for any given grid node point with known coordinates X, Y, the corresponding height value Z can be calculated.

Table 10.3 Polynomial equation used for surface representation.

Individual terms	Order of term	Descriptive term	No. of terms
a_0	Zero	Planar	1
$+ a_1 X + a_2 Y$	First	Linear	2
$+ a_3 X^2 + a_4 Y^2 + a_5 XY$	Second	Quadratic	3
$+ a_6 X^3 + a_7 Y^3 + a_8 X^2 Y + a_9 XY^2$	Third	Cubic	4
$+ a_{10} X^4 + a_{11} Y^4 + a_{12} X^3 Y + a_{13} X^2 Y^2 + a_{14} XY^3$	Fourth	Quadratic	5
$+ a_{15} X^5 + \cdots$ etc.	Fifth	Quintic	6

To make a correct selection of the terms which will best represent or model the terrain surface, the surveyor or civil engineer must keep in mind the shape produced by each term in the polynomial equation (Fig. 10.8). Typical of the simpler types of surface used to model individual grid cells or patches are:

(i) The 4-term bilinear polynomial, as used in the HIFI package (section 10.5.1.1), with the form

$$Z = a_0 + a_1 X + a_2 Y + a_3 XY \quad (10.2)$$

(ii) The 10-term cubic polynomial, as used in the CIP package (section 10.5.2.1)

$$Z = a_0 + a_1 X + a_2 Y + a_3 XY + a_4 X^2 + a_5 Y^2 + a_6 X^2 Y + a_7 XY^2 \\ + a_8 X^3 + a_9 Y^3 \quad (10.3)$$

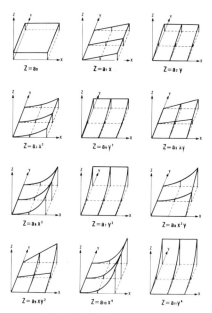

Figure 10.8 Surface shapes produced by individual terms in the general polynomial equation.

(iii) The 16-term bicubic polynomial, as also used in HIFI (section 10.5.1.1)

$$Z = a_0 + a_1 X + a_2 Y + a_3 XY + a_4 X^2 + a_5 Y^2 + a_6 X^2 Y \\
+ a_7 XY^2 + a_8 X^2 Y^2 + a_9 X^3 + a_{10} Y^3 + a_{11} X^3 Y + a_{12} XY^3 \\
+ a_{13} X^3 Y^3 + a_{14} X^3 Y^2 + a_{15} X^2 Y^3 \qquad (10.4)$$

10.4.1.5 *Contouring from grid data.* Considering a single grid cell (Fig. 10.9), a simple linear interpolation is carried out along each of the four sides in turn, based on the values at the nodes. The positions of all the contour values are determined for each side. Next they are connected up, say by straight lines or vectors, since for every entry point there must also be an exit point.

However, ambiguities may also occur, with alternative solutions and impossible situations. Taking the data points given in the example (Fig. 10.10), there are four possible solutions which give quite different positions for the contour and also a fifth (impossible) alternative. A solution to the problem is to revert to a centre point figure, i.e. splitting the grid cell into four triangles, assigning the average of the four grid nodes to the central point and then implementing an arbitrary rule, such as 'keep the high ground to the right of the contour'. This is of course a quite arbitrary solution to overcome the difficulty.

In view of these potential difficulties, in many contouring packages based on regular gridded data no attempt is made to carry out direct threading of the contour. Instead, the interpolation of the contour is based on some type of function fitting, typically using the bicubic polynomial already discussed in the context of random-to-grid interpolation, for example. Thus a new patch is formed based on a group of grid cells. If the full 16-term bicubic polynomial is employed, then the minimum size of patch that can be

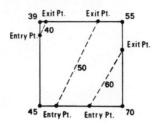

Figure 10.9 Grid contouring: linear contour interpolation in a single cell.

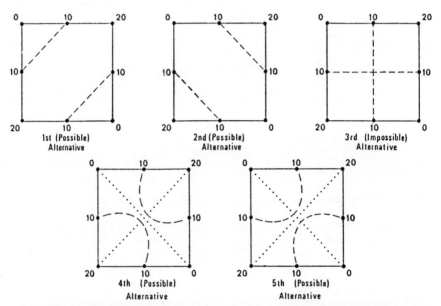

Figure 10.10 Grid contouring: ambiguity in contour threading in a single cell.

used is one made up of 4 × 4 = 16 nodes with 9 cells (Fig. 10.11). If a smaller number of coefficients are used, as in the 9-term biquadratic polynomial, then there will be redundancy, and a least-squares solution will be employed in the determination of the parameters. This can help to give extra smoothing and a better fit or continuity between cells. The interpolation of the contour through an individual grid cell is then carried out with reference to the parameters describing the surface of the whole patch, rather than by simple linear interpolation in each individual grid cell.

Thus, first of all, the values of the coefficients a_0, a_1, a_2,⋯ of the local surface patch are determined using the 16 grid node values. In some contouring packages, such as Calcomp's GPCP (General Purpose Contouring Package), additional height values are then determined using the known values of the coefficients for all the points located on a sub-grid which is generated within each grid cell (Fig. 10.11). The contours are then interpolated linearly and threaded between the points on the sub-grid cell.

The length of each of the straight line vectors will be quite short in each case, and the

DIGITAL TERRAIN MODELLING

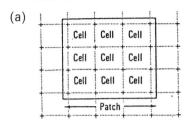

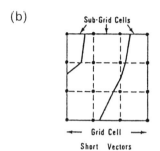

Figure 10.11 Grid contouring: (*a*) patch formed by 16 grid nodes (9 cells); (*b*) contour threading through sub-grid cells.

quite abrupt changes in direction which would be seen with contours derived from linear interpolation on the whole cell are lessened considerably, with evident benefit to the visual appearance of the contours as seen by the user. It is possible to further divide the sub-grid cell itself into centre point triangles, and conduct the contour—threading on the cutting points on the diagonals—in which case, the vectors will be still shorter and the appearance even more acceptable.

Quite a number of contour packages using grids try to get more pleasing contours by an alternative approach, contour smoothing, which involves fitting a series of cubic splines between the entry and exit points. This does produce smooth contours which are visually more acceptable than the straight-line type produced from the same data. However, a separate curve or spline will be fitted to each segment of each contour line to ensure its smoothness. A difficulty which can then arise in areas of steep terrain with close contouring is that the curved smoothed contours may then cross, which of course is not permissible in terms of actual terrain.

10.4.2 Triangle-based terrain modelling

This method, which in North America is frequently termed the TIN (Triangular Irregular Network) method, is being used to an ever-increasing extent in terrain modelling. The reasons for this development are that every measured data point is being used and honoured directly, since they form the vertices of the triangles used to model the terrain, from which the height of additional points may be determined by interpolation and the construction of contours undertaken. Furthermore, the use of triangles offers a relatively easy way of incorporating break-lines, fault lines, etc.

Any triangular-based approach should attempt to produce a unique set of triangles

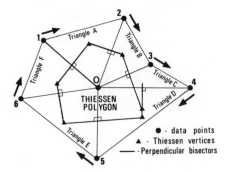

Figure 10.12 Construction of the Thiessen polygon.

that are as equilateral as possible and with minimum side lengths (McCullagh, 1983a). Although other possibilities exist, usually one or other of two main algorithms are used to implement these requirements. These are

(i) The Delaunay triangulation method
(ii) The Radial sweep method.

10.4.2.1 *Delaunay triangulation.* Associated with the Delaunay triangulation (also referred to as the Direchlet Tesselation) is the Thiessen polygon (Fig. 10.12) which endeavours to define geometrically the region of influence of a point on an areal basis. This is done by constructing the series of perpendicular bisectors on each of the triangles formed around that specific point. They intersect at the Thiessen vertices. The polygon so defined is the Thiessen polygon. Because of their distinct pattern when viewed on a computer graphics terminal they are often referred to as *tiles*. Furthermore, the data points surrounding a specific data point (e.g. *O*) are known as its Thiessen neighbours.

Often, a preliminary step before triangulation begins is to define a set of (artificial) boundary points to form a perimeter around the edges of the data set area. This is necessary to create a frame to the terrain model and a set of boundary triangles which will allow contours to be extrapolated outside the area of the data set itself. Once these boundary points (which may have arbitrary values) are defined and have been added to the data set, the whole area can then be triangulated starting with a pair of the artificial boundary points, e.g. those located at the bottom left-hand corner of the area (*A* and *B*) as the initial known neighbours.

The search for the next neighbour is then made by constructing a circle with the base *AB* as diameter and searching to the right (i.e. clockwise) to find if any data point falls within this circle (Fig. 10.13). This search can be carried out quite quickly using a computer. If no data point lies within the circle, it is increased in size to perhaps twice the area of the original circle, with *AB* now a chord in the larger circle (Fig. 10.13). Any data points lying within the new circle are tested to discover which meets the criteria set for the nearest Thiessen neighbour.

Once this has been achieved, the search for the next neighbour then continues to the right (i.e. clockwise), then on to the next neighbour, and so on, till the next boundary

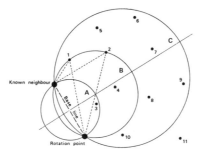

Figure 10.13 Delaunay triangulation: expanding circle search for nearest neighbour (McCullagh, 1983a).

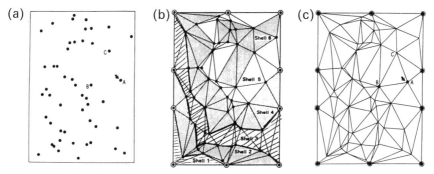

Figure 10.14 Advancing shells of a Delaunay triangulation (McCullagh, 1983a): (a) data points, (b) advancing shells; (c) completed triangulation.

point is reached. The triangles so formed constitute a so-called *shell*. The process of the triangulation then continues with each point in the shell being used in turn as the starting point for the search for the next set of Thiessen neighbours. This continues in a systematic manner until the neighbours for all the points existing in the data set have been found and the corresponding triangles formed (Fig. 10.14).

Now that it is better known and understood, the Delaunay method is that used to form triangles in the majority of terrain modelling packages based on the triangulation method.

10.4.2.2 *Radial sweep method.* The radial sweep algorithm is an alternative to the Delaunay triangulation which was devised and first published by Mirante and Weingarten in 1982. As before, the input data is in the form of randomly located (as distinct from systematically located) points with X, Y and Z coordinates. The points will have been located on summits, along ridges and break-lines, etc., in the usual way.

The point or node which is located nearest to the centroid of the data set is selected as the starting point for the triangulation. From this central point, the distance and bearing of all the other points in the data set are calculated and the points are then sorted and placed in an order by bearing.

Once this has been done, the radiating line to each point is established and a long thin triangle formed by connecting a line between the new point and the previous point

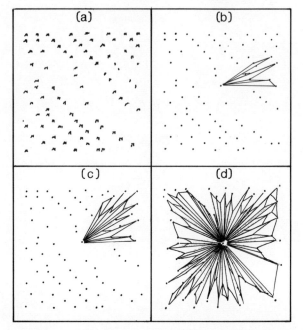

Figure 10.15 Radial sweep method (Mirante and Weingarten, 1982).

(Fig. 10.16). If two points have the same bearing, then they are used to form a pair of triangles on each side of the common line. As each point is accessed, it is added to a linked list which forms the other boundary of the triangulated network. Once the initial radial sweep has been achieved, the concavities created by the initial radial sweep triangulation must be filled by new triangles. Each point or node on the boundary list is combined and compared with the next two nodes on the list and checked to see if an inside triangle can be formed. If so, then the new triangle is added to the database, and the second (inside) node is removed from the list of outer boundary points. After the process has been completed, the list will comprise the points or nodes forming the convex edge of the terrain model.

As will be seen from Fig. 10.15, all the data points are now triangulated with non-overlapping triangles. However, the shapes and connections are far from desirable. To optimize the shapes, each triangle is now tested against each of its neighbours. A quadrilateral is formed by a pair of triangles. This is tested by calculation of the two distances formed by the opposite pairs of nodal points. If the distance between the two common points is greater than the distance between the two unique nodes, then the triangle indices are switched and the database pointers updated. The process is repeated with successive passes through the data set till an entire pass through the database produces no changes.

10.4.2.8 *Contouring from triangulated data.* As with contouring of regular gridded height points, so with randomly located triangulated height data, there are two main options for the contour threading:

(i) Simple linear interpolation of the contours
(ii) Generation of curved smoothed contours using some type of function.

Unlike the situation with gridded data, the use of direct linear interpolation for the contour generation is very common when the terrain model is based on triangulated data. Ambiguities regarding the direction that the contours might take are either not present or can be resolved. So direct linear interpolation gives a simple and robust solution. The contour threading usually begins at the boundary triangles. Using linear interpolation, all the entry points along the perimeter are located and the corresponding exit points found on the interior sides of each boundary triangle. These then of course act as entry points for the next internal triangle.

Just as with the grid-based method, so in the triangle-based method, the area of an individual triangle can be sub-divided into smaller triangles (sub-triangles) in order to get over the difficulty of long vectors with abrupt changes of direction along the common line between two adjacent triangles. The values at the vertices of each sub-triangle can be calculated by simple linear interpolation. Then the cutting points are determined for each required contour along the sides of each sub-triangle. Joining up the cutting points gives the required contour—again, this looks smoother, since it consists of a series of short vectors, instead of long vectors spanning the whole triangle. Obviously the threading decisions in any sub-triangle are also simplified, to a choice between the two remaining sides of the triangle as against the three possible sides within a grid cell.

If curved smoothed contours are required, then, as before, this can be achieved using a series of cubic splines or polynomials fitted through the string of interpolated cutting points along the triangle boundaries. Again, in areas of very close contours in steep areas, crossing contours may occur if this procedure is adopted.

The alternative approach is again to fit some type of curved three-dimensional surface patch to each triangle, so ensuring a smooth transition from one triangle to the next, instead of having a series of planar triangular facets with abrupt changes to the direction of contours between facets along the lines common to adjacent triangles. This will entail forming a polygon patch using the central point and its Thiessen neighbours. There is no difficulty in generating the polygon patch, but it will be irregular in shape (since there is no regular grid), so the data points will have a rather intractable format which is not easy to handle. The distances from the central point will all be unequal, and the fitting function could be weighted inversely with increasing distance. So an inverse distance-weighted, low-order polynomial surface could be fitted through these irregularly located data points which comprise the patch.

10.4.3 *Hybrid approaches to terrain modelling*

Although the discussion till now has been of two different approaches—grid and triangular used independently of one another—there are some packages, such as PANACEA (McCullagh, 1983a, b) and TILE/CONOCON/SOLID (Sibson, 1986), where elements of both approaches are used.

For example, both packages have been used to generate DTM data from digitized contours (Fig. 10.16). In each case, data points are first selected at suitable intervals along each contour. This pre-processing is then followed by the Delaunay triangulation procedure. Thus a triangular network is formed from the randomly located measured data which therefore honours all the data points. Next the elevation value at each

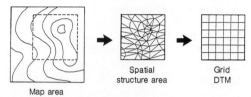

Figure 10.16 Hybrid approach: contours to triangulation to grid DTM (McCullagh, 1983*a*).

specified grid node is determined by reference to the triangular facet in which it falls. Thereafter contours (for checking purposes) and perspective views are formed from the interpolated gridded data rather than the measured triangulation data.

10.5 Software for and applications of terrain modelling in topographic mapping and civil engineering

A wide range of software packages have been written in the past decade for processing DTM data. Although the packages have many common features, they have normally been developed for one specific area of application. Table 10.4 presents a classification scheme based on six alternative application areas. The relative merits of each are considered in more detail in Petrie and Kennie (1987).

For engineering surveying applications, the two categories of particular relevance are those related to topographic mapping and civil engineering design. Two principal factors distinguish packages of these types. First, they offer the ability to model surface discontinuities such as valley floors, ridges, stream channels and other forms of breaklines, and second, and particularly relevant to the latter category, they enable models of design surfaces to be created.

By considering both the distribution of the data points and the geometric shape of the elements forming the DTM, it is possible to identify three distinct sub-categories, and these are defined in Table 10.5.

10.5.1 *Grid-based packages*

The use of a regular cartesian system of referencing elevations is historically one of the oldest forms of terrain modelling. It is particularly attractive because the simple data

Table 10.4 Classification of terrain modelling software packages

Category	Typical examples
1. Modelling for topographic mapping and surveying	SCOP, HIFI, CIP
2. Modelling for civil engineering design	MOSS, HASP ECLIPSE
3. Modelling as part of a computer-aided design (CAD) system	GDS-SITES Pro-Surveyor
4. Modelling geological and geophysical data	Z-Map, UNIMAP, CPS, PANACEA
5. Modelling for military applications	EAMACS, Scicon Viewfinder
6. Modelling general scientific datasets	GINOSURF, GHOST, GPCP

Table 10.5 Classification of terrain modelling software for topographic mapping and civil engineering applications

	Regularly located data points	Irregularly located data points
Gridded elements	Packages designed primarily for processing data obtained photogrammetrically from analytical stereoplotters, or from digitized contour lines. Examples include HIFI and SCOP.	
Triangular elements	Several general purpose modelling/contouring packages adopt this approach, e.g. GHOST	Packages designed primarily for processing field survey data for civil engineering applications. Examples include MOSS and HASP.

structure makes the storage and retrieval of information relatively easy to perform. Two packages which can be used to process data acquired in both forms are HIFI (Height Interpolation by Finite elements) and SCOP (Stuttgart Contour Program).

10.5.1.1 *Height Interpolation by Finite Elements (HIFI).* HIFI (Ebner and Reiss, 1980, 1984) is a suite of programs developed by the Institute for Photogrammetry at the Technical University of Munich. It is also available commercially from Zeiss (Oberkochen) as part of the software for the Planicomp analytical plotter and Orthocomp analytical orthophoto systems. The software was developed especially to deal with the large datasets generated by photogrammetric methods, particularly when progressive sampling techniques are used (Ebner and Reinhardt, 1984) or when profile scanning for orthophotograph generation is taking place.

The programs are based on the use of the finite element method. Finite elements are widely used in civil and mechanical engineering, particularly for structural analysis. The method depends on the generation of a series of finite planar elements or exact fit patches to describe the surface being examined. The patches or elements are normally simple geometric shapes such as triangles, squares or polygons. In terrain modelling terms, such a process may also be referred to as a patchwise polynomial method of interpolation (section 10.4.1.3).

Input to HIFI can be in several forms, ranging from directly measured grid points (which may involve progressive sampling), to irregularly located data. In situations where breaklines are also required to define the ground surface, the program approximates the terrain by a series of bilinear linked patches or elements. Thus a separate surface patch is generated using the heights of the random, or gridded, data points (Fig. 10.17). For every data point, a bilinear equation of the form previously defined in section 10.4.1.4 is created, that is:

$$Z = a_0 + a_1 X + a_2 Y + a_3 XY \qquad (10.2)$$

where Z represents the known measured heights, X and Y represent the known coordinates, and a_0, etc., represent the four unknowns which are solved for each individual patch. Spline functions also ensure that adjacent or surrounding patches fit at common points. When breaklines are available (as a string of height points), the intersection of the breakline with the bounding lines of each patch is calculated. The condition is then imposed that no connections can be formed across the breakline. This then leads to the formation of a sub-grid patch (Fig. 10.18).

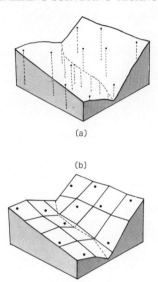

Figure 10.17 Bilinear interpolation from random points to gridded data using HIFI: (*a*) random points and breakline; (*b*) gridded surface elements and breakline.

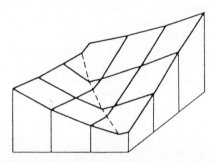

Figure 10.18 HIFI—bilinear sub-grid elements created from inclusion of a breakline into the dataset.

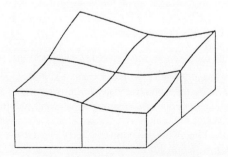

Figure 10.19 HIFI—bicubic elements.

In circumstances where it is appropriate to define the terrain without breaklines, it is possible to increase the size of the finite elements or patches and use a bicubic element to define the characteristics of the gridded patch (Fig. 10.19). In this case, the equation formed is of the bicubic form previously discussed, thus:

$$Z = a_0 + a_1 X + a_2 Y + a_3 XY + a_4 X^2 + a_5 Y^2 + a_6 X^2 Y + a_7 XY^2 \\ + a_8 X^2 Y^2 + a_9 X^3 + a_{10} Y^3 + a_{11} X^3 Y + a_{12} XY^3 + a_{13} X^3 Y^3 \\ + a_{14} X^3 Y^2 + a_{15} X^2 Y^3 \quad (10.4)$$

With a large dataset there will normally be redundant data points in both cases, so that a least-squares solution will be enforced. Once the gridded data points have been interpolated, it is possible to derive secondary products such as contour perspective views and volume reports.

10.5.1.2 *Stuttgart Contour Package (SCOP)*. SCOP is a machine independent series of computer programs written in Fortran IV, which is designed to deal with the large datasets produced by modern analytical stereoplotters. Although originally developed in the early 1970s for large mainframe computers (VAX, IBM, etc.) a more limited version has recently been announced for the IBM Personal Computer. The development of the software has been largely carried out by the Technical Universities of Stuttgart and Vienna, although the software is also marketed by Kern (Aarau) and others.

The overall structure of the software is illustrated by Fig. 10.20 and it can be seen to consist of a series of input/data manipulation facilities, interpolation functions and applications programs for deriving secondary products from the grid DTM. The central element of the system is a random access database consisting of a hierarchical series of directories (Kosli and Sigle, 1986). The main advantage of this structure is the efficiency of correcting or updating the DTM, since interpolation of the grid DTM is restricted only to those parts of the database which have been altered or added.

SCOP interpolates a grid DTM of varying density from irregularly located data points by means of linear prediction or linear least-squares interpolation. In order to apply this technique, it is necessary to define or compute a distance or co-variance function defined by a matrix B (Schut, 1976). Matrix B in equation (10.5) would therefore contain the values of the distance function between the reference points within the area of interest. The vector b in equation (10.5) represents the value of the distance function for the distance from the interpolated point to the reference points. Consequently, if h is the vector whose components are the heights of the reference points, then the height of the interpolated grid node point Z would be given by

$$Z = b^T B^{-1} h \quad (10.5)$$

Limiting the interpolation area to sub-areas containing an average of 70 points helps avoid difficulties with the inversion of matrix B. In situations where breaklines exist, the linear prediction is modified to impose the condition that the points separated by the breakline should not correlate with one another (Assmus, 1976). By this process, the edges of the terrain are clearly indicated; an example of the contours created by SCOP is illustrated by Fig. 10.21.

10.5.2 *Triangular-based packages*
The main limitations of grid-based packages are well documented and have been

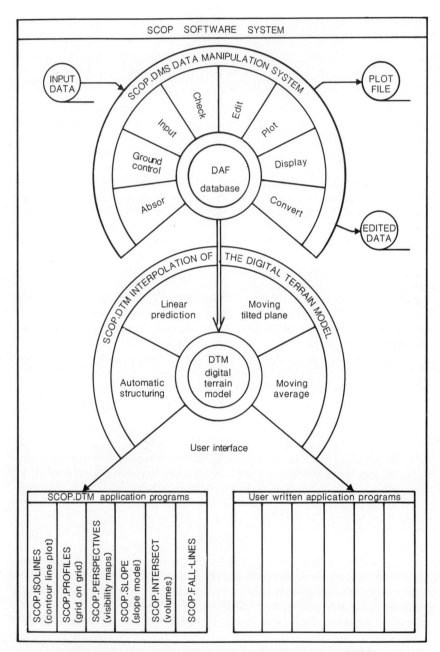

Figure 10.20 Components of the Stuttgart Contour Program (SCOP).

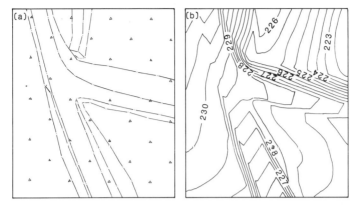

Figure 10.21 (*a*) SCOP—typical plot of input terrain points and breaklines; (*b*) corresponding contour plot generated using SCOP.

widely reported (McCullagh, 1987). Among the more significant disadvantages are:

(i) The need for considerable computer time for interpolation of a regular grid
(ii) The lack of flexibility in responding to variations in the slope of the ground (unless progressive sampling is used)
(iii) Non-honouring of the datapoints
(iv) The need for relatively complex techniques to represent cliffs and breaklines adequately.

Triangular modelling techniques offer an alternative to grid techniques which overcome most of the deficiencies described above. The most common methods used to form a triangular mesh are the Delaunay technique (section 10.4.2.1) and the radial sweep method (section 10.4.2.2). The essential characteristic of both is that they generate the unique set of triangles many sides of which are nearly equilateral. Also, the density of the triangular mesh is directly related to number of survey points and hence to the rate of change of slope over the ground.

Having generated the triangular mesh, most packages assume that the terrain can be approximated by simple planar triangular facets. Others, for example, the Contour Interpolation Package (CIP) from Wild Heerbrugg, use more sophisticated techniques to improve the surface definition and also help ensure continuity between adjacent triangles.

10.5.2.1 *Contour Interpolation Program (CIP)*. CIP (Steidler *et al.*, 1984) is one of two terrain modelling/contouring packages currently offered by Wild Heerbrugg. Like its companion product, System 9 DTM (Steidler *et al.*, 1986), it uses triangular finite elements to define the terrain. Originally developed in Fortran to run on Data General Nova and DEC VAX 11/700 series computers, it can also be run on the IBM PC/AT microcomputer.

Data input can be in either a random or regular grid form, together with information about the location of breaklines and faultlines. The model is created using a Delaunay triangulation, ensuring that lines or polygons which have known three-dimensional

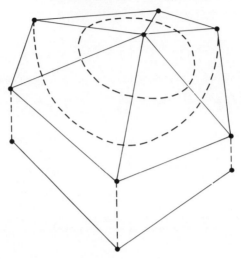

Figure 10.22 CIP—10-term cubic polynomial surface patch.

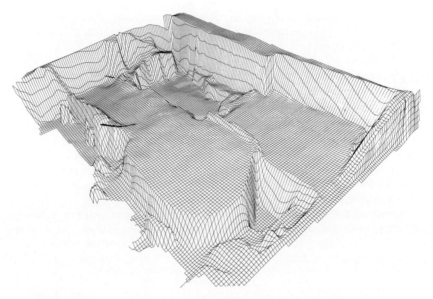

Figure 10.23 CIP—perspective view of opencast mine. Courtesy Wild Heerbrugg.

coordinates always form the edges of triangles. The surface normals at each node are then computed (Fig. 10.23) and used to define the surface by a series of Zienkiewicz functions (Bazely *et al.*, 1965) which consist of 10-term cubic polynomials of the form previously described, thus:

$$Z = a_0 + a_1 X + a_2 Y + a_3 XY + a_4 X^2 + a_5 Y^2 + a_6 X^2 Y + a_7 XY^2 \\ + a_8 X^3 + a_9 Y^3 \tag{10.3}$$

The use of surface patches of this form ensures continuity from one triangular element to another. Contouring is carried out using 16 sub-triangles. The cutting points along the edges of the sub-triangles are then determined. Other options include facilities for generating the heights of single points, a regular grid DTM, profiles, perspective views and volumes relative to a reference surface. Figure 10.24 represents a perspective view of an opencast mine generated using CIP.

10.5.2.2 *MOSS*. MOSS (Craine, 1985; Houlton, 1985) is another triangular-based package, and is probably the most widely used software for terrain modelling and design in the civil engineering industry in the UK. It currently has over 250 users worldwide, and has become established as the industry standard for terrain and design modelling. The system was initially developed in the early 1970s by a consortium of UK County Councils (Durham, Northamptonshire and West Sussex), and continued to be developed by this consortium until 1983, when the marketing and development rights were taken over by MOSS Systems Ltd.

Although several systems, such as the British Integrated Program System for highway design (BIPS), existed before the development of MOSS, these early systems were limited in several respects. First, the data structure was normally restricted to either a cross-section or square grid format. Consequently, terrain definition was poor. Second, it was often difficult to deal with complex junctions and interchanges. Although this was not a major limitation when dealing with major highway projects of constant road cross-section, it became a much greater constraint when concerned with congested and more complex urban interchanges. Third, it was not possible with these early packages to merge models and generate composite models. Current packages such as MOSS offer the facility to derive data from these composite models; for example, the ability to generate isopachytes or contours of equal thickness is often used, particularly for resurfacing projects. Fourth, the drafting facilities of many of the early systems were very limited and the output required considerable manual cartographic additions before it became acceptable as a final contract drawing. Finally, these early packages were generally limited to batch processing on large mainframe computers with no interactive graphics capabilities.

Many of the fundamental procedures developed for terrain and design modelling by the MOSS system are now in common use by other systems. An understanding of some of the basic principles is therefore of importance.

Surface representation in the MOSS system is carried out by storing data in either point or string form. *Point data* simply refers to single three-dimensional coordinate sets which represent discrete features in the terrain or on a design surface (Fig. 10.24a). *String data* refers to a linked series of points (Fig. 10.24b). Strings may be of several forms, and Fig. 10.25 summarizes the characteristics of some of the more common string types. The combination of points and strings in the computer is referred to as a MOSS model.

MOSS software is structured into a series of option commands, and the options are categorized into either major or minor options. Major options define the primary functions of the software and are used to generate models, design surfaces and create drawings. Some of the more common major options are illustrated by Fig. 10.26.

Minor options exist within each major option, and are used to perform operations on the strings within the model specified by the major option. Over 150 minor options can be invoked by the operator, and are identified by a three-digit number. Option 941,

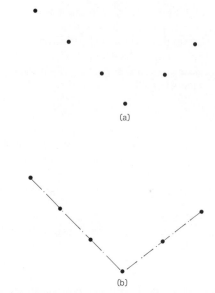

Figure 10.24 Forms of data used to create a MOSS model; (*a*) points; (*b*) strings.

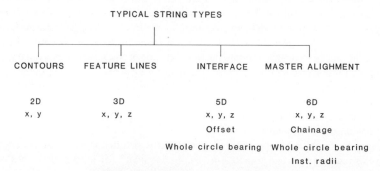

Figure 10.25 MOSS—typical string types.

for example, creates a triangulated surface from the string data. The formation of the triangulated surface is performed by a unique algorithm (exact details of which have not been published). The algorithm not only derives the most equilateral series of triangles, but also ensures that all elements on a string line form vertices of the triangles (Fig. 10.27).

MOSS models can be created, interrogated and processed using two distinct modes of operation. Batch processing is the most common, and normally involves the creation of a short job-control file to control the execution of the necessary operations. This is generally the preferred mode when large design projects are being processed, since even on fast mainframe computers such operations may take several hours to process. For users unfamiliar with the MOSS command structure, it is possible to use screen menus which offer default values for the various menu choices. While batch processing is

DIGITAL TERRAIN MODELLING

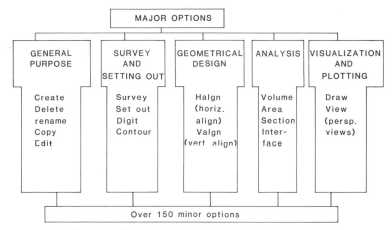

Figure 10.26 MOSS—relationship between major and minor options.

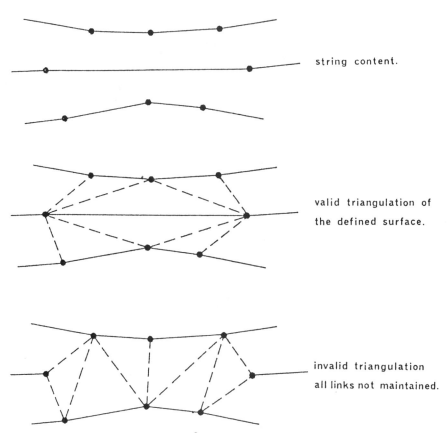

Figure 10.27 MOSS—triangulation is used to generate the surface model ensuring that all string elements form vertices of the triangles. Courtesy MOSS Systems.

suitable for large projects, it can be cumbersome to use on small projects or for editing drawings. The development of interactive graphics workstations, such as the Apollo Domain, and fast microcomputers, such as the MicroVAX II, has led to the development of Interactive MOSS or I-MOSS as an alternative to batch processing and is most suited to editing drawings created by MOSS. The technique makes use of on-screen scrolling menus, and a mouse for editing and choosing particular options. Such hardware also enables sophisticated visualizations to be produced, two examples of which are shown in Fig. 10.28.

The original impetus for the development of MOSS was the expansion of the

(a)

(b)

Figure 10.28 (*a*) and (*b*): Perspective visualizations of highway projects.

highway network in the United Kingdom during the 1970s. Consequently, much of the early development work was concerned with methods of automating the traditional manual design techniques used by highway engineers. In addition to replicating the traditional manual techniques (such as horizontal alignment using fixed intersection points and radii), MOSS also offers the design engineer the ability to design using elements. This latter technique allows greater flexibility in the design phase.

It would be inappropriate, however, to give the impression that the use of MOSS is restricted to highway engineering. It has now been developed into a more general-purpose civil engineering modelling and design system. The list of projects where MOSS has been used is impressive, and ranges from opencast mining in Australia (Fig. 10.29), to the design of complex highway interchanges (Fig. 10.30).

10.5.2.3 *Personal-computer (PC) and desktop-computer-based triangular packages.* During the past five years or so, a variety of terrain modelling packages have become available on PC and desktop computers. Milne and Motlagh (1986) recently reported details of many of the packages available in the UK which run on IBM PCs and Apricot and Hewlett Packard microcomputers. A typical example of such a system is HASP.

The HASP-DIGICAL IIS system consists of a series of program modules developed for surface modelling, volumetric calculations, road and building design, digital mapping and cadastral surveying (Hogan, 1984).

The modules, although originally written in HPL (a programming language specific to Hewlett Packard), are now available in Pascal. They are designed to run on the Hewlett Packard (HP) 200 and 300 series of desktop computers and the HP 9816 and 9825 series of personal computers (Fig. 10.31).

Modelling in HASP is carried out by the Triangular Irregular Network (TIN) technique using the radial sweep method described in detail in section 10.4.2.2. An important feature of the triangulation technique is the ability to incorporate breakline information, so avoiding triangle sides crossing known surface discontinuities such as valley bottoms, tops of banks and so on. The procedure involves checking each triangle which crosses a breakline and optimizing the triangulation, so that the breakline forms one side of the final triangulated surface. The consequence of not including such features is illustrated by Fig. 10.32, where the omission of the breakline between the coal stock piles has resulted in a model forming across the valley floor.

A recent addition to the HASP software has been a module for performing visual impact analysis. This technique is becoming an increasingly important requirement at the planning and feasibility stages of a civil engineering project or for landscape design (Ketterman, 1985). In addition to providing perspective views of the terrain, the package also enables areas of intervisibility to be examined.

10.5.3 *Hybrid packages*

Hybrid packages which combine the attributes of the grid and triangular techniques are also available. Although this combined approach is uncommon, it nevertheless offers the user some of the benefits of the triangular technique, such as the honouring of the data points, together with the display benefits of the grid-based techniques.

10.5.3.1 *PANACEA.* The PANACEA software suite is a typical example of a hybrid triangular/grid-based package. It has been devised and written in Fortran 77 by Siren

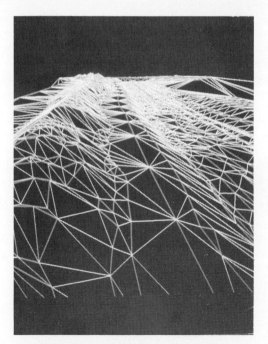

Figure 10.29 (top) MOSS is used by many operators of opencast mines to provide accurate estimates of extraction rates; (bottom) perspective view of a triangulated surface model of an opencast mine.

DIGITAL TERRAIN MODELLING 421

Figure 10.30 Applications of MOSS Modelling System: highway design and earthworks measurement on the M25 London orbital motorway. (*a*) Oblique aerial photograph of Leatherhead interchange; (*b*) plan view of main horizontal alignment; (*c*) perspective.

Figure 10.31 HASP Digital IIS System. Courtesy HASP Inc.

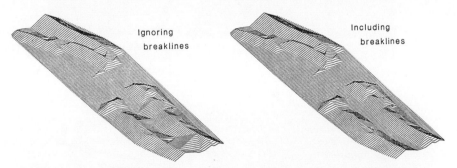

Figure 10.32 The effect of breaklines on a triangular model (Hodges and Alderson, 1985).

Systems, and is also available from LaserScan Laboratories Ltd. The software was designed primarily for the processing of small-scale terrain models created from digitized contours or geophysical records.

PANACEA differs from the previous two categories in that it first generates a triangular terrain model and then produces a grid DTM based on the triangular structure (Fig. 10.16). The various modules which carry out these operations and the relationship between them are illustrated in Fig. 10.33.

(i) The first module, PAN, is a pre-processing program which reduces the number of points which were originally digitized, using a variation of the Douglas Peucker algorithm (Douglas and Peucker, 1973).

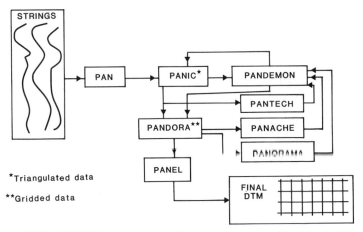

Figure 10.33 PANACEA—interrelationship between programs (McCullagh, 1983a).

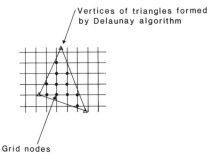

Figure 10.34 PANDORA—height determination of grid nodes by linear interpolation from the vertices of the triangles.

(ii) Using the reduced dataset created by PAN, a second module, PANIC, generates a Delaunay triangulation over the entire dataset.

(iii) The third module, PANDEMON, allows the user to graphically edit the dataset to eliminate any erroneous points, or to correct points which cross breaklines.

(iv) The interpolation of the grid DTM is then performed using PANDORA. The grid heights can be determined either by linear interpolation within each triangular facet (Fig. 10.34), or by fitting smooth surface patches between the vertices of the triangles. The latter technique requires considerably more computational time, and furthermore needs an estimation to be made of the derivatives at each data point to ensure continuity between adjacent triangular patches (McCullagh, 1981).

(v) Display of the gridded data can then be performed using either PANACHE to generate contour plots from the gridded data (Fig. 10.35), PANORAMA to create perspective views, or PANTECH for direct contouring of the triangular structure. Each can be very useful when checking the input data for gross errors.

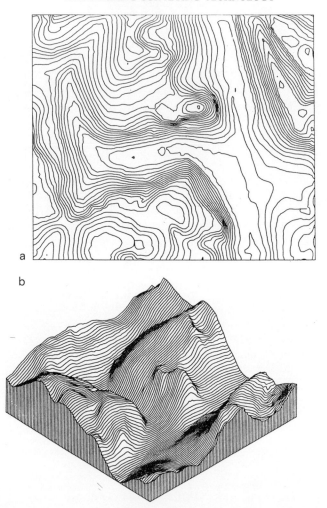

Figure 10.35 (*a*) Contour plot generated using PANACHE, (*b*) isometric view of same dataset using PANORAMA (McCullagh, 1987).

(vi) PANEL may be used in order to combine grids of different areas, or of different resolutions to form a single model.

Finally, PANTHEON is a complementary suite of programs which can be used to derive and display other information from the dataset created by PANACEA. Figure 10.36 illustrates the facilities available as part of the PANTHEON sub-system. Typical applications of the system have included the production of hazard maps for landslide prediction (McCullagh *et al.*, 1985).

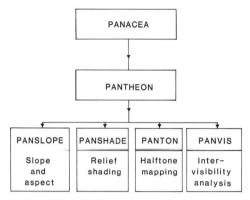

Figure 10.36 Relationships between PANACEA and PANTHEON.

10.6 Conclusions

Digital terrain modelling is now a commonly used technique both in topographic mapping and civil engineering design. It is also used widely in other fields such as landscape planning, flight simulation and geological/geophysical exploration where generally the accuracy requirements of the elevation data are lower than for surveying and engineering applications.

For the future, the use of terrain modelling methods will undoubtedly continue to develop and expand, particularly with continued improvements in the price performance ratio of computer systems. National and regional terrain databases based on existing topographic maps, are now being developed in many parts of the world, and these will play an increasingly important role in terrain visualization during the preliminary planning stages of engineering projects. For small site or route projects, however, the primary source of terrain data is likely to continue to be directly measured spot heights. Other issues which will become increasingly important in the future include the integrity and quality assurance of the original data, the accuracy of the algorithms used for interpolation and derivation of information such as volumes, the creation of comparative benchmarks for computer packages and the continued development of national and international standards for the transfer of terrain modelling data.

References

Assmus, E. (1976) Extension of Stuttgart Contour Program to treating terrain breaklines. *Proc. XIIIth Congr. Int. Soc. Photogramm., Commision III*, Helsinki.

Bazely, G.P., Cheung, Y.K., Irons, Y.K. and Zienkiewicz, B.M. (1965) Triangular elements in bending-conforming and non-conforming solutions. *Proc. 1st Conf. on Matrix Methods in Structural Mechanics.* Air Force Institute of Technology, Wright Patterson A.F.B.

Craine, G.S. (1985) MOSS. *Civil Engineering Surveyor* **10** (10) 17–21.

Douglas, D.H. and Peucker, T.K. (1973) Algorithms for the reduction of the number of points required to represent a digitised line or its caricature. *Canadian Cartographer* **10** (2) 112–122.

Ebner, H. and Reiss, P. (1980) HIFI–a minicomputer program package for height interpolation by finite elements. *Proc. 14th Congr. Int. Soc. Photogramm., Commission IV, WG/1*, Hamburg, 14 pp.

Ebner, H. and Reinhardt, W. (1984) Progressive sampling and DEM interpolation by finite elements. *Int. Arch. Photogramm. and Remote Sensing* **25** (A4) 125–134.

Ebner, H. and Reiss, P. (1984) Experience with height interpolation by finite elements. *Photogramm. Eng. and Remote Sensing* **50** (2) 177–182.

Hodges, D.J. and Alderson, J.S. (1985) Automated volumetric surveys. *University of Nottingham, Mining Dept. Mag.* **37**, 71–76.

Hogan, R.E. (1984) HASP Digital survey design system. *Proc. ASP-ACSM Convention*, San Antonio, 468–475.

Houlton, J.M. (1985) Computer aided drafting in highway engineering. In *Current Issues in Highway Design, Proc. Conf. on Highway Design, Inst. of Civil Engs*, Thomas Telford, London, 149–162.

Ketterman, M.R. (1987) The applications of terrain modelling for mine surveying. *Proc. Short Course on Terrain Modelling in Surveying and Civil Engineering*, University of Surrey, 13 pp.

Kosli, A. and Sigle, M. (1986) The random access data structure of the DTM program SCOP. *ISPRS Commission III Symp., Mapping from Modern Imagery*, Edinburgh, 45–52.

Leberl, F.W. and Olsen, D. (1982) Raster scanning for operational digitising of graphic data. *Photogramm. Eng. and Remote Sensing* **48** (4) 615–627.

Makarovic, B. (1973) Progressive sampling for digital terrain models. *ITC Journal* **1973–3**, 397–416.

Makarovic, B. (1975) Amended strategy for progressive sampling. *ITC Journal* **1975–1**, 117–128.

Makarovic, B. (1977) Composite sampling for DTM's. *ITC Journal* **1977–3**, 406–433.

McCullagh, M.J. (1981) Creation of smooth surface patches over irregularly distributed data using local surface patches. *Geogr. Analysis* **13** (1) 51–63.

McCullagh, M.J. (1983*a*) 'If you're sitting comfortably we'll begin'. *Workshop Notes on Terrain Modelling*, Australian Computing Society, Siren Systems, 158 pp.

McCullagh, M.J. (1983*b*) Transformation of contour strings to a rectangular grid based elevation model. *EURO-CARTO II*, 18 pp.

McCullagh, M.J. (1987) Digital terrain modelling and visualisation. *Proc. Short Course on Terrain Modelling in Surveying and Civil Engineering*, University of Surrey, 27 pp.

McCullagh, M.J., Cross, M. and Trigg, A.D. (1985) New technology and super-micros in hazard map production. *Survey and Mapping '85*, Paper D4, 16 pp.

Miller, C. and LaFlamme, R.A. (1958) The digital terrain model–theory and applications. *Photogramm. Eng.*, **24** (3) 433–442.

Milne, P.H. and Motlagh, K.C. (1986) Land surveying software review. *Civil Engineering Surveyor* **11**(7) 14–21.

Mirante, A. and Weingarten, N. (1982) The radial sweep algorithm for constructing triangulated irregular networks. *IEEE Computer Graphics and Applications* **2** (3) 11–21.

Petrie, G. and Kennie, T.J.M. (1987) Terrain modelling in surveying and civil engineering. *Computer Aided Design* **19** (4) 170–187.

Schut, G.H. (1976) Review of interpolation methods for digital terrain models. *Proc. XIIIth Congr. Int. Soc. Photogramm.*, Helsinki, 23 pp.

Sibson, R. (1986) Terrain modelling using quadratics. *British Computer Society Displays Group Meeting on State of the Art in Stereo and Terrain Modelling*, London, 8 pp.

Steidler, F., Zumofen, G. and Haitzmann, H. (1984) CIP: a program package for interpolation and plotting of digital height models. *ACSM-ASP Annual Convention, Washington* (also available from Wild Heerbrugg), 10 pp.

Steidler, F., Dupont, C. Funcke, G., Vuattoux, C. and Wyatt, A. (1986) Digital terrain models and their applications in a database system. Wild Heerbrugg Publication, 10 pp.

Turnbull, W.M. and Gourlay, I. (1987) Visual impact analysis: a case study of a computer based system. *Computer Aided Design* **19** (4) 197–202.

11 Land information databases

D. PARKER

11.1 Introduction

A computerized land information system is a computer database of spatially referenced land-related data. Within the database, graphical information of topographic mapping and spatially referenced textual information are held in a carefully structured form. The functions of such a system are to make land-related data more accessible and to assist improved management of land resources.

Society is demanding more effective use of land. There is, for instance, a need to re-use land previously made derelict. Planning land use requires knowledge of a wide range of information, relating not only to the immediate area of interest but also to surrounding land. This may include, for example, information on past, current and proposed land use, on land ownership and rights of way, on soil types, pollution control restrictions, or financial incentive schemes. Coordination of the efforts of the public utilities and local authorities in the installation and repair of underground services is an example where land-related data from a wide variety of sources has to be integrated. The public utilities, local authorities and the Land Registry make the greatest demands upon traditional large-scale mapping. There are approximately 220 000 large-scale maps (1:1250 and 1:2500) covering the built-up and lowland rural areas in the UK. These are used to record the routes of underground cables and pipelines, the location of plants and substations, and administrative and ownership information. There are also many other organizations, ranging through administration, commerce and engineering, which depend on smaller scale map coverage.

The intense and complex demands upon land use are stimulating the growth of computerized information systems in those organizations responsible for collecting, recording, supplying and interpreting land-related data. A further stimulus has come from computer manufacturers, software houses and system suppliers, who have seen the growth of computerized land information systems as an opportunity to increase their outlets. There is an overlap between the spatial data handling requirements of computer aided drafting and design systems (CADD) and those required for map-based queries. Further, the methods of management of the textual data are the same as those employed with the numerous large data sets maintained in many administrative and commercial environments. Textual data management in many organizations involves the control of input, upkeep, security and integrity of the data and the provision of access methods. All these requirements are shared in common with mapping-based information systems.

The advantages of computerizing land information systems are similar to the advantages claimed in many other fields. The volume of data available from a wide variety of sources may be too great to be assimilated by any one person unaided. Through computerization, the data can be integrated and made available to many users rapidly and simultaneously. Given suitable data retrieval and analysis procedures, the data can be used for purposes previously considered impossible. The advantage to the data organizers is that the integrity of data can be maintained much more easily.

There are of course disadvantages in the computerization of land information systems. Most significant is the requirement to transform all data to digital form. A number of trial computerized land information systems have been established recently; most have identified problems which have still to be overcome. The initial expense of computerization is high and it is yet to be proven that this is outweighed by subsequent savings in cost. Such proof perhaps must wait until major organizations have gained experience with systems for an extended period of time.

There are problems in coordinating the activities and standards of those organizations responsible for maintaining the existing non-computerized data sets. Even the simplest spatial information system requires both graphical and tabular data from several sources. Where a large multi-faceted system is implemented for a regional area, the data will be shared between several institutions. The data will come from many sources ranging from mapping organizations, the land registry, the public utilities and the census office, to national and local administrations. Collaboration is required for such enterprises. All too often this collaboration is not forthcoming. Institutional problems can be more difficult to overcome than technical.

The availability of suitable computer hardware has, until recently, been a problem, but computers with the required power are now available. Data storage devices with the necessary capacity for very large data sets are under development. The technology for high speed and reliable data transfer between computers is becoming available, though it is not yet in general use. When it is, data sets which are limited in both content and coverage will be held in separate locations. They will be accessed by many users along with many other data sets held in different locations, thus forming a distributed database. (McLaughlin, 1986).

Land information must be structured in such a way that it satisfies the demands currently made of it and will be able to meet new needs. This chapter discusses the ways in which data may be organized within a computer system. It considers digital data sets where the data relate only to two-dimensional space. It ignores video and raster scanned images and also data relating to the third dimension, since this has been dealt with to a large extent in Chapter 10.

11.2 Spatial databases

A computerized land information system is not intended for cartographic i.e. digital mapping purposes alone. It must answer many queries, some of which cannot be identified when the data set is first created. The answer to some of these queries may be presented in the form of a map display on a screen or as a hard-copy map, but many analytical processes may be required to compile the result. Spatial information held about any particular feature must detail its absolute position, its geometric shape and

LAND INFORMATION DATABASES

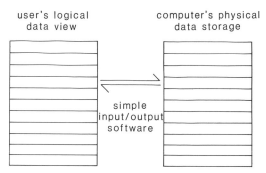

Figure 11.1 Elementary data storage.

show its relationship to neighbouring features. It must also show the nature of the feature and its attributes.

Map-type data may be stored in a computer *database*. A user may have access to data stored in a database, without having to know how the data are stored in the computer. Access is achieved through a special *query language* or by application programs. Application programs can be changed or new ones created without modification to the data structure. Further, the methods and hardware of data storage can be altered without the need to rewrite the application programs or to modify the query language. The query language allows a user to retrieve and analyze data independently from other users. It is often incorporated into a host programming language, such as Pascal, Fortran or Cobol, or into the computer operating system so that calls can be made directly from application programs.

In the early days of computer development, it was normal for a user to store and access data records serially (Fig. 11.1). The way in which a user viewed the data was essentially the same as the way in which it was physically stored in the computer. Application programs had to be modified whenever there were changes in the physical data structure. With early databases, the *database management system* (DBMS) software removed the need for data to be physically stored in the same way as it was logically structured (Fig. 11.2). With this approach, different logical views required by various users can be derived from the same physical data set. Since individual data elements can be shared between diverse applications, the necessity to have multiple copies of the same data is reduced and sometimes eliminated. Complex forms of physical data organization can be used without complicating application programs, which no longer need to be changed when the physical organization is altered. The physical storage methods can be optimized for each particular computer hardware configuration used. Many levels of indexing on the data can be incorporated to speed up access. To assist the management of large databases it is necessary for a global view of the complete data set to be maintained. This so called *conceptual view* is a representation of the entire information content, in a form which is somewhat abstract compared with the physical storage methods.

Current data storage systems for both spatial and non-spatial data referred to as databases are actually some subset of the ideal database described. Future designs for data storage system will incorporate many of the principles outlined above, but not

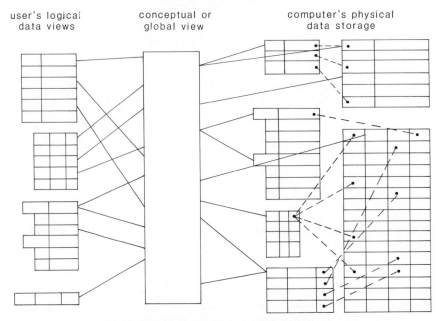

Figure 11.2 Database approach to storage.

necessarily all. Many applications may never warrant the additional effort required to incorporate all possible improvements.

11.3 User's spatial data model

Exact representation of every detail of the land surface requires records of all positional and factual attributes of every point on the surface. This would result in an infinite number of data points, each with many attributes, and possibly with infinite relationships between them. To limit the data quantities, it is necessary to define spatial units and to associate these with the appropriate attributes—a model of reality has to be constructed. The conventional model of the land surface is the paper plan or map. This is made by the surveyor or photogrammetrist, who measures the features of the land surface to a density and accuracy sufficient for representation at the scale of plan being produced. A *feature* is an item of direct interest, such as a house, kerb, pipeline or manhole. Features such as administrative land parcels and regions delimiting soil types are represented as *areas*. Property boundaries and the cable and pipeline routes of public utility services such as gas and electricity authorities are described as *lines*. The positions of such features as trees and lamp standards may be referred to as *points*. Thus, topographic and thematic information shown on a map is described by features, areas, lines and points.

One major problem is that the unit level of description can change with map scale. An example of this is a town, which when it covers many large-scale map sheets will detail every property and side-road, but when it is part of an application such as long-distance route planning, will be represented by a single point on a small-scale map.

Computerized land information systems therefore contain data which are not scale-free.

It is essential that there should be a common spatial reference between all data held in the system. Often, this is based on the geographic coordinate system of latitude, longitude and height (ϕ, λ, H), or on the OS National Grid in terms of easting, northing and height (E, N, H). A postcode, street address or simply a unique number can also be used to cross-reference one data set to another.

The thematic information which is associated with each feature may be represented on the map by the use of colours, symbols or by cross-references to an associated list of data. In the computer, the area, line and point features of the map are represented spatially by some combination of polygons, links and nodes. Each feature has *attributes* associated with it. These attributes describe the characteristics or properties of the feature such as the name of the owner of a house, the diameter of a pipe or the reference number of a manhole. In database terminology, the concept of an *entity* is used. The definition of an entity is much wider than that of a feature. An entity is any element or object. A *relationship* is an association or link between one entity and another or between an entity and an attribute (NTF, 1986).

11.3.1 *Topological data model*

A topological data model is a method of structuring land-related information normally shown on maps and plans. Topology is the study of spatial relationships such as connectivity between entities. Topological data structures are close analogies of land-related data. They allow the user to make queries similar to those normally resolved by reference to conventional paper maps. In addition, they are readily processed using advanced computer graphics and database management software.

11.3.1.1 *Polygons, links and nodes.* As an example of a topological data structure, consider a set of mutually exclusive area units such as that illustrated in Fig. 11.3. Each area unit is delimited by a *polygon*. The boundary lines of these polygons are called *links*

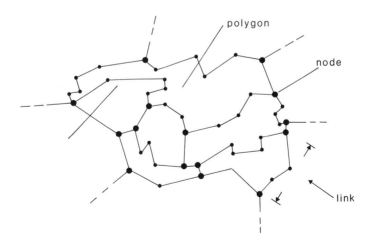

Figure 11.3 Polygons, links and nodes.

and the places where these links intersect are called *nodes*. (Broadbent, 1980). Such a set of mutually exclusive polygons has a number of characteristics:

(i) each polygon is completely bounded by a set of links
(ii) no two polygons overlap
(iii) each link has exactly one polygon on each side of it
(iv) no two links cross
(v) each link runs from one node to another
(vi) each node has at least three links leading from it.

The spatial interrelationships or *connectivity* between the polygons, links and nodes are completely described by this set of characteristics. Hence:

(a) each polygon is defined by an ordered sequence of boundary links;
(b) each link is defined by i) the nodes at the beginning and end, their order defining the direction in which the link is traversed, ii) references to the two polygons it divides, one to the left, the other to the right, iii) information describing the spatial geometry of the link;
(c) each node is defined by i) information describing the position of the node, ii) an ordered list of links beginning or ending at that node.

The example of Fig. 11.3 does not cover all types of polygons. In the example shown in Fig. 11.4 there are also:

(i) Areal units with a single boundary link which form *islands*. These can be

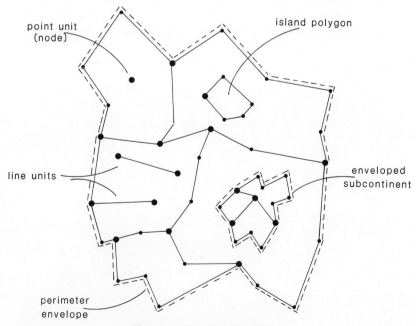

Figure 11.4 Topological problems.

described by selecting a single point on the boundary link as a node and starting and ending the link at this point.

(ii) Groups of polygons totally contained within another polygon which are termed *subcontinents*. These have to be flagged as belonging to the enclosing polygon, hence creating a hierarchy of polygon levels.

(iii) Boundaries of the mapping area and of subcontinents, best considered by the term *envelope polygons*.

Some lines require special consideration, such as those line units which project into, or lie entirely within, a polygon; these may be considered to be links which have the same polygon on either side.

The topological data structure can also incorporate *points* with the following characteristics:

(i) the structure described caters for a point unit coinciding with an existing node;
(ii) a point unit lying on an existing link can be considered to split the link at this

```
Polygon
    ├ unique identifier
    ├ positional information
    │      ┌ list of bounding links
    │      └ (or) polygon extent (raster only)
    ├ list of enclosed polygons
    └ list of enclosed nodes

Link
    ├ unique identifier
    ├ node
    │   ┌ at start
    │   └ at end
    ├ polygon division
    │   ┌ left
    │   └ right
    └ positional information

Node
    ├ unique identifier
    ├ type
    │   ┌ full node
    │   └ isolated
    ├ positional information
    └ (if full node)
        └ list of joining links
```

Figure 11.5

position and thereby create a new node; and the two new links created take on the topological characteristics of the divided one;

(iii) a point unit lying isolated within a polygon must be considered as a special case, as a node without any attached links, and a list of these should be included in the record of any polygon containing such point units.

The notes above have set down the important concepts of topological data models. These are summarized diagrammatically in Fig. 11.5. Even though this presentation has not detailed all eventualities, its structure can be adapted to cope with them. Considerable computing power and complex software are required to construct topologically sound polygon networks, maintain them and perform analyses based on them.

11.3.2 *Representation and description of features*

The distribution and interconnection of any mutually exclusive configuration of area, line and point units can be described in the form of polygons, links and nodes. In reality, however, topography consists of a jumble of intersecting, overlapping and coincident features or entities. To model these, consider the example in Fig. 11.6. The simple topology illustrated with the aid of symbols and names can be reduced to a mutually exclusive collection of polygons, links and nodes. Any collection of features can be reduced in such a way. Each area feature corresponds to an exact number of polygons; each line feature to an exact number of links; and each point feature to a single node. The position, shape and topological relationships of real surface features can therefore be recorded by referencing them to one or more of these basic units. In the example, the area feature 'wood' consists of polygons 9, 11 and 12, the linear feature 'road' comprises links 111 and 110 and the point feature 'railway station' is node 266. This topological data model models only the geometrical properties and the connectivity between features. The data set must also record the characteristics or *descriptive attributes* of each feature. The total data required to be associated with each feature is shown diagrammatically in Fig. 11.7.

It has been common to use a single coded value to describe the properties of each feature. This practice arose when creating digital spatial data sets for the computer-assisted production of conventional hard-copy maps on paper or film, i.e. for digital mapping. Most of the descriptive information on a map is implicit from the line style and geometric shape of the lines. At large scales, this information includes such features as houses, road edges, manholes and, at small scales, features such as reservoirs, roads or hamlets are described. These can be represented easily in terms of a single coded value (Parker, 1986). Any feature with the same coded value, or *feature code*, can be represented on the plan with the same symbol. The feature codes are, in effect, display codes. Feature codes can be expressed as a simple numeric value (46 = motorway) or a single alpha value (FE = fence). In more complex coding, the order of the individual characters follows a hierarchical system (UGPP = utility, gas, pipe, plastic).

For a computerized land information system, much more data are needed than can be represented by a single value. The association of a feature's characteristics or attributes with its spatial data, combined with the facilities to analyze them together, are the essence of the geographically referenced information system. Not all the information needs to be held in the same data set, provided there is some method of cross-referencing. A unique reference number for each feature can be used for this

LAND INFORMATION DATABASES

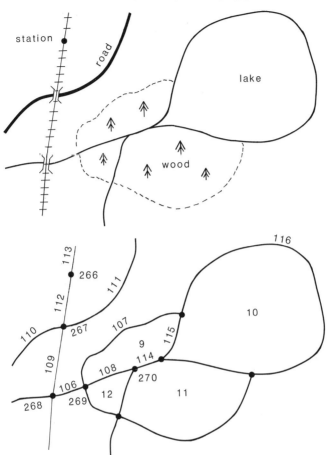

Figure 11.6 Representation of features.

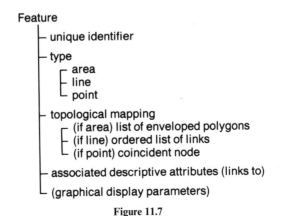

Figure 11.7

purpose. In this way, the descriptive data can be held in many separate locations in a variety of formats. The importance of linkage mechanisms cannot be overstressed. It is the ability to query both the spatial and the descriptive data together that makes land-related databases potentially such powerful tools.

11.3.3 Simplified data models

The data model set out so far can record any combination of topographic features and their interconnections. Any enquiries that can be satisfied by visual inspection of a conventional map can be replicated from such a data set. For many purposes, however, the topological model as described may be too complex, either because the necessary data collection, manipulation and validation processes to create and maintain it are too demanding, or because the range of queries it considers can be answered using a more simplified approach.

There are two ways to create simpler data models. The number of types of spatial unit within the model (areas, lines and points) can be reduced, perhaps considering only links and nodes or simply areas alone for example. Alternatively, the level of connectivity recorded within the model can be reduced. With this second approach, however, the reduced effort in data capture and maintenance results in increased search times and greater processing effort during subsequent data analysis (Peuquet, 1984b; Van Lamsweerde, 1983).

11.3.3.1 *Link and node model.* For many applications of information systems, especially those based on large-scale topographic plans, there is a simpler data model. This recognizes only links and nodes, but can still satisfy a large range of enquiries. The main data source currently in use in the UK is conventional monochrome plans and maps, and these can be collated into the format of the simpler data model relatively easily.

Fig. 11.8 shows a section of a typical large-scale urban plan. It includes an indication of the position of nodes within the link and node model. (The building stipple has been left in for clarity but does not contribute to the model.) Linear features are formed by a series of links; point features occur at nodes; an area feature can be considered by delimiting its boundaries with a loop of links. No polygon units have been defined, and therefore the links do not record the polygons they divide. Hence, there is no connectivity between adjacent areas.

Link and node structures can cope with the majority of retrieval demands made of spatial data in computerized land information systems. The structure allows the relative location of features possessing certain characteristics or groups of characteristics to be computed or displayed. It also permits network analysis. However, not all processing packages can perform these operations, even though the data structure makes it theoretically possible.

11.3.3.2 *Areal model.* A polygon structure may be sufficient for applications involving only areal information, such as defining areas with a particular soil type or vegetation, or designating administration units or catchment areas. Since a topological data model records polygons by reference to their boundary links and records these links using their terminating nodes, there might not appear to be much of an advantage in this simplification. However, simpler ways of representing the geometrical properties of polygons when their boundaries have no significance, may be useful.

In data sets where areal features alone are considered, their connectivity commonly

LAND INFORMATION DATABASES

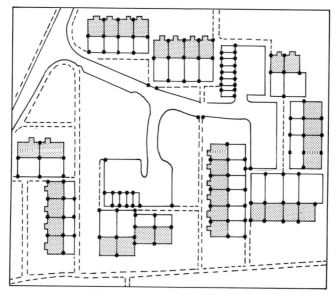

Figure 11.8 Link and node model.

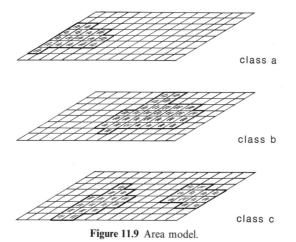

Figure 11.9 Area model.

is not recorded. Each classification of an areal feature is considered as a separate *layer* (Fig. 11.9). The analysis software answers queries by computing the limits of areas where two or more specified classes of feature overlap (intersect), or by computing areas from which certain classifications of features are excluded (difference).

A special case of an area model is one in which the basic unit of area is a grid square. In some statistical information systems, data are often collated for large grid cells (1 km squares for example). Definition of the grid then automatically provides the absolute location, extent and connectivity of one area unit to another.

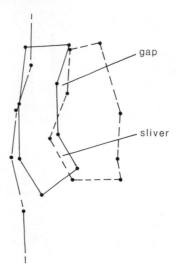

Figure 11.10 Spaghetti model.

11.3.3.3 *Spaghetti model.* In a *spaghetti model*, all reference to connectivity can be omitted so that features (closed loop, linear, point) are independent of each other. Data sets such as these are sometimes used in digital mapping systems. They are extremely easy to capture and maintain, but have many serious disadvantages. These latter all stem from the problem of having to search the geographical information for common positions in order to ascertain whether one feature overlaps, adjoins or is coincident with another. Since positional data is neither absolutely nor infinitely accurate, problems arise from *gaps* and *slivers* in the information (Fig. 11.10). These are difficult for computer software to resolve. Geometric data in this form may not be suitable even for graphical display since the gaps and slivers produce linework which generally is unacceptable in quality. The positional discrepancies of gaps and slivers are best resolved during a postprocessing stage, immediately following data capture.

Data sets without connectivity may exist during data capture even when creating more topologically complex models. Such a situation may arise when a CADD system has been adapted as a mapping information system. Several overlays of the same base map are digitized and the resulting data may be held as separate units, forming a series of *layers* or *levels*. Depending on the structure within a specific level, it may be possible to perform complex analyses involving the data on a single level, but to cross-reference between levels of data, where no connectivity has been recorded, would involve much search and processing time.

11.3.3.4 *Centroids model.* Specifying the centroid position of an area feature by its geographical position provides a measure of spatial referencing. No information describing the connectivity or shape and extent of features is held, but queries relating to the separation and distribution of features can be answered, from computations based on the geographical data.

11.3.4 *Representation of position–vector or raster*

So far, the discussion has not considered the method of representation of position (location and extent) of polygons, links or nodes. Ideally, spatial data users of a database do not need to know the method adopted to record the geometric data. In practice, knowledge of the method employed is required, since the possible modes of data retrieval depend heavily on this (Burrough, 1986). There are two contrasting ways of representing the location and extent of polygons, links and nodes: the first is by vectors, using coordinates, and the second is by the raster-based approach, using grid cells.

11.3.4.1 *The vector format.*
This mimics the representation used in conventional cartography. The boundaries of polygons are formed by links, links are formed by a series of interconnecting points and these points and any nodes are described by coordinated positions. There are two main methods by which the coordinates can be specified. With the *absolute mode*, the X, Y coordinate pair (with respect to the system origin, or sometimes a false origin in the south-west corner of the area of interest) is stored. In the relative or *incremental mode*, only the first point on a vector is defined in absolute terms; all subsequent points are then described by an ordered sequence of relative coordinates. These relative coordinates may be either rectangular (dX, dY) or polar (angle, distance).

In the majority of cases, the lines connecting pairs of points in vector format are straight. The use of straight lines is simple but many more segments and hence points are necessary to achieve a fine resolution. The lines can be better represented as curves by mathematical functions, from simple circular arcs to high order polynomials and splines. The savings in coordinated points by using mathematical functions are usually offset, however, by the need to store the functional coefficients and the additional processing required to interpret the data.

11.3.4.2 *Raster format.*
In raster format, the location and extent of polygons, links and nodes are built up from a set of squares (pixels) on a grid or raster. The square is either filled or empty, depending on whether the polygon, line or node occupies that square.

Several important differences between the two data formats can be demonstrated with reference to the example in Fig. 11.11. The vector representation requires less data than that of the raster to record the locational details of the simple water features illustrated. Sixteen coordinated pairs are sufficient to define the location and extent of the three features in the vector representation. The coarser raster structure requires data describing forty cells. To improve the resolution of the raster format to that of the vector at this scale, the grid interval would have to be reduced to a value nearer the plotting accuracy of the absolute coordinates ($\simeq 0.5$ mm). In the simple raster representation, it is apparent that there is no clear point of connection (node) between the linear and area features. This is one of the reasons why the raster format is seldom used with full implementations of a topological data structure.

There are problems in recording the *accuracy* with which data have been captured. The accuracy in the position of a point or line represented in raster format cannot exceed the grid resolution; conversely, with vector data it is possible to record a coordinate pair more accurately than it has been measured. During data capture, a balance has to be made between the accuracy achieved and the cost and effort required

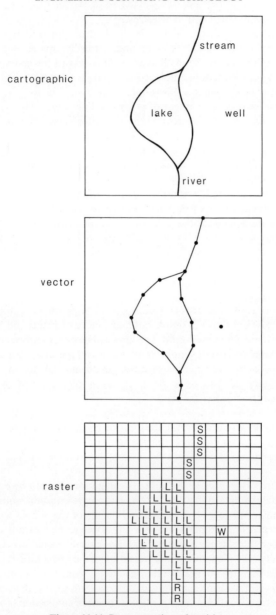

Figure 11.11 Representation of position.

to improve accuracy. When spatial data were collected solely for map or plan production, the scale of the plan determined the significance of any errors and hence the accuracy required. There are no such criteria for the spatial information held in a database. Such data may be used for a variety of purposes and can be displayed at a range of scales. It is essential that some measure of accuracy in position should be associated with coordinate pairs.

Many algorithms exist for the conversion of raster data to vector data and vice versa. The conversion from raster to vector can be very demanding on computer processing time. These operations may become more commonplace as processing power increases.

11.3.5 *Information retrieval and analysis*

One fundamental reason to use a spatial database management system (DBMS) is to facilitate a wide range of applications of data. Theoretically, it is unnecessary for the designer of the system to know all the retrieval and analysis procedures likely to be required. The database designer is, however, responsible for the provision of a set of fundamental *algorithms* or *tools* to be used by the application's programmer. These should consist of a set of simple commands in which a user can easily formulate any request for information, and a set of output tools, either graphical or tabular, for the presentation of the results.

11.3.5.1 *Information retrieval.* There are two general ways in which information can be directly retrieved from a topographic database. These are

i) by locating features of a specific *type* e.g. all manholes;
ii) by retrieving the information of all features which match a specific *location* e.g. occur within specific area or pass through a specific line or point.

The locational data and descriptive data recorded for a feature are normally linked by a unique identifier. To retrieve the location or description of a feature for which a unique identifier is known requires only a direct access to the database. Where the location of a feature of a specific type is required, the unique identifier must first be found. This entails a search of the descriptive attributes associated with each feature. The descriptive attribute data must be organised in a manner which allows maximum freedom in the requests that may be made and efficient implementation of search methods.

Retrieval by location is required every time a display of a *window* of data is requested. The limits of windows on data sets can be defined in several ways: rectangular areas to fit display screens or map sheet lines; circles delimiting a region around a point of interest; or buffer zones, which are bands of land defining a pathway either side of a linear feature. Many other requests for data or information require access by location, such as details of nearest neighbours. Retrieval by location therefore must be highly efficient. To this end, it is common for each feature to have associated with it additional data elements which describe location and overall extent (possibly indexed). These can be scanned rapidly and assist efficient retrieval.

11.3.5.2 *Data analysis.* The adaptability of the data model is reflected in the range of analyses possible and the retrieval methods available. For the comparison of areal features, most requirements can be formulated using combinations of Boolean operations such as 'and' and 'or'. 'And' defines an intersection where two area features overlap, whilst 'or' defines an amalgamation of areal features. Analytical tools to test for adjacency can be combined with these operators. They can be used to define contiguous areas where certain pre-set conditions are true.

There are many cases where it is inappropriate to store information on all classes of features in a single, topologically modelled database. The number of times it is necessary to cross-reference underground utility information with local administrative boundaries, for example, may be extremely limited. In such circumstances, it may be preferable to hold the two types of data in different databases. To analyse information from more than one database at a time requires the overlaying and comparison of polygons, lines and nodes in one data set with polygons, lines and nodes in another. The user may need to know, for example, whether a line feature crosses a defined area or a point feature falls within a pre-set area. Similarly there is a need for *network analysis*. For example, it is often necessary to trace pathways through networks of linear features when planning vehicle routes or preparing flow analyses through river and pipe networks. Specific operations include *minimum path analysis*–finding the shortest routes, and *critical path analysis*—identifying bottlenecks.

11.4 Computer data structures

From the foregoing discussion, it is apparent that spatially referenced information can be modelled to reflect the user's view of the data. This model must be structured in a form suitable for storage within a computer system. The main memories of computers are far too small to be able to store permanently all the data describing the spatial information, even for a small institution. Data have to be organized into units of files and held on some external storage media, usually a disk of some kind.

11.4.1 *Computer records and files*

A *file* consists of a number of records. A *record* consists of a number of fields. Each *field* holds a data element of some predefined type: e.g. a string, real number, integer number, etc. Records can be of fixed or variable length. *Fixed length records* are easier to process, but are wasteful of space since blank fillers may be needed if a record does not fill the allotted space. A *variable length record* must indicate where it ends, either by using a unique end-of-record symbol or by having a field count at the start of the record. Updating data often requires a change in the length of the stored record.

Data records usually are not accessed in the order in which they are physically stored. Access is controlled by the use of pointers. The term *pointer* is used here to describe a data element held in a specific field in one data record. It gives the starting address, that is, it points to a related record. There are three main ways to access records in files: serial, ordered and indexed access.

11.4.1.1 *Simple serial files.* With the simplest form of file organisation, records are held in *unordered lists*. This approach was in common use when data had to be stored on punched cards or paper tape. The only method of access is to go from one data item to the next in the physical order in which they are stored. This access method is normally very slow. The only procedure for new records to be included is for them to be added at the end of the file.

11.4.1.2 *Ordered sequential files.* In this case, one field is nominated in each record as a *key value*. The physical record order is based on this key. Typically, key values are arranged in numerical or alphabetical order. The disadvantage is that adding a new item means that the data must be reorganized to create the appropriate space for its insertion. The advantage is that records can be accessed much faster using processes such as 'skip' or binary searches. A *skip search* is one where, for instance, every 100th record in the ascending order key sequence is read until the first key value is found that exceeds the search value. From this point, the previous 99 points that were skipped would be searched. With a *binary search*, the range of records is halved with each read. The search key is first compared to the appropriate data element in the record midway in the ordered list. Based on the result, either the upper or the lower half becomes the new search area. This process is repeated until the required record is found.

The skip search was appropriate when it was common to have records stored on serial access devices such as tapes. The binary search is more efficient when it is used to search items in the main computer memory. It can be less efficient however for data held on disk where large and frequent physical movements of the reading head may be required. It is often used to search indexes.

11.4.1.3 *Indexed files*. An *index* is an additional file created both to speed access and minimize reorganisation of records in a data file (Date, 1986; Fig. 11.12a). A *dense index* can be made for a data file on any key value not in sequential order. The key values are sorted into sequential order and records created in the index file for each value. These identify the address in the main file where the appropriate data record is held. There will be as many entries in the index file as there are in the main file (Fig. 11.12b). A non-dense or *sparse index* can be made on any key where the values are held in sequential order. Only selected values of the key are entered into the index. Typically, the values selected would be the first occurrence of a new starting letter, if indexing an alphabetic

(a) Data file

record address	unique feature number	description	
1	F001	house	.
2	F039	fence	.
3	F042	house	
4	F043	house	
5	F106	wall	
6	F109	wall	
7	F203	road	
8	F267	house	
9	F333	wall	

(b) Dense index on description

description	record address
fence	2
house	1
house	3
house	4
house	8
road	7
wall	5
wall	6
wall	9

(c) Non-dense index on feature number

unique feature number	record address
F000	1
F100	5
F200	7
F300	9

Figure 11.12 Index files.

sequence, or the start of a new range of numbers if indexing a numeric sequence (Fig. 11.12c). In the example, the record for feature number F109 is found by first searching the index table for the address in the main data file of the first feature with a number equal to or greater than F100. The search for the required record then continues in the main file. Even though the process still requires serial file scanning, the number of items to be scanned is greatly reduced. The techniques of skip searching and binary searching can also be applied.

Any file, whether it be a main data file or an index file, can have at most a single sparse index, since such an index relies on a unique physical sequence. However, there can be any number of dense indexes.

11.4.1.4 *B-Tree index.* The use of multi-level or tree-structured indexes, usually termed B-trees, is probably the most common indexing method employed in general database systems. The other indexing systems mentioned so far require considerable physical scanning of files, but the B-tree approach reduces this requirement by building indexes on top of indexes, commonly up to three levels.

An example of one variation is shown in Fig. 11.13a. The lowest level of the tree is a dense sequential index providing fast access to the data via pointers. At the levels above this, the tree structure provides fast direct access to the sequential index. The numbers 2, 7, 11... 99 are values of the field indexed in the main file. The single top node, called the *root*, holds two of these values (51 and 76) and three *pointers*. When searching for a value s, for records with s less than or equal to 51, the left-hand pointer is followed. For values of s greater than 51 but less than or equal to 76, the middle pointer is followed. For the remaining values, those with s values greater than 76, the right-hand pointer is followed. This process continues down the tree until the appropriate value in the dense sequential index is found. Fig. 11.13b demonstrates the worst case that has to be dealt with when inserting a new record. To insert a pointer to record 38 in (a) where the tree is already full, an additional level has to be introduced to create space whilst still maintaining a balanced structure.

To implement the B-tree index on the computer, each node in the structure is represented by one record of five fields. Two fields hold the indexed numbers, and three hold the pointers to other nodes.

11.4.1.5 *Pointer chains.* A *pointer chain* is a series of records that are linked together by pointers which are data elements within the records. Pointer chains are also known as *link lists*. Pointer chains are ideal for handling data sets with many interconnecting lists. The principal advantage is that the insert and delete processes are comparatively simple and efficient (Prince, 1971).

Fig. 11.14a illustrates an independent list of records. A start pointer holds the address of the first record. The first data element in this record holds the address of the second record and so on through the list until some end of sequence marker is encountered. Figure 11.14b illustrates the relative ease of editing the list. This involves only manipulation of pointer values. The second data record is eliminated from the logical list (it does, however, remain in the same space in physical storage) whilst a new record, physically stored at the end of the file, has been logically inserted. The final diagram demonstrates that records of two or more logical lists can be held in the same physical file, with the start of each list being accessed from an index file.

There are many variations and additions that can be made to the simple structure

LAND INFORMATION DATABASES 445

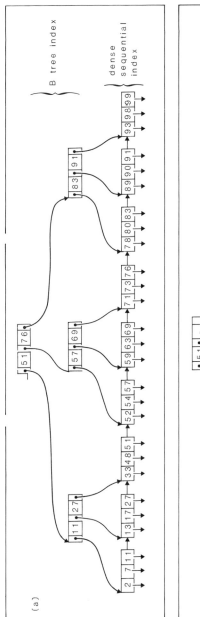
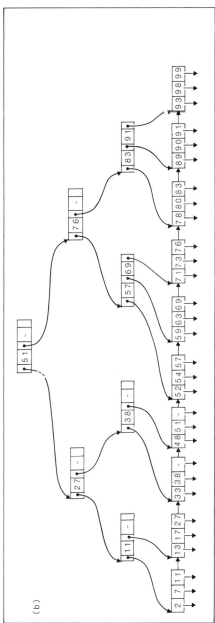

Figure 11.13 B-tree index.

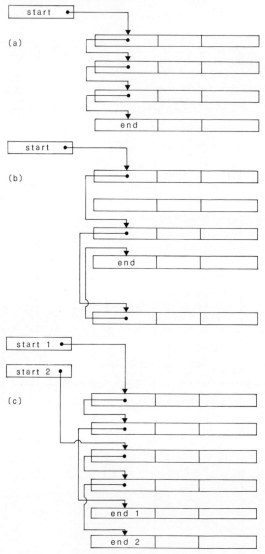

Figure 11.14 Pointer chains.

shown. It is very common to form pointer chains into *rings* with the last item referring back to the first. In this case, if any one item has been found, any other item in the list can be accessed by stepping forwards or backwards. A further modification might incorporate both forward and reverse pointers to allow retrieval from either direction. The use of these extra pointers results in improved access but demands additional storage space and, more significantly, additional effort in editing. Despite these disadvantages, ring pointer chains are a popular and successful method of data organisation, especially for graphical data.

11.4.1.6 *Indexing by spatial location.* The methods of record indexing presented so far cannot treat more than one field as the key simultaneously. Spatial data should be organized according to its spatial location, to allow searching in both the east–west and the north–south directions during the same operation for instance. This requires reference to two, and in some cases three, key fields defining the coordinate values.

Many database systems are designed to deal with data sets of limited coverage. In general, these do not utilize any spatial indexing. They rely on very high speed storage access and possibly single-key indexing to ensure acceptable response times to queries. As spatial database usage becomes more sophisticated, data set coverage will increase and so will the necessity for spatial indexing. There is already a demand when viewing data and zooming or panning across an area for the same response time to be achieved no matter what the spatial extent (Ritchie, 1987).

Both individual coordinated points and complete features can be indexed. Point-in-polygon tests can be used to determine whether an individual point lies within or outside a window. It is possible for a large linear or areal feature to cross or impinge on a selected window of interest, without being demarcated by any coordinates which fall within it. When indexing features, therefore, the index must be provided on coordinate values which delimit the maximum extent in each direction.

Some possible strategies for two dimensional spatial indexing values are given in Fig. 11.15. In example (a), space is regularly segmented and a sparse index maintained to point to the starting address of the data for each segment. This method has the advantage of being relatively easy to implement. It can work well for evenly distributed data sets; small-scale data often exhibit this characteristic. Large-scale data, however, tend to be more clustered. The method then becomes wasteful of storage since the space allotted for some area segments is unused whilst the storage allotted for other segments is overflowing and requires additional allocations.

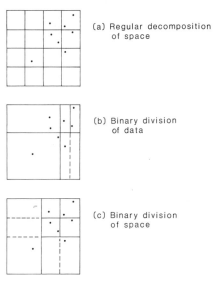

Figure 11.15 Spatial indexing.

The other two approaches shown in Fig. 11.15 attempt to organize data into regions containing the same number of data values. In example (b), the *data set* is repeatedly divided into two, until regions contain a number of points suitable for an indexed group; in example (c), the *object space* is repeatedly divided to achieve an equivalent result. In this example, a maximum of two data points per segment is allowed. Sparse indexes are maintained to point to the starting addresses. By dividing the data set, boundaries are dependent on the data distribution; when dividing the object space, divisions always occur at predefined locations. With both approaches, when the original data set is indexed by subdivision into segments, storage blocks are incompletely filled to leave some room for additions. With subsequent editing, when a block is filled, a new division is introduced and the segments are re-indexed.

11.4.1.7. *Data clustering.* Using present technology, the quantity of data required for a spatial database means it can be held only on disk type storage devices. All disk devices depend on a reading head which physically traverses a rotating disk in a radial direction, to read the optically or magnetically stored data. The characteristics of this physical movement determine the performance of any data retrieval system. Minimizing the number of disk transactions can improve retrieval performance significantly. This has some important implications for the ordering and indexing of data sets. (Date, 1986).

The database has so far been viewed as a collection of stored records in files. The organization of these files is supported on the computer by the *file manager*. This treats the database as a collection of units of disk storage referred to as *pages*. A disk page is a temporary store or buffer of data read from disk and held in the main computer memory. These pages are controlled by the *disk* manager, a component of the underlying computer operating system which is responsible for all disk input and output transactions. A *bucket* is the term used to refer to the quantity of data read from the disk in a single operation. A disk page consists of many buckets. Disk space can be used most economically by ensuring that the bucket size holds an integer number of records. The choice of item to index is also important. It is not efficient to index individual records; to be most efficient, the blocks of records that fill a bucket should be indexed, since these have to be read from the disk at one time. It is advantageous in terms of speed if records that are logically and spatially related, and hence frequently used together, are stored physically close together on the disk. This clustering will help minimize the reading head movement. The final consideration is the choice of page size. A set of records might need to be read several times for one particular type of spatial analysis. If this set can be kept together in a page of the main computer's memory, several disk transactions can be saved.

11.4.2 *Database techniques*
A definition of a *database* is:

> 'A collection of interrelated data stored together with controlled redundancy to serve one or more applications in an optimal fashion; the data are stored so that they are independent of programs which use the data; a common and controlled approach is used in adding new data and modifying and retrieving existing data within the data set' (Martin, 1976).

The all embracing advantage of using a database approach is that it provides an organization with central control over all its operational data (Date, 1986). The more specific benefits include:

(i) *data independence.* The form of application programs is not directly related to the way in which the data are stored. This means that existing programs can be modified and new ones written without changing the stored data; conversely, the physical storage of the data can be altered to take advantage of new storage techniques and hardware without having to change applications programs.
(ii) *controlled data redundancy.* Since each data element is in general stored only once, the problem of inconsistency, created by several users having their own individual but slightly modified versions of the same data set, is reduced.
(iii) *data sharing.* Data can be shared not only between users but also between a range of applications, some not envisaged when the data set was created.
(iv) *uniform standards.* The maintenance of common standards, agreed at local, national or international level, aids the exchange of data between organizations.
(v) *security control.* Although it is essential that data sets be shared, for security reasons, there will always be some data for which access cannot be permitted to all users; databases can incorporate such selective restrictions of data access.
(vi) *improved integrity.* It is inherent in the database approach that validity checks, such as whether a data value is within acceptable limits, are included.

The internal level of a database consists of many files of records many with indexes. There is no one file structure or indexing system that is optimum for all data. Many different file types are used when storing a large data set. To access data from one or more files, it is essential to have some form of structure integrating them. There are three recognized structures termed hierarchical, network and relational. Each of these is discussed below.

11.4.2.1 *Hierarchical structure.* A hierarchical database consists of an ordered set of trees (Fig. 11.16a). Each level in a tree consists of one or more records of a particular type with each record possibly containing many fields. Access up or down the tree from one level to another is via key fields in each record.

This method of structuring provides rapid access and is convenient when the data have a one-to-many or parent-child relationship. Its two great advantages are that it is both easy to understand and easy to update and expand. It suffers the disadvantage, however, that it is difficult to directly access data which are not the key in the record. This restriction makes it difficult to answer some speculative enquiries, such as those not conceived at the time the structure was designed. Another significant disadvantage is that some record field values have to be repeated many times, leading to data redundancy.

11.4.2.2 *Network structure.* The network structure can be regarded as an extended form of the hierarchical structure. In the latter case, a child record has exactly one parent record. Access is via a route which passes only up and down through the various levels of a tree. In many situations, such as those involving graphics features, this lack of freedom in access routes is too restrictive. In network structures, a child record can have any number of parent records (Fig. 11.16b). This arrangement can overcome one of the main shortcomings of the hierarchical approach in that it makes good use of all data and hence reduces redundancy. Two major disadvantages still exist however.

(a)

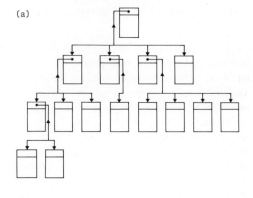

(b)

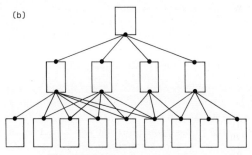

Figure 11.16 Database structures.

Firstly, any required linkages between records must be directly specified. Secondly, because linkages are normally implemented in terms of ring pointer structures, the database size is greatly enlarged by the overhead of pointers.

An example of the implementation of a network type structure to a link and node data model is given later. Here, as is common in graphic related data sets, it is implemented using a ring pointer structure.

11.4.2.3 *Relational structure.* A relational database is one that is perceived by its users as a collection of tables, or relations, of information. Data are stored in records, similar types of records being grouped together in two-dimensional tables normally in a single file. The relational structure has no hierarchy and stores no pointers. Unique key values for each type of record are used to link one set of records to another.

Relational databases have the advantage that their structure is very flexible. Any queries that can be formulated using relational algebra can be answered. In answer to any query, a new table is created from those present in the data set. Editing data is easy since this simply involves adding, amending or removing records from one or more tables. The disadvantage of the relational structure is that because the data has no preconceived linkages, most operations involve many searches through files to find appropriate data elements. This can consume much computer time. It is becoming less of a problem as computational speeds increase with improved hardware. Relational databases and their management software have to be skilfully designed to ensure that systems operate at reasonable speeds.

Almost all commercial database systems developed for general purposes over the past few years are relational. The majority of these support a version of the relational database language *Structured Query Language* (SQL). Almost all current database research is based on relational ideas. There is little doubt that for the majority of applications in the foreseeable future, a relational approach will be adopted. This is certainly true for the storage and manipulation of the descriptive information associated with spatial data. There are also several applications for relational databases in spatial data structuring. For this reason, the concepts of a fully relational database will be considered further, using a simple example.

11.4.2.4 *An example of a relational database.* The details of three simple relations recording descriptive attribute data are tabulated in Fig. 11.17. Relations U (utility feature) and B (building feature) are similar. Each row or *tuple* in the table details information about an individual feature. There could be as many tables as there are broad categories of feature. Each feature has a unique reference number. This is used as the *primary key*. Further descriptive attributes for each feature are listed in the remaining columns or fields: displaycode, date, etc.

In this example, the feature reference number is unique not just to the relation in

Relation U: Utility feature

featurenum	displaycode	date	service	depth	diameter	material
.	.	.	.	.	.	.
16	96	8706	gas	1.6	0.2	plastic
17	96	8706	gas	1.5	0.2	plastic
18	23	8701	elec	0.5	0.08	sheath
19	24	8701	elec	0.7	0.1	sheath
20	.	.	.	.	.	.

Relation B: Building feature

featurenum	displaycode	date	owner	use	construction
.	.	.	.	.	.
45	03	8609	Smith	house	brick/tile
46	04	8609	Smith	office	brick/tile
47	04	8512	Dixon	shop	stone/slate
48	.	.	.	.	.

Relation D: Display code

displaycode	kind	type	colour	size
.	.	.	.	.
03	15	02	R	4.3
04	15	02	G	5.2
05	16	02	R	5.2
06	.	.	.	.

Figure 11.17

which it falls, but to all the relations detailing features. Using such a unique reference number, a cross-reference can be made to the spatial data held in a separate database.

The third table (Relation *D*) is used to define the display code that describes the way in which lines and symbols are to be presented on graphic display screens and plotters. The primary key is the display code itself. A primary key does not have to be based on a single field; a unique combination of fields can equally well be used if required.

Three aspects of the relational model are apparent from Table 11.17. Firstly, there is only ever one data value in each row and column position. If it were required to list several items in one position, a new table would be introduced. Secondly, every non-key field must be fully dependent on the primary key. As an example, although it is possible to combine tables B and D by adding the display parameters from D alongside the appropriate display code listed in B, this is not permitted by this particular rule. It would result in a table where fields were not only dependent on the primary key, but also on the display code. Any such dependencies must be removed by the use of additional tables. Lastly, no pointers exist between tables. The only method available to represent a *relationship* between two values is to specify them in the same row of a table. For example, the network connectivity of the linear utility features in table U could be represented by Relation J: Junction

Junctionnum	endfeature	startfeature
45	16	17
46	17	32
47		

The relation J considers a single linear utility feature starting where another has ended—Utility feature 16 ends at Junction 45 where Utility feature 17 starts, etc. This idea can be extended to relate any arrangements of interconnections.

Information within the database is entered, retrieved and edited using the commands of a *query language*, part of the DBMS. Such commands can be issued directly to answer individual unique demands. Alternatively they can be accessed through a programming language to allow the development of application programs. These can involve complicated combinations of commands.

There is a group of commands referred to as *data definition statements*. An example of one of these is CREATE. A CREATE TABLE statement must specify not only the table name, but also the name and data type (string, integer etc.) of all its fields. Further, it must specify the range of values acceptable in any field (e.g. negative values are not permitted).

The most commonly used data retrieval operation is SELECT. This, along with its many qualifiers, can specify any of the relational algebraic functions shown diagrammatically in Fig. 11.18.

The result of a SELECT command is always another table:

Query:	SELECT	featurenum, use
	FROM	B
	WHERE	owner = 'Smith'
Result:	featurenum	use
	45	house
	46	office

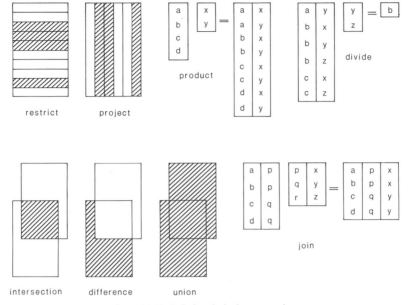

Figure 11.18 Relational algebra operations.

An asterisk (∗) can be used at any time as a shorthand statement to refer to all items. The expression LIKE 'C%' will access all items starting in 'C'. Preceding an attribute by a table name limits its scope to just that table. Terms such as DISTINCT and ORDER BY are available to eliminate duplicates from the result and present items in numerical or alphabetic order. A range of conditions can be set using qualifiers such as $> = <$, and Boolean operators like AND, OR and NOT. For example,

Query: SELECT   ∗
        FROM   U
        WHERE   depth < 1.0
        ORDER BY diameter

Result:

featurenum	displaycode	date	service	depth	diameter	material
19	24	8701	elec	0.7	0.1	sheath
18	23	8701	elec	0.5	0.08	sheath

The query language incorporates a number of built in functions to enhance its analytical power: COUNT, SUM, AVERAGE, MAX, and MIN are some examples. These functions operate on a collection of values in one column of a table. This table may be either permanent or derived as a result of a query. The sum and average functions must of course act on numeric values. They result in a single value.

One of the more powerful features of relational systems is the ability to 'join' two or more tables. The example below results in a table associating each feature number from B with the appropriate display colour from D:

Query:	SELECT	featurenum, colour
	FROM	B, D
	WHERE	B. displaycode = D. displaycode
Result:	featurenum	colour
	45	R
	46	G
	47	G
	48	

Another powerful feature is that multiple levels of nesting of SELECT statements are acceptable. For example, the argument of a WHERE expression can itself contain a SELECT statement. The query below would tabulate the possible colours in which Building features could be displayed. The nested query is computed first. It lists the distinct display codes referenced in relation B. The main query then lists the distinct colours referenced within the reduced list of display codes:

Query:	SELECT	DISTINCT colour
	FROM	D
	WHERE	displaycode IN
		[SELECT DISTINCT displaycode FROM B]

A data manipulation language must also support update operations such as UPDATE, DELETE and INSERT. The operation UPDATE acts on fields within rows whilst the other two act only on complete rows. The data definition as set in CREATE must be adhered to at all times.

The inherent simplicity of the relational data structure is demonstrated by the requirement to use so few fundamental operators to manipulate the data set. In general, the more complex the data structure, the more operators are needed. The software within the DBMS is not however straightforward. To execute a multi-nested series of commands might require the temporary creation of many tables of differing sizes. The DBMS must, at all times, optimize all sequences of commands to be efficient in use of both computer storage space and time.

11.5 Vector data storage

As an example of computer storage techniques applied to spatial data models, Fig. 11.19 tabulates the contents of six files of a hypothetical implementation of a simple link and node model. Each record in the features (F) file details data for an individual feature in the data set. Each feature has associated with it a unique reference number, a classification code and coordinate values defining its maximum extent. Depending on whether it is a linear or point feature, a pointer (represented throughout the example by P) indicates the records in the link list (L) file or the node (N) file where the appropriate spatial data are first referenced.

Access to features is most often by classification, so the records throughout the feature file are ordered by the classification used and a B-tree index is maintained to ensure rapid access. This indexing, and the further cross-referencing between files described later, requires a serial record number, shown in the first column for this and other files. With any indexed file, insertion is difficult so new entries are added to the end of the file until a periodic sort re-orders the records.

Feature (F) records file

	record	unique ref	feature type	class	coord extent	^node or link list	----
	.	.	.	.	.	.	
index on class	F036	192	linear	Building	xy/xy	^L060	
	F037	396	point	Transport	xy/xy	^N154	
	F038	021	linear	Utility	xy/xy	^L065	
	F039	022	point	Utility	xy/xy	^N153	
	.	.	.	.	.	.	

Link list (L) records file

record	^link detail	^next link in feature	^previous link in feature	----
.	.	.	.	
L060	^D096	^L061	---	
L061	^D097	^L069	^L060	
.				
L069	^D106	---	^L061	
.	.	.	.	

link Details (D) records file

record	^start node	^end node	^start coord	^end coord	^usage	----
.	.	.	.	.	.	
D096	^N161	^N162	^P368	^P374	^U256	
D097	^N162	^N163	^P580	^P604	^U256	
.						
D106	.					
.	.	.	.	.	.	

Node (N) records file

record	^coordinate	^usage	----
.	.	.	
N153	^P367	^U204	
N154	^P914	^U202	
.	.	.	
N161	^P...	---	
N162	^P...	---	
N163	^P...	---	
.	.	.	

coordinated Point (P) records file

	record	^type	^next point	^previous point	X	Y	^link or node	----
	.	.	.	.	.	.	.	
index on coords	P367	node	---	---	xxx	yyy	^N153	
	P368	link	P369	---	xxx	yyy	^D096	
	P369	link	P370	P368	xxx	yyy	^D096	
	P370	.						
	.	.	.	.	.	.	.	

Figure 11.19

```
link and node Usage (U) records file
  record    | ^feature  | ^next   |  -----
            |           | usage   |
     .           .           .
   U202       ^F037      ^U203
   U203       ^F067       ---
   U204       ^F039       ---
     .           .           .
   U256       ^F036       ---
     .           .           .
```

Figure 11.19 Cont'd.

It is not essential to maintain data elements detailing the maximum coordinate extent since this information is held elsewhere, but this facility does speed up spatial searching. When searching by classification, it eliminates the need to access any other files to establish whether a feature falls within a chosen window of interest. Repeating data such as coordinate limits in more than one location goes against the design philosophy of databases. Without careful software design, it can cause problems when data values are edited.

Linear features can be composed of any number of links taken in order. References to the links defining each feature are held in pointer chains in the link list file. Take as an example feature reference number 192 in file F. The list of its defining links starts with record L060. Record L060 points to L061 and this on to L069. These three records form a pointer chain. Each record in the chain contains one further data element which points to a record in the *link detail (D) file* containing link information. This file structure, where link details are held separately from the link lists, permits a link to be used with any number of features. Within each record in the link detail (D) file there are pointers to both the start and end nodes and to the start and end of the coordinate list defining the link shape. All the spatial data required for a link can be extracted from this information. The link can be traversed from either direction. A further data element points to a record in the usage (U) file. This is discussed later.

Two record types have contained pointers to the *node file:* records of point features in the feature file and records in the link details file. Node records are very simple. They contain only pointers to a coordinate record and a usage record. The coordinates of the node could have been included but there are advantages in incorporating such data in the coordinate point (P) file.

References to the *coordinate file* are made from the two record types—node records and link detail records. Each record in the coordinate file contains the coordinates of either a node or a single point occurring along a link. In the case of a node, the records in the file are independent of all the others. In the case of a link detail point, the records form part of a chain pointing both forward and backward. Mixing these two types of records ensures that all coordinate values are contained in this one file. This benefits spatial searches and computations such as coordinate transformations. To find the coordinated point nearest to a selected location, it is necessary to scan only this file. To speed up access, records can be indexed on coordinate values. The index could be based

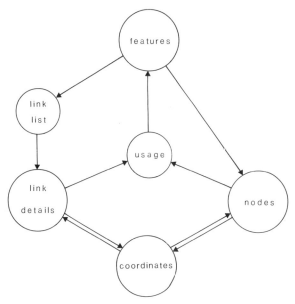

Figure 11.20 Data linkages.

either on a single key value or on a dual key using one of the spatial location approaches described earlier.

The record structure described so far has shown a hierarchy where features point to lists of links or single nodes. These in turn point to coordinates. There is also a need for an access route through the data in the opposite direction. This will enable the user to locate the one or more features associated with a particular coordinated point. Access starts with a field in each coordinate record which points back to the node or link detail record that referenced it. The node or link detail record found contains pointers to records in the usage (U) file. Each of these usage records points back to a feature record. If a link or node forms part of more than one feature, a pointer chain of usage records is formed. This forward and backward network of links are shown diagrammatically in Fig. 11.20.

11.6 Raster data storage

The raster method of position representation does not lend itself to efficient storage in computer records. The rows and columns of grid cells (pixels) representing an area can be recorded using a two-dimensional array. Such arrays are commonly used within high-level programming languages such as Pascal and Fortran. The problem with this approach is that a two-dimensional array spanning the complete area would be required to record each and every spatial unit. This would in effect create a three-dimensional array (Fig. 11.9). The resolution in the third direction would be determined by the number of units to be recorded. To overcome this problem, many spatial units are recorded on the same 'layer'. If this is done, all units on the same layer possess the same characteristics. Alternatively, a code is associated with each grid cell

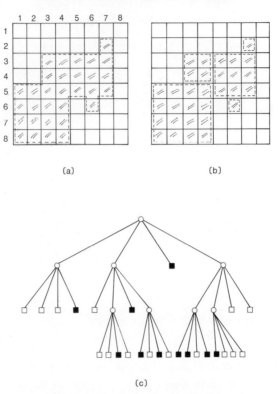

Figure 11.21 Raster data compaction.

so that a variety of characteristics can be mixed on each layer. These solutions decrease storage requirements, but increase subsequent computer search and processing times.

11.6.1 *Data compaction methods*
There are several simple approaches to compacting raster data. They all result in a structure suitable for holding in computer records. Unfortunately, most of the simplest approaches suffer from the need to recreate a grid structure when procedures of area analysis are performed (Burrough, 1986).

Chain coding is a compact way to record the boundary of a region. It can be used directly to calculate areas and perimeters and to evaluate boundary shapes. Using the digits 0, 1, 2, 3 to represent the directions east, north, west, south respectively and a superscript to represent a number of grid cells, the boundary of the example area unit in Fig. 11.21a, starting from cell row 5 column 1 is

$$0^2, 1^2, 0^4, 1, 0, 3^4, 2, 3, 2, 1, 2, 3^3, 2^4, 1^4.$$

This sequence, along with its start position, is sufficient to describe the area for a complete data set of any total extent. Even if the total area of the data set occupied many thousand by many thousand grid cells, the sequence is still sufficient to define the area. This property is true of other simple methods of data compaction.

Run-length encoding can record any area by listing for each row the columns enveloped by the area:

row	column
2	7
3	3–7
4	3–7
5	1–7
6	1–4, 6
7	1–4
8	1–4

A further technique, *block coding*, subdivides the area into a minimum number of square blocks of any size. The position and size of each block is described in terms of just three data elements: the row and column position of an origin and the dimension of the square. The shape in the example shown in Fig. 11.21b can be defined in terms of 5 blocks: one 4∗4, one 3∗3, one 2∗2 and two 1∗1. This procedure is of particular advantage when recording large, regularly shaped areas.

11.6.2 Quadtrees

Quadtrees are a further means of compacting data and result in a code that is easily held in computer records and can be processed directly for spatial analysis. A *quadtree* successively divides an area of 2^n by 2^n grid squares into quadrants to create a hierarchical structure or tree of n levels. At each level, the four divisions of the unit area are usually considered in the order NW, NE, SW, SE (Fig. 11.21c). If the quadrant is contained totally within the area unit, the tree is terminated with a *black square node*. If the quadrant is totally outside the area unit, the tree is terminated with a *white square node*. If the quadrant is neither totally within nor without the area, it is represented by a *circle node*. It is then further subdivided to create a new level in the tree.

In the computer, quadtrees can be held with each node represented by a single record of five pointers and a single data element. Of the five pointers, four point down and one points up. If it is a terminating node, the data element indicates whether it is inside or outside the area. This approach allows efficient processing but requires a large amount of computer storage for the pointers.

Alternative approaches, termed *linear quadtrees*, allow efficient processing whilst keeping storage requirements small (Hogg, 1986). One such method associates the digits 0, 1, 2, 3 to quadrants at each level of the tree. A multi-digit number, which is essentially an address in base 4, can then be used to represent any quadrant at any level; the lower the level, the greater the number of digits. Since only quadrants within the shape need to be recorded i.e. those represented with black squares at nodes, the example of the area unit can be fully described by the linear list

2
03
12
112
130
132
300
301
303
310.

Spatial analysis requires operations such as union, intersection or adjacency of areas to be performed by arithmetic operations. This can be done directly on quadtree addresses. Another benefit of the quadtree approach is that it readily lends itself to display at variable levels of resolution. Considered very simply, at coarse resolutions only large blocks of cells represented by addresses with the least number of digits need to be displayed.

11.7 Land and geographic information systems

11.7.1 *Slimpac*
SLIMPAC (Surveyor's Land Information Management Package) is a very simple mapping information system developed by the author and his collaborators at the University of Newcastle. It is designed to handle large-scale spatial data of limited areal extent and any amount of thematic information associated with it. It is designed primarily to run on powerful microcomputers. SLIMPAC is based around two widely used applications software packages—a computer aided drafting (CAD) package and an alphanumeric database management system.

The *drafting package* used is AutoCad. The spatial database within AutoCad stores entities. The most widely used entities for mapping applications are points, polylines and blocks. A polyline is a series of lines, and a block is a collection of other entities. Spatial data can be input from a range of sources: from field survey, photogrammetry, existing mapping, or from another computer system. All data collection software produces data in the same standard format. At this stage, the land related features which are recognized are topographic details such as boundary fences, road edges and building walls. The spatial data defining these features are stored in the AutoCad database as points and polylines. The resulting data set can be considered to be a spaghetti model as described in section 11.3.3.3.

Routines assist the user to group together the topographical features into more useful units of information. The units of information depend on the application. They can be point, line or area units. A common area unit is the *land-parcel*. The polyline entities representing the boundary are grouped together into block entities. Since adjacent land parcels share common boundaries, the problems of gaps and slivers are avoided. Any other entities representing details within the land-parcel can be included once the perimeter has been defined. A unique reference number associated with each block is used as a key to access associated records in the alphanumeric database.

The *alphanumeric database* used is Superfile. Its structure is close to that of a relational database. A user who wishes to make a query first selects one of a range of screen forms. One or more items on the form must be entered. The data entered defines the model. Examples of such models are 'features with an address of 23 Acadia Avenue' and 'features that are private residences with more than four bedrooms'. The database management system finds the feature or features that match the model and presents them to the user in one or more screen forms. The user does not know of the relations within the database or the command language by which they are accessed.

The graphic and alphanumeric databases can be accessed independently but more often they are used together in one of two ways. One or a group of features are selected by pointing at them on the screen. Any associated details are then extracted from the alphanumeric database and displayed on screen forms. Alternatively, features which match a particular descriptive model are automatically found from the textual database and displayed graphically. A wide range of functions are available within AutoCad and Superfile's DBMS for operations such as plotting and report generation.

SLIMPAC is representative of a fairly simple mapping-oriented information system, developed by the staff and students of a small university department. At the other extreme are the expensive, highly sophisticated GIS/LIS systems which are offered commercially by large software houses such as Environmental Systems Research Institute (ESRI), Synercom Technology, Geovision, etc. which specialize in this area. These firms integrate the data capture equipment, computer hardware, and display and plotter devices from a wide range of suppliers together with their own specially written software into a single system. Another source of large, sophisticated GIS/LIS systems are the well-known manufacturers of surveying and photogrammetric instruments— Wild, Kern, Intergraph, etc.—who have also chosen to enter this market. Usually certain items from their own hardware catalogues, such as electronic tacheometers, analytical stereo-plotting instruments and flatbed plotters, form integral parts of the input to and output from the system. An example of a GIS/LIS system from each of these two main groups of system suppliers is the ARC/INFO system from ESRI and System 9 from Wild.

11.7.2 ARC/INFO

The ARC/INFO system has probably the largest installed base of GIS/LIS systems at the present time. It is already in use by many organisations in the USA and worldwide. Users include local and national administrations, local and regional planning and development authorities, public utilities and oil, gas and mineral exploration companies.

The ARC/INFO system has been implemented on a wide range of computers ranging from: mainframes (e.g. IBM 3090); multi-user minicomputers (e.g. Prime, Dec VAX and Data General); graphics workstations (e.g. Sun, Apollo, Tektronix, Hewlett Packard and IBM 6150 or PC/RT); and microcomputers (e.g. IBM PC/AT and PS/2 series and compatibles). A large number of the digitizers, displays and plotters described in Chapter 9 can be interfaced, handled and driven by the ARC/INFO software as represented by Fig. 11.22.

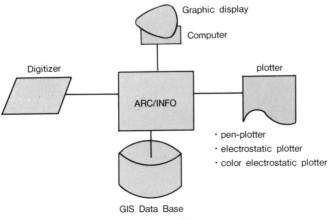

Figure 11.22

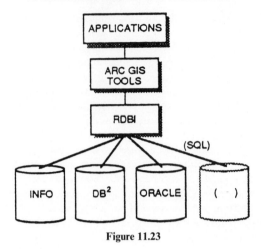

Figure 11.23

While the ARC/INFO part of the software suite has been written by ESRI, the INFO part of the suite is a *relational database management system* (RDBMS), which has been produced by Henco and is marketed and sold separately by the latter company for use with non-graphic data. The integrated ARC/INFO system is produced and sold by ESRI. It should be noted, however, that ESRI has recently produced a few GIS systems in which the ARC part of the system has been integrated with other non-graphic RDBMSs such as Oracle, RdB, INGRESS and dBASE II. This has been achieved via a new interface tool: the *Relational Data Base Interface* (RDBI), produced by ESRI (Fig. 11.23). The ARC/INFO combination, however, is by far the predominant combination which has been sold to date.

ARC/INFO is a modular software suite. *ARC* is the main program environment and controls access to all the other modules. It is used to manage the graphic data, and to provide the cartographic front- and back-end to the overall system. Most of its commands can run in a batch environment without the use of a graphics terminal. The data editing commands include all those to merge data sets from many sources. Analytical operations include feature buffering, map overlay and nearest-neighbour analysis. The INFO software is a complete RDBMS which manages the tabular data describing the attributes of the cartographic features. All the facilities of the DBMS can be applied to both the spatial and attribute data which are stored in relations in the database.

ARCEDIT is the interactive graphics editor. It has two functions: it interactively edits the spatial database, and it places text and symbols in high quality graphics displays. ARCPLOT is used to compose maps for output from graphic displays. All additional information such as legends, borders and titles can be added to the displays and the result sent to a plotter. The MAP LIBRARIAN module aids the control of many different data sets.

ARC/INFO supports a full polygon, link and node topological data model. The vector spatial data are stored in the relational database as are the feature attribute data. The COGO (coordinate geometry) program supports direct land-survey data input and ADS (ARC digitising system) supports digitizer input from cartographic sources.

The creation of the topological data model is fully automatic on input. A unique reference number links a feature's spatial data with its descriptive data.

The Network software performs analyses on networks modelled using ARC/INFO. The major functions are routing and allocation. *Routing* determines optimum paths for the movement of resources through a network; *allocation* finds the nearest centre (minimum travel cost) for each link in a network to best serve the network. These two functions can be used on networks such as city streets, waterways and telephone lines for operations such as vehicle routing, optimum facilities siting and time–distance flow analysis.

The TIN (triangulated irregular network) software, which can be provided as an additional module to ARC/INFO, is essentially a triangularly-based digital terrain modelling package which can store, manage and analyze the data describing 3D surfaces. Besides terrain, the TIN surface models can be applied to rainfall, population census data and geophysical data, etc. Additional modelling capabilities include: the calculation of slope, aspect, volume and surface length; the implementation of profiling; and the determination of stream networks and ridgelines. Display facilities include perspective views with hidden line removal.

Related software can manage, analyze and display spatial data in raster format. For efficient storage, the raster data can be stored in a run-length encoded format. A wide range of analyses are possible including proximity calculations and terrain analysis. Remotely-sensed image sets, such as those produced by the Landsat MSS and TM scanners can also be handled, although it is more usual for such data to be used in collaboration with a digital image processing system such as ERDAS. Special links have been devised to allow the interchange of data between the two systems. Through these, the map images generated after analysis of spatial data in an ARC/INFO system can be displayed as an overlay in register with the coloured images produced after digital image processing and analysis in the ERDAS system. A comprehensive overview of all the functions, operations, manipulations which can be carried out on a large sophisticated GIS/LIS system such as ARC/INFO is presented in Fig. 11.24.

11.7.3 *System 9*

System 9 is a new geographical information system developed by Wild of Switzerland intended to replace the Wildmap system described in Section 9.7.2. It is based upon the family of Sun-3 graphics workstations which offer considerable local processing power, data storage and graphics display facilities. The Sun workstations used in System 9 utilize both the Unix operating system and the Sun window management system. A range of workstations is available. All have the database management system installed but additional hardware and software allow data input by field survey, stereo-plotting instruments and tablet digitizing (Fig. 11.25). Thus the workstations can be coupled directly to the tablet digitizer, in the S9-D digitizing station, or integrated into the design of the stereo-plotting instrument in the S9-AP analytical plotter. An editing station, S9-E, offers the capability of changing graphic and non-graphic data held within the database. Each workstation can have its own project database, but if many are linked together using an Ethernet network (Fig. 11.26) the database can be distributed over many machines. Any workstation can extract a partition from the project of a connected workstation.

The software needed for the efficient implementation and operation of a large system such as Wild's System 9 is extremely complex (Fig. 11.27). Besides the kernel toolbox

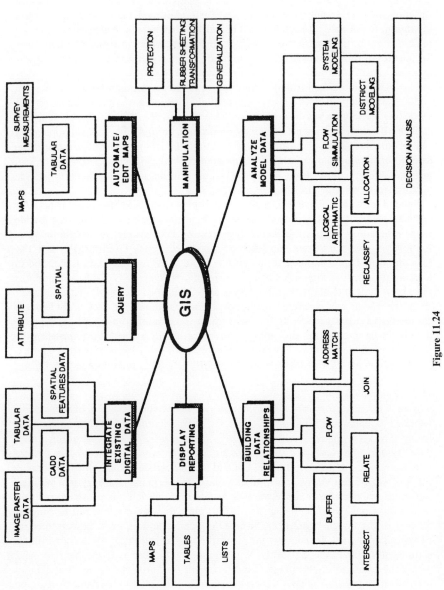

Figure 11.24

LAND INFORMATION DATABASES 465

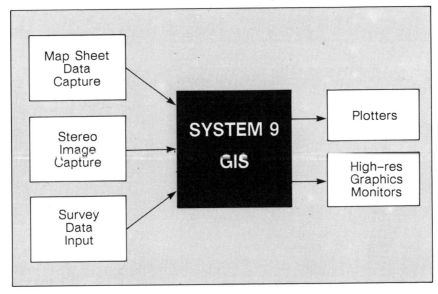

Figure 11.25

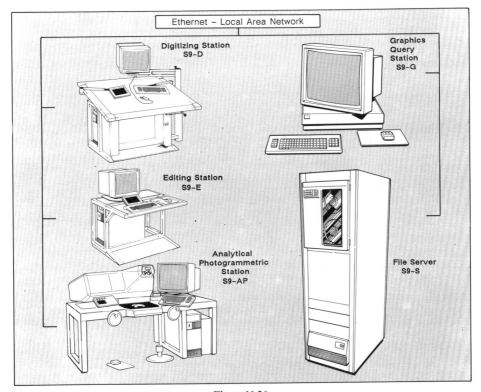

Figure 11.26

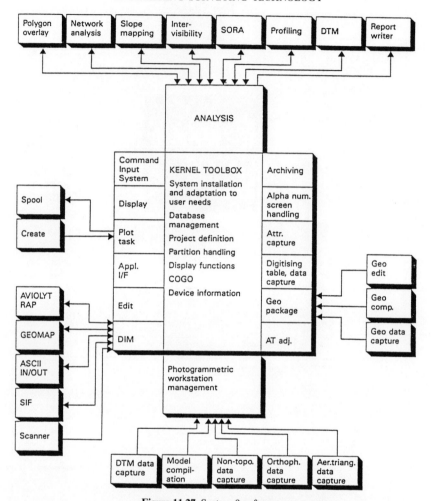

Figure 11.27 System 9 software.

which comprises the DBMS and the various software tools already discussed, there are numerous other software modules; these are shown distributed around the periphery of Fig. 11.27. The modules shown at the foot of this figure are related to the management of the photogrammetric data capture operations, discussed in Chapter 7. The data interchange modules (DIM) shown in Fig. 11.27 are designed specifically to allow rapid transfer of files of data between existing digital mapping systems, such as Wild's RAP (photogrammetric-based), Geomap (field survey-based), and Wildmap systems and those CADD-based systems which utilize the SIF (Standard Interface Format) for data exchange. At the top of Fig. 11.27, a number of analysis programs are shown; these act as modules for standard tasks such as polygon overlay, network analysis, slope mapping, etc. Finally, at the right hand side of the figure, the Geo modules for the capture and editing of data held on existing maps are shown.

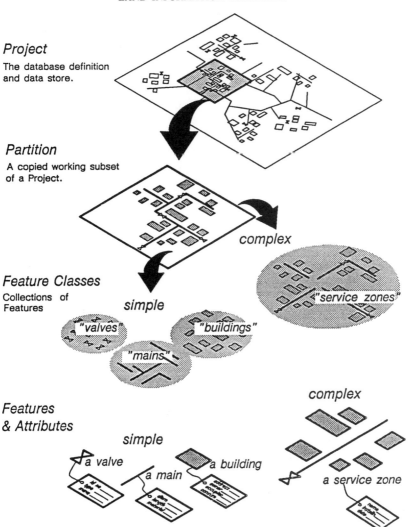

Figure 11.28 Hierarchical organisation of data in System 9.

System 9 supports a full topological data model of polygons, links and nodes. The elements of the model and all attribute data are stored using the EMPRESS relational database management system. The software design differs from other GIS packages in being 'object orientated'. The software is divided into modules not solely by operation but also by *object*, which comprises the data structure plus the procedures to operate on it (Fig. 11.28). For example, individual drains or sewers are considered as *simple features*. A logical group of simple features such as a drainage area is a *complex feature*. A complex feature can contain any number of other complex features. All complex features of a particular type comprise a *complex feature class*. Complex features can have attributes associated with them in the same way as simple features. The attributes of each of the complex features apply to all the simple features which it references, hence obviating duplication.

Within System 9, three organisational groups of information are recognised: projects, partitions and themes. A *project* is the highest level of data organization, representing the entire database for a geographical area. It comprises two components—the spatial/attribute database, and the database definition which specifies the project structure such as feature classes and themes. The *theme* definition determines how features such as colour, symbols, line types, etc. are to be displayed. Since many theme definitions are permitted, any data set can be differently displayed to suit different users and requirements.

The database is created and updated by means of *partitions*, working subsets of a project which are extracted according to the type of work to be done. When editing is complete, the partition is merged back into the project database to effect the update. A partition definition describes spatial extent, contents and their representation. The merit of the partition structure is that it makes it safe for different sections within an organization to work on data from the same common project.

It will be extremely interesting to see the impact on the market of this new GIS/LIS system from one of the leading survey and photogrammetric system suppliers with such a large, established customer base. That Wild should go in this direction and make the large investment required to develop System 9 is an acknowledgement of the importance now attached to GIS/LIS systems, within the surveying and mapping community.

11.8 Applications of land and geographic information systems

Three very different applications are described in this section to illustrate the scope and diversity of the operations that can be carried out by a land or geographic information system. The first concerns the application of such a system within a large public utility in the UK; the second is an example of an Australian land information system concerned principally with land ownership and tenure and its associated cadastral mapping; the third describes a project carried out in the United States in which a GIS system has been used to provide the basis for environmental management and decision making.

11.8.1 *British Telecom Line Plant Graphics (LPG)*

British Telecom (BT) has begun to implement an extensive network system known as its *Line Plant Facilities Management System*, which is designed to manage the spatially related data of its entire UK inland telecommunications network. The term AM/FM (Automated Mapping/Facilities Management) has recently come into use to describe this particular type of geographically referenced information system, used increasingly by public utilities. These AM/FM systems are concerned especially with the description and mapping of a physical network of cables, pipelines, etc. and its associated plant, and with the provision of spatially referenced information used in the management of these very complex networks which extend over large geographic areas.

The existing BT *plant record* is a complex representation of its underground and overhead transmission lines, contained in four main types of graphical record all of which are geographically or spatially referenced (Cole and Voller 1988). These comprise the following types of plan and diagram (Fig. 11.29):

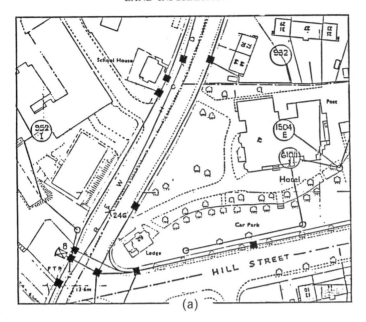

(a)

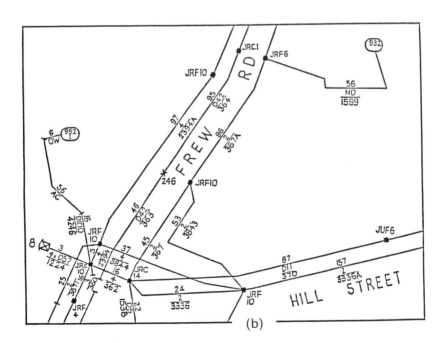

(b)

470 ENGINEERING SURVEYING TECHNOLOGY

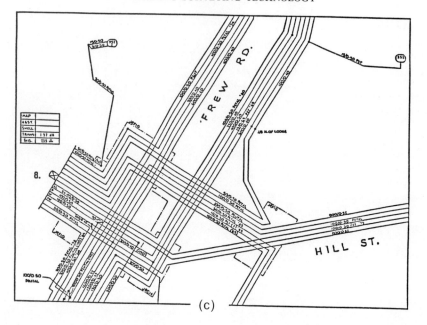

Figure 11.29

(a) plant-on-map;
(b) duct plan;
(c) cable diagram;
(d) duct space record.

The *plant-on-map* series (Fig. 11.29a) shows the positions of buried ducts and overhead cables represented by lines drawn or overlaid on existing OS plans and maps at 1:1250, 1:2500 and 1:10000 scales. Symbols drawn on these sheets represent nodes on the network, surface boxes, manholes, poles, etc. The *duct plan* (Fig. 11.29b) shows the detail of the duct network and is semi-geographical in character in that correct directions and approximate locations are retained, but most of the detail of the OS map is omitted, with only major roads and other very prominent terrain features being retained to provide a geographical reference. The *cable diagram* (Fig. 11.29c) goes still further towards a schematic diagram, with many details of cable characteristics being appended to the lines but with correct directions and retention of major streets to provide the spatial orientation and location of the cables. Finally, the *duct space record* (Fig. 11.29d) is a detailed alphanumeric table of information about a duct and its actual occupancy by cables.

The total extent of this record amounts to 150 000 plant-on-map sheets, 300 000 duct plans and cable diagrams and a host of duct space records and indexes. These form an essential element in the maintenance and repair of the network and in the planning of its extension, 1800 staff are required to maintain this record manually. Since the maintenance of the record in hardcopy form is a highly labour-intensive activity and the record is incompatible with other computer-based information systems used by BT at both local and national level, the decision has been taken to convert the existing record to a computer-based information system.

The task of converting the existing graphical records to digital form is an enormous one—1200 man years of effort have been estimated (Cole and Voller 1988). As discussed in Section 9.7.3, the OS is already engaged in digitizing its large-scale series, but the rate at which BT requires the OS digitized map data for its new system far exceeds the Survey's capacity to produce the data. Thus an agreement has been made with the OS that BT will employ contractors to undertake this work to an OS specification and quality assurance standard, and that the data produced will become part of the OS National Archive. The conversion of the BT duct plans and cable diagrams will be undertaken locally by each BT district with the support of contractors if necessary. Both the OS map data and the overlay data will be made available in the National Transfer Format (NTF) to enable it to be used by other public utilities.

Initially, the system will largely replicate the existing manual record since the two methods must coexist for a period. When it is completed, the digital record will form a 2D model of BT's telephone network; it will be capable of supporting most sections of BT's telephone activities and of being interrogated to produce management, marketing and commercial information, as well as underpinning the technical and engineering aspects of BT's telecommunications network.

The hardware/software system is being supplied by Intergraph and is based on a series of computing nodes or centres with large Digital VAX 8800 processors at each node. The first five of these will be located at Belfast, Glasgow, Leeds, Wolverhampton and Bristol, connected by fast kilostream communication links (Fig. 11.30). A further five centres will be installed in due course to complete a national network. When

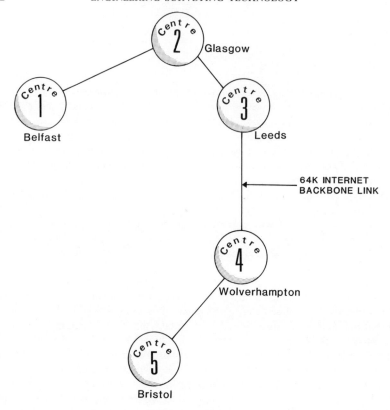

Figure 11.30

complete, BT will have one of the largest geographically based and spatially related information systems in Europe.

The basic software environment consists of Intergraph's well-established Interactive Graphics Design System (IGDS) and Data Management and Retrieval System (DMRS), with the facilities management (FM) software called FRAMME (Facilities Rule-Based and Application Model Management Environment). The overall arrangement is shown in Fig. 11.31.

The map and diagram features are recorded as the *graphic* elements to form the vector graphics data base, within the IGDS part of the system. The *non-graphic* elements include, in this case, the duct space records and all the statistical, economic and other numeric and textual information which are related to the spatial information. These are recorded and held in the DMRS part of the system. DMRS employs a network database structure which, together with storage techniques that substantially compress the data, facilitates the rapid processing of large volumes of textual information. The closely interwoven links between IGDS and DMRS produce

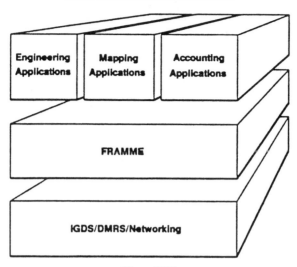

Figure 11.31

integrated files that result in a comprehensive digitally-based graphics database with linkages to the DMRS descriptors and text. The dual structure allows bidirectional access to graphics or text from any Intergraph graphics work station.

The superimposed FRAMME software supplies the FM component of the BT system. It is a so called rules-based 4GL (4th Generation Language) which contains the standard rules and specifications for the company's records, including its own drawing symbologies, feature definitions, design menus, etc. A *data dictionary* accommodates this *rule base*, ensuring: data entry, access and processing procedures are controlled and verified; proper relationships are established between facilities; and the data are structured to give the highest degree of flexibility in using the stored data. FRAMME also includes a feature definition language which defines the specific features to be accommodated in the system. For example, a pole can be represented graphically by a simple circle, while its associated non-graphic description would include its height, class, material and any other attributes pertaining to it. The *knowledge base* of FRAMME provides a model that includes:

(i) the location of each item of plant;
(ii) the relationship between one item of plant and another;
(iii) related information about each item of plant.

Each item (or feature) in the model has both a graphics and non-graphics representation which is handled as a logical unit of information, in this case, a logical unit of plant (LUP). The graphics items, maps, plans and diagrams etc., are stored in the IGDS part of the database. The non-graphics attributes of these items, e.g. the connectivity of plant, are stored in the DMRS database. Data can be stored or modified in the facilities model according to the definitions contained in the data dictionary. The model analysis and the collection of attribute information is performed through database tracing. The final output from the FRAMME can be in the form of either a graphics (map or diagram) or a text-based report.

A very similar system to that described above has been implemented at the Southern Bell telecommunications company in the United States.

11.8.2 *South Australia Land Information System*
The South Australia Land Information System was conceived in 1974 as a corporate data resource for the state. A nodal approach was taken, with the main land information system comprising four major databases with a myriad of peripheral systems attached. The databases and peripheral systems are operated independently, but a series of procedures and standards have been adopted to allow the transfer and integration of land-related data between nodes. The four *primary nodes* consist of a legal/fiscal database; a geographical database; an environmental database; and a socio-economic database (yet to be implemented). These are considered in greater detail below.

(i) The *land ownership and tenure system* (LOTS) is the major part of the legal/fiscal database. It is an alphanumeric database of land parcel information. It contains no spatial data. The LOTS database itself comprises a number of separate but integrated systems, each with their own function, data input, update systems and associated authorities. The individual systems document such information as title, valuation, sales history and land tax. LOTS is the best developed and most heavily used part of the complete land information system.

(ii) The main element of the geographical node is the *digital cadastral data base* (DCDB). This is currently being implemented. For each land-parcel, the information on the large-scale cadastral plan recording its shape and position with respect to other parcels and permanent ground reference marks is captured digitally. A unique coordinate reference system for all the permanent ground reference marks is being established so that the limits of all parcels will be uniquely defined in such a system. The addition of large-scale topographic data to the cadastral data set is under consideration.

An example of a secondary node is the utilisation of the DCDB by the Engineering and Water Supply Department as a reference base for the spatial location of its water and sewerage networks. The Department has brought in computer hardware and software to capture and maintain its own facilities records. It has created links between these and the DCDB. The local gas supply organisation is taking similar action.

(iii) The *environmental node* of the information system is a topologically structured spatial database created from small-scale map data and other data sources. Currently it exists only for limited areas of the state. The ARC/INFO geographical information system software is used. It is intended that a comprehensive database and set of analysis procedures for all environmetal planning purposes will be established.

11.8.3 *Connecticut Geographic Information System Project*
In 1984, the United States Geological Survey (USGS) and the National Resources Centre (NRC) of the State of Connecticut began a joint GIS project. The main goal was to demonstrate how a GIS could be incorporated into the ongoing data collection processes of the NRC and to test its effectiveness in environmental management. The ARC/INFO GIS software is used. The USGS had been using the complete ARC/INFO package for some time and NRC had adopted the relational database package INFO for alphanumeric data handling some years before this.

An area of approximately 400 km^2 was selected for the project. The area had a

Data set	Resolution/scale	Data type
Land use	1:24 000	polygon
Hydrography	1:24 000	polygon/line
Transportation network	1:24 000	line
Political boundaries	1:24 000	line
State land	1:24 000	line
Drainage basin limits	1:24 000	line
Geographical name	1:24 000	point
Digital elevation model	30 m grid	grid
Census boundaries	1:24 000	line
Wells	1:24 000	point
Pollution sources	1:24 000	point
Sewered areas	1:24 000	polygon
Water supply areas	1:24 000	polygon/line
Water quality class	1:24 000	polygon/line/point
Municipal zones	1:24 000	polygon
Biophysical land class	1:50 000	polygon
Landsat scenes	80 m grid	grid
Surface materials	1:24 000	polygon
Wetland inventory	1:24 000	polygon
Bedrock elevations	1:24 000	line
Soils	1:15 840	polygon
Endangered species	1:24 000	polygon
Bedrock geology	1:24 000	polygon
Water table elevations	1:24 000	line
100-year flood zones	1:24 000	polygon
Mining inventory	1:24 000	polygon
Water utility lands	1:24 000	polygon
Aeromagnetic survey	1:24 000	line

Figure 11.32

variety of landscapes and supported a wide range of land uses. A total of 28 primary data sets were identified. These are listed in Fig. 11.32 along with some details of the data resolution and the topological structure used for each individual data set. A great deal of preprocessing of some material was necessary before the data could be digitized. Each data set was captured separately; paper maps that contained more than one type of data were either recompiled as separate layers or were colour-coded according to feature type. Features had to be labelled with unique identifiers to provide the link between the spatial data and the existing INFO attribute database.

Four trial applications were tested. The first two required the extraction of spatial and attribute data for direct input into existing waterflow simulation software. In both cases, it was demonstrated that the required data could be extracted rapidly and automatically and permitted frequent simulations. The third application also involved ground water. The final product from this test was a map depicting all suitable sites where water for domestic use could be extracted from shallow wells. Only sites within a 800 m 'buffer area' around the water utility service area were selected and these had to satisfy several criteria: only wooded areas within suitable water-quality zones further than 500 m from polluting sources and 100 m from waste-receiving streams were accepted. The final map was overlayed with standard topographic details.

The final application was also a site-selection exercise. The product of this assessment was a map identifying potential sites for industrial development. Unsuitable factors included excessive slope, poor soil characteristics, seasonal flooding, insufficient size and any controversy associated with 'environmentally sensitive' areas. Data of water supply, sewer and transportation facilities were combined with these negative factors to determine the availability of these public services and to provide an overall assessment of site suitability.

The Connecticut trial resulted in a national strategy being devised to encourage the creation and exchange of digital spatial databases and their use in all environmental investigations.

11.9 Conclusion

Spatially referenced land information databases and systems are powerful tools that offer a wide range of facilities and possibilities in the supply of land-related information. Such information will guide the decision-making of a large number of users, including administrators, planners, field scientists and lawyers, as well as the surveyors and engineers for whom this book is primarily intended. No single database or system alone is capable of providing the myriad information required by such a large and diverse community of users in such a large area of terrain, especially in highly developed countries. It will be extremely interesting to see the way in which land-related information systems develop over the next decade. Assuredly it will be a rapid, if confused, period of development and one in which many surveyors and engineers will be heavily involved. Hopefully the introduction given in this chapter will assist them in their understanding of the underlying concepts of land information databases and the ways in which they may be implemented in practice.

References

Broadbent, C.K. (1980) Topological data structures. (*unpublished*) 36 pp.
Burrough, P.A. (1986) *Principles of geographic information systems for land resources assessment*. Clarendon Press, Oxford. 193 pp.
Dangermond, J. (1983) ARC/INFO, a modern GIS system for large spatial data bases, *Technical Papers ACSM-ASP*, Fall Convention (Virginia), 81–89.
Date, C.J. (1986) *An introduction to database systems*, Volume 1. Addison-Wesley, Massachusetts, 639 pp.
Encarnacao, J. (1980) Computer aided design modelling, systems engineering, CAD-systems. *Lecture notes in computer science*, Springer-Verlag, Berlin. 459 pp.
Feuchtwanger, M. and Lodwick, C.D. (1986) Position-based land information systems. (*unpublished*) 192 pp.
Hogg, J. (1986) Evaluation of regional land resources using geographic information systems based on linear quadtrees. *Int. Soc. for Photogrammetry and Remote Sensing* (Holland), **Commission VII**.
Lodwick, G.D., and Feuchtwanger, M. (1987) Land-related information systems (*unpublished*), 233 pp.
McLaughlin, J. (1986) Land information management: a review of selected concepts and issues. *Survey Review* 28 (222).
Martin, J. (1976) *Principles of database management*. Prentice-Hall, New Jersey. 352 pp.
NTF Working Party (1986) National Transfer Format User Manual.
Parker, D. (1986) Prospects for digital mapping in the North East: System requirements. *Proc. RICS Digital Mapping Seminar*, (Newcastle) 14 pp.
Peuquet, D.J. (1984a) Data structures for a knowledge-based geographic informtion system. *Proc. Int. Symp. on Spatial Data Handling* (Zurich) **2**, 372–391.

Peuquet, D.J. (1984b) A conceptual framework and comparison of spatial data models. *Cartographica* **21** (4), 66–113.
Prince, M.D. (1971) *Interactive graphics for computer aided design*. Addison-Wesley, Massachussetts, 301 pp.
Ritchie, A.J. (1987) The capture and storage of locational data: an exercise in integration. MSc. thesis, University of Newcastle-upon-Tyne, 111 pp.
Van Lamsweerde, A.A.P.J.M. (1983) Spatial data structure for land information systems. *Proc. FIG XVI International Congress*, **Commission 3**, paper 301.5, 12 pp.

Glossary of abbreviations

A considerable number of acronyms have been used throughout the book, and the following list summarizes those used.

ATM	Airborne Thematic Mapper
AIS	Airborne Imaging Spectrometer
APPS	Analytical Point Positioning System
BE	Broadcast Ephemeris
BIPS	British Integrated Program Suite
BLUE	Best Linear Unbiased Estimate
B:H	Base:Height (ratio)
CAD	Computer Aided Design (or Draughting)
CADD	Computer Aided Design and Draughting
CCD	Charge Coupled Device
CCTV	Closed Circuit TeleVision
CIO	Conventional International Origin
CIP	Contour Interpolation Package
CRT	Cathode Ray Tube
DBMS	DataBase Management System
DCCS	Digital Comparator Correlator System
DCDB	Digital Cadastral DataBase
DEM	Digital Elevation Model
DFUS	Digital Field Update Station
DGM	Digital Ground Model
DHM	Digital Height Model
DIP	Digital Image Processing
DISCOS	DISturbance COmpensation System
DLT	Direct Linear Transformation
DTED	Digital Terrain Elevation Model
DTM	Digital Terrain Model
ECDS	Electronic Coordinate Determination System
EDM	Electronic (or Electromagnetic) Distance Measurement
EM	ElectroMagnetic
EPF	Electronic Position Fixing
ERTS	Earth Resources Technology Satellite
FILS	Ferranti Inertial Land Surveyor
FINDS	Ferranti Inertial Navigation Direct Surveyor
FRAMME	Facilities Rule based and Application Model Management Environment
GAST	Greenwich Apparent Sidereal Time
GCP	Ground Control Point
GIS	Geographic Information System
GPCP	General Purpose Contouring Package

GLOSSARY OF ABBREVIATIONS

GPM	Gestalt PhotoMapper
GPS	Global Positioning System
HDDT	High Density Digital Tape
HIFI	Height Interpolation by FInite Elements
HIRIS	HIgh Resolution Imaging Spectrometer
IERS	International Earth Rotation Service
IFOV	Instantaneous Field of View
IMC	Image Motion (or Movement) Compensation
INS	Inertial Navigation System
IR	InfraRed
ISS	Inertial Surveying System
JPL	Jet Propulsion Laboratory
LASER	Light Amplification by Stimulated Emission of Radiation
LCD	Liquid Crystal Display
LED	Light Emitting Diode
LIS	Land Information System
LOD	Length of Day
LOTS	Land Ownership and Tenure System
MCE	Mapping and Charting Establishment
MOMS	Modular Optoelectronic Multispectral Scanner
MSS	MultiSpectral Scanner
NASA	National Aeronautical and Space Administration
NAVSTAR	NAVigation Satellite Timing And Ranging
NNSS	Navy Navigation Satellite System
NTF	National Transfer Format
OS	Ordnance Survey
OSTF	Ordnance Survey Transfer Format
PADS	Position and Azimuth Determination System
PE	Precise Ephemeris
PICS	Photogrammetric Integrated Control System
PMI	Programmable Multispectral Imager
PROM	Programmable Read Only Memory
PUSWA	Public Utilities Street Works Act
RBV	Return Beam Vidicon
RDBSI	Relational DataBase Interface
RDBMS	Relational DataBase Management System
ROV	Remotely Operated Vehicle
RMS	Remote Measuring System
SAR	Synthetic Aperture Radar
SCOP	Stuttgart COntouring Package
SIF	Standard Interchange File
SIR	Shuttle Imaging Radar
SLAR	Side-Looking Airborne Radar
SLIMPAC	Surveyors Land Information PACkage
SLR	Side Looking Radar
SLR	Satellite Laser Ranging
SMART	StereoMapping with Radar Techniques
SPACE	System for Positioning and Automated Coordinate Evaluation
SPOT	Le Système Probatoire d'observation de la Terre

SPRITE	Signal Processing In The Element
SQL	Standard Query Language
STARS	Simultaneous Triangulation and Resection
TIN	Triangular Irregular Network
TM	Thematic Mapper
TVFS	Thermal Video Frame Scanner
VCC	Video Correlator Controller
VLBI	Very Long Baseline Interferometry
VLL	Vertical Line Locus

Index

absolute orientation 242, 293
acceleration 96
accelerometer 96
accuracy 149, 163
aerial camera 185
aeronautics 318
alignment lasers 65
altimetry 141
AM/FM 468
analogue stereoplotters 239, 303, 392, 394
analytical stereoplotters 255, 266, 306
architecture 314
areal model 436
Arc/Info 462
atmospheric errors
 − EDM 18
 − GPS 130, 137
 − TRANSIT 130
automobile engineering 319
autoplumbing 68
Azimuth
 − transfer 84

B-tree index 444
best linear unbiased estimate 150
binary research 443
Boolean operations 441
breaklines 395, 411, 413
British Telecom line plant graphics 468
broadcast ephemeris 125
bundle method
 − close range photogrammetry 295
 − theodolite data 41

cadastral mapping 106, 420
calibration
 − EDM 20
cameras
 − aerial 185
 − CCD 191
 − metric 298
 − non metric 301
 − stereometric 300
centrifugal force 104
centroids model 438

cartographic plotters 363
 − raster 376
 − vector 364
carrier phase measurement 132
cathode ray tube 357
 − colour 361
CERN 57, 95, 140, 171
Charge Coupled Devices (CCD) 37, 191, 280, 321
CIP 414
classification 218
colinearity equations 259, 293
comparators
 − mono 260, 305
 − stereo 263, 305
compass
 − datum 84
 − prismatic 85
 − theodolite 86
 − trough 86
 − tubular 86
Computer Aided Design (CAD) 319
computer data structures
 − files 442
 − pointers 442
 − records 442
composite sampling 397
contouring
 − grid 401
 − triangles 406
contrast stretching 215
coplanarity equations 259, 293
correlation
 − area 274
 − close range photogrammetry 321
 − epipolar 278
correlator 253, 273
covariance 153, 163
cycle slips 138
cyclic error 20, 24

database
 − hierarchical 430, 449
 − network 449
 − relational 450

481

data compaction
- block coding 458
- chain coding 458
- run length encoding 459
Database Management System (DBMS) 331, 429
data models
- areal model 436
- centroids model 438
- link and node 436
- spaghetti model 438
- topological 431
data structures 431
- raster 352, 439
- vector 352, 439
data recorders 31
dead reckoning 96
deflection measurement
- laser 70
deformation 165
delay lock loop 132
Delaunay triangulation 404, 413, 423
density slicing 218
design
- networks 164
Digital Image Processing (DIP) 214
- classification 218
- contrast stretching 215
- density slicing 218
- principal components analysis 218
- ratioing 217
- spatial filtering 216
Digital Terrain Modelling (DTM) 391
- grid 393
- triangular 403
digitizer 244
- automatic line following 342
- automatic raster scanning 347
- Lasertrak/Fastrak 343
- line following 334
- mechanically based 335
- raster scanning 334
- tablet 338, 393
digitizing 332, 393
- accuracy 342
- 'blind' 339
- line data 332
- point data 332
Digital Field Update System (DFUS) 389
digital monoplotter 257, 261
digital photogrammetry 279, 321
digital tape recorders 32
dimensional control 45
Direchlet triangulation 404, 413, 423

display devices 356
- raster refresh 360
- raster storage 363
- vector refresh 358
- vector storage 359
display memory 361
disk storage devices 448
- buffer 448
- bucket 448
- disc page 448
- file manager 448
Doppler technique 126
drift 105
drop line plot 255
drum
- plotters 369, 379
- scanners 347

earth rotation 118
earth spin 88
ECLIPSE 381, 408
Electronic Coordinate Determination Systems (ECDS) 35
- accuracy 42
- hardware 35
- principles 38
Electronic Distance Measurement (EDM) 13
- calibration 20
- errors 17
- principles 13
electronic tacheometers 28, 392
- integrated 28
- laser 57, 60
- modular 28
electronic theodolites 7
- circle measuring 8
- tilt sensors 12
electronic transfer 252
electrostatic plotters 376
- colour 376
ephemeris 122
- broadcast 125
- precise 126
error ellipse 160
errors 147

Fastrak 393
feature coding 34, 434
fields 442
files (computer) 442
- B-tree index 444
- pointer chains 444
- sequential 442
- serial 442
finite elements 171, 409
frames 358
frame buffer 361

frame imaging systems 184
frame scanners 197
  – Barr and Stroud IR18 198
free adjustment 156, 297, 312
Fresnel zone plate 70

gas lasers 51
geoid 117, 143
Geographic Information Systems (GIS) 331
geostationary 120
geotechnics 309
Gestalt PhotoMapper (GPM) 274
global congruency test 167
global interpolation methods 398
Global Positioning System (GPS) 130
graphics workstations 340, 356, 418
  – analogue stereoplotters 247
  – analytical stereoplotters 269
  – digitizing 340
gravity 98, 115
grid north 85
grid sampling 393
gyro attachment 90
gyroscopes 87
  – laser 99

hand-held computers 34
HASP 420
Helmert transformation 124, 138
hydraulics 314
hierarchical structure 450
HIFI 408, 409

indexed files 443
  – B-tree index 444
Image Movement (Motion) Compensation (IMC) 188
image processing 214
  – classification 218
  – contrast stretching 215
  – density slicing 218
  – principal components analysis 218
  – ratioing 217
  – spatial filtering 216
image space plotter 264
imaging spectrometers 204
inertial
  – platforms 101
  – space 87
  – survey systems 95
industrial metrology 43
inner orientation 240, 292, 304
input axis 88
integration 99
interferometry
  – laser 24, 55
  – very long baseline 143
intersection 38

Kalman filter 104
Kepler's laws 119, 120

LAGEOS 64
Land Information Systems (LIS) 331
Landsat
  – multispectral scanner 194
  – thematic mapper 194
laser
  – alignment 65
  – detectors 78
  – digitizers 343
  – gas 51
  – gyroscopes 99
  – interferometers 24, 55
  – levels 71
  – liquid 54
  – plotters 375
  – rangers 58, 60
  – satellite ranging 112, 140
  – semi-conductor 53
  – solid state 52
Laser Scan
  – Fastrak 393
  – HRD-1 360
  – Laser Plot 369, 370
  – Lasertrak 343, 393
  – LITES 340
light spot projector 375
line extraction 351
line thinning 351
linear array scanners 198
  – digitizing 350
  – MOMS 203
  – SPOT 200
link 432
link and node model 437
link lists 445
local level system 99

National Joint Utilities Group (NJUG) 387
National Transfer Format (NTF) 387
navigation
  – offshore 108
NAVSTAR 130
nearest neighbour search 397
network design 164
network structure 449
nodes 431
normal equations 152

magnetic
  – attraction 85
  – compass 85
  – meridian 85
  – variation 85
marine engineering 320

INDEX

measurement patterns (DTM) 392
mechanical projection stereoplotters 242, 304
Mekometer 17, 57
metrology 43
microwave sensors 206
 − side looking radar 206
 − synthetic aperture radar 211
minimum constraint 156
modulation 15
Moiré fringe 9
MOMS 203
MOSS 408, 415
 − major options 415
 − minor options 415
 − strings 415
multispectral line scanners 193
 − Daedalus ATM 197
 − Landsat MSS 194
 − Landsat TM 194
multistation photogrammetry 296

observation equations 151
 − GPS 132, 135
 − TRANSIT 130
optical projection stereoplotters 240
optical transfer 251
orbit relaxation 138
orbits 119
Ordnance Survey
 − digital mapping 385
 − transfer format 387
orientation
 − absolute 242, 293, 304
 − inner 240, 292, 304
 − relative 241, 293, 304
orthometric heights 117
orthophotography 248, 410
orthophotoprinter 272
OSCAR 125
oscillation 92
overlap (photographic) 186
outliers 157
output axis 88

PANACEA 424
parabolic antennae 43
patchwise interpolation methods
 − exact fit 398, 407, 409
 − overlapping 399
phase measurement 15, 55
photographic systems 185
 − aerial cameras 185
 − close range cameras 298
phototriangulation 295
pipe-laying laser 65
plotters
 − cartographic 356, 363
 − stereo 239

pointer chains 444
pointers 442
pointwise interpolation methods 397
polygon 432
polynomials 400
 − bicubic 401, 411
 − bilinear 400, 409
 − biquadratic 402
 − cubic 400, 414
post processing 105
precession 89
precise ephemeris 126
precision 149, 159
process engineering 320
progressive sampling 394, 409
prismatic compass 85
projection
 − mechanical 242, 304
pseudo ranging 113
Public Utilities Street Works Act (PUSWA) 388
pushbroom scanners 198
 − MOMS 203
 − SPOT 200

Q-switching 52
quadtrees 459
query language 429, 452
 − *see also* Structured Query Language

radial sweep method 404, 405
random sampling 396
rank deficiency 154
range data 129
raster
 − data structure 352, 439
 − digitizers 347
 − displays 360
 − plotters 376
raster format 457
 − quadtrees 459
raster scanners 347
 − areal arrays 348
 − drum scanners 347
 − flatbed scanners 348
 − linear arrays 348
raster to vector conversion 351
 − line extraction 351
 − line thinning 351
ratioing 217
real time photogrammetry 321
 see also digital photogrammetry
rectifying 248
reference frame 98
refractive index 18
relational structure 450
relative orientation 241, 293
reliability 149, 157